Quantum Communication and Information Technologies

NATO Science Series

A Series presenting the results of scientific meetings supported under the NATO Science Programme.

The Series is published by IOS Press, Amsterdam, and Kluwer Academic Publishers in conjunction with the NATO Scientific Affairs Division

Sub-Series

I. **Life and Behavioural Sciences**	IOS Press
II. **Mathematics, Physics and Chemistry**	Kluwer Academic Publishers
III. **Computer and Systems Science**	IOS Press
IV. **Earth and Environmental Sciences**	Kluwer Academic Publishers
V. **Science and Technology Policy**	IOS Press

The NATO Science Series continues the series of books published formerly as the NATO ASI Series.

The NATO Science Programme offers support for collaboration in civil science between scientists of countries of the Euro-Atlantic Partnership Council. The types of scientific meeting generally supported are "Advanced Study Institutes" and "Advanced Research Workshops", although other types of meeting are supported from time to time. The NATO Science Series collects together the results of these meetings. The meetings are co-organized bij scientists from NATO countries and scientists from NATO's Partner countries – countries of the CIS and Central and Eastern Europe.

Advanced Study Institutes are high-level tutorial courses offering in-depth study of latest advances in a field.
Advanced Research Workshops are expert meetings aimed at critical assessment of a field, and identification of directions for future action.

As a consequence of the restructuring of the NATO Science Programme in 1999, the NATO Science Series has been re-organised and there are currently Five Sub-series as noted above. Please consult the following web sites for information on previous volumes published in the Series, as well as details of earlier Sub-series.

http://www.nato.int/science
http://www.wkap.nl
http://www.iospress.nl
http://www.wtv-books.de/nato-pco.htm

Series II: Mathematics, Physics and Chemistry – Vol. 113

Quantum Communication and Information Technologies

edited by

Alexander S. Shumovsky

Department of Physics,
Bilkent University, Bilkent, Ankara, Turkey

and

Valery I. Rupasov

Landau Institute for Theoretical Physics,
Moscow, Russia and
ALTAIR Center,
Massachusetts, U.S.A.

Kluwer Academic Publishers

Dordrecht / Boston / London

Published in cooperation with NATO Scientific Affairs Division

Proceedings of the NATO Advanced Study Institute on
Quantum Communication and Information Technologies
Ankara, Turkey
3–14 June 2003

A C.I.P. Catalogue record for this book is available from the Library of Congress.

ISBN 140201452X (hb)
ISBN 1402014538 (pb)

Published by Kluwer Academic Publishers,
P.O. Box 17, 3300 AA Dordrecht, The Netherlands.

Sold and distributed in North, Central and South America
by Kluwer Academic Publishers,
101 Philip Drive, Norwell, MA 02061, U.S.A.

In all other countries, sold and distributed
by Kluwer Academic Publishers,
P.O. Box 322, 3300 AH Dordrecht, The Netherlands.

Printed on acid-free paper

Proceedings of the NATO-ASI
Quantum Communication and Information Technologies
Ankara - Antalya, Turkey, June 3-14, 2002

Director

Alexander S. Shumovsky *Bilkent University*

Co-Director

Valery Rupasov *Landau Institute*

International Organizing Committee

Vladimír Bužek *National University of Ireland, Slovak Academy of Sciences*
Joe. H. Eberly *University of Rochester*
Herbert Walther *Max-Planck-Institut für Quantenoptik, Universität München*

Local Organizing Committee

Erdal Arıkan *Bilkent University*
Ekmel Özbay *Bilkent University*
Bilal Tanatar *Bilkent University*
M. Ali Can *Bilkent University*

Contents

Contents

Preface

This volume contains contributions based on the lectures delivered at the NATO Advanced Study Institute on Quantum Communication and Information Technologies (NATO-ASI QCIT) held in Turkey, 3-14 June, 2002. This Institute, hosted by Bilkent University, was attended by about 70 participants from all over the world. There was also a high level of participation by graduate students and post-docs, who greatly benefited from the opportunity to attend a world-class meeting. There was very active audience participation. Lectures were conducted in the morning and in the afternoon and this left the midday and evening free for informal discussions. The lectures were delivered by 15 invited lecturers, representing the major experimental and theoretical groups working in the field. The Conference was hosted in Bilkent University (Ankara) during the first week of activity and then in Club Hotel Sera (Antalya). The cultural program included excursions to Cappadocia and to the ancient cities of Aspendos and Perge on the Mediterranean Coast.

The importance of investigations in the field of quantum communications and quantum information processing and their implementations is well known. Remarkable recent developments in the field, in particular realization of quantum teleportation teleportation and quantum communication channels based on the use of entangled states, entangled states and discovery of quantum computing Quantum computing algorithms, have made quantum information one of the most important branches of modern physics and technologies. Every year, there are a number of Workshops, Conferences, and Summer Schools on quantum information. The distinguishing feature of NATO-ASI QCIT is that the discussion of fundamentals was tightly interlaced with technological problems.

The editors would like to thank the lecturers who provided notes for publication within the deadline. We especially thank the secretaries of the meeting Emine Benderlioğlu and Ayşe Bulut and on behalf of the participants we thank Banu Sabrioğlu who organized almost everything during the Antalya part of the meeting. We wish to thank the various organizations that have provided either financial or technical support. These include: Office of the President of Bilkent University, the Scientific and Technical Research Council of Turkey (TÜBITAK) and Özcivelek Tourism.

Finally, editor greatly acknowledge the collaboration of M.Ali Can from Bilkent University. Without his help this volume would never appear.

ALEXANDER SHUMOVSKY

Contributing Authors

Joseph H. Eberly Center for Quantum Information and Department of Physics and Astronomy, University of Rochester.

Alexander V. Sergienko Quantum Imaging Laboratory, Department of Electrical and Computer Engineering and Department of Physics, Boston University.

Vladimír Bužek Department of Mathematical Physics, National University of Ireland, and Research Center for Quantum Information, Slovak Academy of Sciences.

Luigi A. Lugiato Dipartimento di Scienze Chimiche Fisiche e Matematiche, Universita dell'Insubria.

Victor V. Kozlov Abteilung für Quantenphysik, Universität Ulm and Fock Institute of Physics, St.-Petersburg University

Herbert Walther Max-Planck-Institut für Quantenoptik and Sektion Physik der Universität München

Alexander S. Shumovsky Department of Physics, Bilkent University

Anatoly V. Andreev Department of Physics, M.V.Lomonosov Moscow State University

Apostol Vourdas Department of Computing, University of Bradford.

Valery I. Rupasov Landau Institute for Theoretical Physics.

Thomas F. Krauss School of Physics and Astronomy, University of St. Andrews.

Ekmel Özbay Department of Physics, Bilkent University

Mario P. Tosi NEST-INFM

Bilal Tanatar Department of Physics, Bilkent University

Erdal Arıkan Department of Electrical-Electronics Engineering , Bilkent University

Introduction

Opening Address by Professor İhsan Doğramacı
President of Bilkent University and Chairman of the Board of Trustees

Distinguished lecturers and participants in the NATO Advanced Study Institute on Quantum Communication and Information Technologies, ladies and gentlemen.

On behalf of Bilkent University I extend to you a warm welcome and express our happiness at hosting such an elite group of scientists here on our campus.

It is well known that we live in the Information Age and that the progress of humanity now depends increasingly on our capabilities to develop and implement information technologies. Optics, in this respect, has a privileged place. As you well know, the American Research Council has recently declared Optics the information technology of the 21st century, and in turn, the famous physicist Michael Berry declared that the 21st century will be shaped by quantum optics and quantum information.

In spite of the fact that Albert Einstein and Ervin Schrödinger introduced a number of basic principles in the early days of quantum mechanics, quantum information is a very young branch of science and technology. It was born only 10 or 12 years ago. Nevertheless, remarkable recent developments in the field herald a new era of scientific and technological capabilities. There is great success in the realization of quantum communication lines completely secured from eavesdropping. eavesdropping Teleportation teleportation, which has been the subject of popular movies like Star Trek, is now successfully realized in physics laboratories.

Rapid progress in the field makes urgent the need for providing advanced training and offering orientation to young scientists and engineers. This is the aim of the present NATO Advanced Study Institute organized by Bilkent University and supported in part by the Scientific and Technical Research Council of Turkey (TÜBITAK).

Bilkent University is one of the leading centers of higher education and science in Turkey. Our scientists work with success in the field of quantum information. Our students from the Faculty of Science and Faculty of Engineering take courses in this important field.

I deeply appreciate the presence of the prominent scientists from Canada, Germany, Italy, Russia, Slovakia, the United Kingdom and the United States who have come to join our Turkish scientists in giving advanced courses in this important area. I extend to them my gratitude and wish them an enjoyable stay here in Turkey. I welcome the participants from all over the world and hope that this NATO-ASI will be beneficial to all of you.

Your meeting will take place here at Bilkent, and in Antalya, one of the most famous Mediterranean resorts.

I wish you all continued success in the further development of Quantum Information as a science and technology.

SCHMIDT-MODE ANALYSIS OF ENTANGLEMENT FOR QUANTUM INFORMATION STUDIES

J.H. Eberly

Center for Quantum Information and Department of Physics and Astronomy,
University of Rochester, Rochester, NY 14627

K. W. Chan

Center for Quantum Information and Department of Physics and Astronomy,
University of Rochester, Rochester, NY 14627

C. K. Law

Center for Quantum Information and Department of Physics and Astronomy,
University of Rochester, Rochester, NY 14627
Department of Physics, The Chinese University of Hong Kong, NT,
Hong Kong SAR, China

Abstract   We present examples of the analysis of quantum entanglement entanglement as an introduction to the fundamental basis for quantum computing and information technology. Pure non-separable two-particle states are analysed using the Schmidt decomposition and we introduce the number of effective eigenmodes as a measure of entanglement. We give an elementary illustration, as well as an overview of more complex calculations we have carried out in situations involving bipartite states in continuous Hilbert space.

1.   Introduction

In quantum mechanics the non-separability of states implies the existence of some kind of correlation. One can utilize such correlation in terms of nonlocal control of one quantum system in relation to another. For example, it is well-known by now that the nonlocality associated with non-separable quantum states is the key to performing computational tasks that cannot be realized classically [1, 2, 3]. One of the most well-studied physical system possessing such behaviour is the spontaneous parametric down-conversion, in which a pair

1

A.S. Shumovsky and V.I. Rupasov (eds.), Quantum Communication and Information Technologies, 1–12.
© 2003 *Kluwer Academic Publishers.*

of entangled photons with the same or different polarizations polarization are emitted. It has been examined in the contexts of photon localization [4] and quantum information [5, 6], and a combination of these in quantum pattern formation and two-mode squeezing [7, 8].

2.   The Schmidt Theorem

As a reminder, we review here the Schmidt theorem [9], which allows one to characterize any bipartite pure state as a unique pairwise association of eigenstates of the subsystem density matrices. This is a matter of relatively simple matrix analysis if the state space is small enough, but a sketch of the theorem indicates its greater generality.

A general state in the composite Hilbert space $\mathcal{H}_a \times \mathcal{H}_b$ can, in some orthonormal bases $\{|a_i\rangle\}$ and $\{|b_j\rangle\}$, be written as

$$|\Psi\rangle = \sum_i \sum_j C_{ij} |a_i\rangle \otimes |b_j\rangle, \tag{1}$$

which is assumed to be normalized. The partial density matrix density matrix of the subspace $\mathcal{H}_a$ is

$$\hat{\rho}_a = \mathrm{Tr}_b \, |\Psi\rangle\langle\Psi| = \sum_{ii'} W_{ii'} |a_i\rangle\langle a_{i'}|, \tag{2}$$

where $W_{ii'} \equiv \sum_j C_{ij} C_{i'j}^*$. We see that the matrix W has three properties, namely,

1   W is bounded:

$$1 = (\mathrm{Tr}_a \, \hat{\rho}_a)^2 \geq \mathrm{Tr}_a \, \hat{\rho}_a^2 \;=\; \sum_{ii'} W_{ii'} \sum_{ll'} W_{ll'} \, \mathrm{Tr}_a \left(|a_i\rangle\langle a_{i'}|a_l\rangle\langle a_{l'}| \right)$$

$$= \sum_{ii'} |W_{ii'}|^2; \tag{3}$$

2   $W_{ii'} = W_{i'i}^*$, i.e., W is hermitian;

3   W is non-negative definite in the sense that, given any complex number f_i, we have

$$\sum_{ii'} W_{ii'} f_i^* f_{i'} = \sum_j \left(\sum_i C_{ij} f_i^* \right) \left(\sum_i C_{i'j}^* f_{i'} \right) = \sum_j \left| \sum_i C_{ij} f_i^* \right|^2 \geq 0. \tag{4}$$

The first property ensures that the partial density matrix density matrix $\hat{\rho}_a$ can be diagonalized (the Singular-Value Decomposition Theorem in matrix theory or the Mercer's theorem for functions of continuous variables [10]) and the remaining two properties imply that the eigenvalues of $\hat{\rho}_a$ in the new basis are real and non-negative.

Let $\{|\alpha_n\rangle\}$ be the basis which diagonalizes $\hat{\rho}_a$ with the corresponding set of eigenvalues $\{\lambda_n\}$. The three properties mentioned above ensure that $\langle \alpha_n|\alpha_{n'}\rangle = \delta_{nn'}$, $\sum_n \lambda_n = 1$, and the set $\{|\alpha_n\rangle\}$ is discrete (its importance is on continuous Hilbert space, see Ref. [11]). In terms of the matrix elements,

$$\langle \alpha_n|\hat{\rho}_a|\alpha_{n'}\rangle = \sum_{ii'} W_{ii'}\langle \alpha_n|a_i\rangle\langle a_{i'}|\alpha_{n'}\rangle \equiv \lambda_n\delta_{nn'}. \tag{5}$$

Now, with the use of this new basis, the general two-system wave function can be rewritten as:

$$|\Psi\rangle = \sum_{ij} C_{ij}\,|a_i\rangle \otimes |b_j\rangle = \sum_n \sqrt{\lambda_n}\,|\alpha_n\rangle \otimes |\beta_n\rangle, \tag{6}$$

where

$$|\beta_n\rangle \equiv \frac{1}{\sqrt{\lambda_n}}\sum_{ij} C_{ij}\langle \alpha_n|a_i\rangle|b_j\rangle. \tag{7}$$

It is easy to check that $\{|\beta_{n'}\rangle\}$ also forms an orthonormal basis in $\mathcal{H}_b$. More interestingly, the above decomposition works in finite as well as infinite, discrete or continuous, Hilbert space. The new bases $\{\alpha_n\}$ and $\{\beta_n\}$ are still countable [11].

In the second writing of Eq. (6) the desired result has finally been achieved. That is, $|\Psi\rangle$ is expressed as a single sum with a unique pairwise relation of the the states of the two subsystems. We call these the "information eigenstates", reflecting their origin as eigenstates of density matrices. Note that they can be ordered according to $\lambda_1 \geq \lambda_2 \geq \lambda_3 \geq ...$, and the degree of entanglement entanglement of the two systems is obviously related to the number of λ's that are "important". As a numerical measure of entanglement we will use the so-called participation ratio denoted by K [12]:

$$K \equiv \frac{1}{\sum_n \lambda_n^2} \geq 1. \tag{8}$$

As an example, if $K = 1$, the Schmidt sum has only one term and the state is not entangled. On the other hand, if there are N states all with $\lambda_n = 1/N$, then $K = N$. Consequently we can usually refer to K loosely as the number of Schmidt states that are significant in making up $|\Psi\rangle$.

3.   Finite Hilbert space

It is implicit in the Schmidt formalism that there will be only as many non-zero Schmidt eigenvalues as the smaller of the two dimensions involved. Here we give an example in which a two-state system (say two quantized one-photon radiation modes) is coupled to a four-state system (say a four-level atom). There are only two non-zero Schmidt eigenvalues in this case.

We designate the two one-photon states of "radiation" by $|A\rangle$ and $|B\rangle$, and use Greek letters to designate the states of the "atom". For the state to be analyzed we shall take the following complicated combination of tensor products of these states:

$$
\begin{aligned}
|\Psi_0\rangle &= |\alpha\rangle\left(2|A\rangle + |B\rangle\right) + |\beta\rangle\left(|A\rangle + 2|B\rangle\right) + |\gamma\rangle\left(|A\rangle - |B\rangle\right) \quad (9)\\
&\quad - |\delta\rangle\left(|A\rangle - |B\rangle\right)\\
&= |A\rangle\left(2|\alpha\rangle + |\beta\rangle + |\gamma\rangle - |\delta\rangle\right) + |B\rangle\left(|\alpha\rangle + 2|\beta\rangle - |\gamma\rangle + |\delta\rangle\right) (10)
\end{aligned}
$$

It may be difficult to be sure that a complicated two-particle state is actually non-separable, but the Schmidt analysis will answer this question.

The total density matrix density matrix is of course $\rho_{\text{tot}} \equiv |\Psi_0\rangle\langle\Psi_0|$, and a first step is to find the reduced density matrix for atom or radiation. This is easily done for either one, which we give here in unnormalized form:

$$
\rho_{\text{at}} = \begin{pmatrix} 5 & 4 & 1 & -1 \\ 4 & 5 & -1 & 1 \\ 1 & -1 & 2 & -2 \\ -1 & 1 & -2 & 2 \end{pmatrix}, \quad (11)
$$

and

$$
\rho_{\text{rad}} = \begin{pmatrix} 7 & 2 \\ 2 & 7 \end{pmatrix}. \quad (12)
$$

The eigenvalues of the normalized reduced density matrices can be checked to be the same: 5/14 and 9/14. The non-zero value of at least two eigenvalues shows immediately that more than a single term in the Schmidt sum is needed, so the initial state was indeed non-separable. Knowing the eigenvalues allows easy calculation of the participation ratio as

$$
K = \frac{1}{(5/14)^2 + (9/14)^2} \approx 1.85.
$$

If we designate the Schmidt eigenvectors as $|F_m\rangle$ for the states of the radiation modes and $|\phi_m\rangle$ for the atomic energy levels, then the Schmidt decomposition gives:

$$|\Psi_0\rangle = \sum_{m=1}^{4} \sqrt{\lambda_m} |F_m\rangle \otimes |\phi_m\rangle$$

$$= \sqrt{\frac{9}{14}}|F_1\rangle \otimes |\phi_1\rangle + \sqrt{\frac{5}{14}}|F_2\rangle \otimes |\phi_2\rangle$$
$$+ (0)|F_3\rangle \otimes |\phi_3\rangle + (0)|F_4\rangle \otimes |\phi_4\rangle. \tag{13}$$

For completeness, the corresponding eigenvectors are:

$$|F_1\rangle = \frac{1}{\sqrt{2}}(|A\rangle + |B\rangle),$$

$$|F_2\rangle = \frac{1}{\sqrt{2}}(|A\rangle - |B\rangle),$$

$$|F_3\rangle = p|A\rangle + q|B\rangle, \quad \text{(arbitrary)}$$
$$|F_4\rangle = r|A\rangle + s|B\rangle, \quad \text{(arbitrary)}$$
$$|\phi_1\rangle = \frac{1}{\sqrt{2}}(|\alpha\rangle + |\beta\rangle),$$

$$|\phi_2\rangle = \frac{1}{\sqrt{10}}(|\alpha\rangle - |\beta\rangle + 2|\gamma\rangle - 2|\delta\rangle),$$

$$|\phi_3\rangle = \frac{1}{\sqrt{3}}(|\alpha\rangle - |\beta\rangle + |\delta\rangle),$$

$$|\phi_4\rangle = \frac{1}{\sqrt{3}}(|\alpha\rangle - |\beta\rangle - |\gamma\rangle).$$

Thus the state can be written, using the Schmidt pairing, but the original basis states:

$$|\Psi_0\rangle = \sqrt{\frac{9}{14}}\left[\frac{1}{\sqrt{2}}(|A\rangle + |B\rangle) \otimes \frac{1}{\sqrt{2}}(|\alpha\rangle + |\beta\rangle)\right]$$
$$+ \sqrt{\frac{5}{14}}\left[\frac{1}{\sqrt{2}}(|A\rangle - |B\rangle) \otimes \frac{1}{\sqrt{10}}(|\alpha\rangle - |\beta\rangle + 2|\gamma\rangle - 2|\delta\rangle)\right] \tag{14}$$

4. Quantum Entanglement in Very Large State Spaces

Many physical situations involve entanglement entanglement of pairs of particles or systems. When the state space for the participating quantum systems increases, the information-carrying potential increases as well. Systems described by continuous variables have infinite state spaces, and these offer the greatest potential. We have studied a number of examples, shown in Fig. 1, and have found values of K as high as 4 or 5. In order to explain how the continuum case is analysed we will deal here with the entanglement that arises from

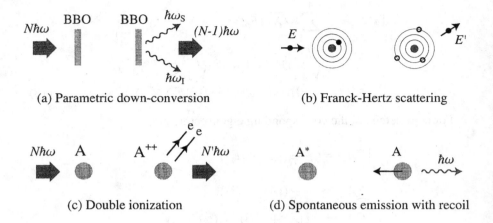

(a) Parametric down-conversion   (b) Franck-Hertz scattering

(c) Double ionization   (d) Spontaneous emission with recoil

Figure 1.   Examples of entangled systems with continuous variables: (a) parametric down-conversion [5, 6], (b) Franck-Hertz scattering [12], (c) double ionization [13], and (d) spontaneous emission [14].

momentum conservation in the recoil of an atom undergoing spontaneous emission [14]. The experimental set-up, requiring detection of photon and atom, is sketched in Fig. 2, similar to the experiments reported by Kurtsiefer *et al.* [15].

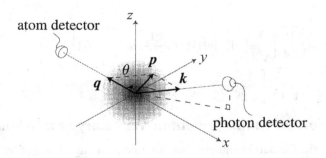

Figure 2.   Experimental set-up for the photon and atom detection.

The Hamiltonian Hamiltonian that describes the free-space spontaneous emission of a two-level atom two-level atom with mass M, transition frequency

ω_0, and ground state ground state and excited state $|g\rangle$ and $|e\rangle$ can be written as

$$H = \frac{\mathbf{P}^2}{2M} + \hbar\omega_0|e\rangle\langle e| + \sum_s \int d^3k \; \hbar\omega_k a_{\mathbf{k}s}^\dagger a_{\mathbf{k}s} \tag{15}$$

$$+ \sum_s \int d^3k \left[\hbar g_s(\mathbf{k})|g\rangle\langle a|e_{\mathbf{k}s}^\dagger e^{-i\mathbf{k}\cdot\mathbf{R}} + \text{h.c.} \right], \tag{16}$$

where $\mathbf{P}$ and $\mathbf{R}$ are the momentum and the center of mass position operators of the atom, and $a_{\mathbf{k}s}$ and $a_{\mathbf{k}s}^\dagger$ are the annihilation and creation operators of the quantized electromagnetic field with photon wavevector $\mathbf{k}$ and polarization s. The first two terms in Eq. (16) are the kinetic energy and the internal electronic state of the atom. The third term is the free field whereas the last term is the interaction in dipole approximation. Note that the ground state ground state energy of the atom is set to zero. The coupling strength is expressed as

$$\hbar g_s(\mathbf{k}) = \sqrt{\frac{\hbar\omega_k}{2\epsilon_0(2\pi)^3}} \mathbf{e}_{\mathbf{k}s} \cdot \mathbf{d}_{ge}, \tag{17}$$

with $\mathbf{e}_{\mathbf{k}s}$ and $\mathbf{d}_{ge}$ denoting the polarization of the photon and the dipole moment of the atom.

We would like to study the momentum entanglement entanglement between the photon and the atom after the emission through their Hilbert space structures. The most convenient way is to look at the joint atom-photon state after the emission process. At the time $t \gg \gamma^{-1}$ where γ is the natural decay rate of the upper atomic level, the state of the system can be written as

$$|\Psi(t)\rangle \rightarrow \int d^3q \int d^3k \; C(\mathbf{q},\mathbf{k}) \, e^{-i(E_q/\hbar + kc)t} |\mathbf{q}, g\rangle \otimes |1_\mathbf{k}\rangle. \tag{18}$$

Here we denote the state of the decayed atom with final momentum $\hbar\mathbf{q}$ by $|\mathbf{q}, g\rangle$ and that of the emitted photon with wavevector $\mathbf{k}$ by $|1_\mathbf{k}\rangle$. The joint amplitude $C(\mathbf{q},\mathbf{k})$ gives us the correlation between the two particles. When $C(\mathbf{q},\mathbf{k})$ is not factorable, $C(\mathbf{q},\mathbf{k}) \neq f_1(\mathbf{q})f_2(\mathbf{k})$, then the state is entangled.

According to Rzążewski and Zakowicz [16], the amplitude is given by

$$C(\mathbf{q},\mathbf{k}) = \frac{g(\mathbf{k}) \, a_0(\mathbf{q}+\mathbf{k})}{(E_q - E_{q+k})/\hbar + kc - \omega_0 + i\gamma}, \tag{19}$$

which is indeed not factorable. Here

$$a_0(\mathbf{p}) = N \exp\left[-\left(\frac{p}{\sigma_p}\right)^2 \right], \tag{20}$$

is the initial momentum distribution of the center of mass of the atom in its excited state. It is assumed to have a Gaussian shape of width σ_p. Such a momentum distribution corresponds to an atom prepared in a thermal state. The symbol $E_p \equiv \hbar \mathbf{p}^2 / 2M$ is a shorthand for the kinetic energy of the atom with momentum $\hbar \mathbf{p}$ in frequency unit.

To conform with the experimental set-up depicted in Fig. 2, we designate a particular angle θ between $\mathbf{q}$ and $\mathbf{k}$. This is accomplished by fixing the positions of the atom and photon detectors. A convenient choice is to set $\theta = \pi$. Physically we expect the atom suffers the strongest recoil in such a configuration. As a result, the joint atom-photon amplitude can be written explicitly as

$$C(q, k) = \frac{N \, e^{-(q-k)^2/\sigma_p^2}}{\frac{\hbar}{2M}(2qk - k^2) + kc - \omega_0 + i\gamma}. \tag{21}$$

The notation for the angle $\theta = \pi$ has been omitted for convenience. The constant N is a normalization factor which incorporates the coupling parameter $g_s(k)$. When we consider an optical transition such as spontaneous emission, this parameter is a slowly-varying function of k as compared to the resonant Lorentzian factor in Eq. (21). Thus we can safely treat it essentially as a constant which is to be fixed by the normalization condition.

Now we may perform the Schmidt decomposition on the non-factorable amplitude. It can be written as

$$C(q, k) = \sum_n \sqrt{\lambda_n} \, \psi_n(q) \, \phi_n(k). \tag{22}$$

The eigenvalues λ_n obtained above can determine the participation ratio K. Also note that the pairing of the atom mode $\psi(q)$ with the photon mode $\phi(k)$ allows a projective measurement of one to determine the other without detecting it directly.

Before using realistic numbers in Eq. (21) to do the Schmidt analysis, one may obtain more insight upon the behaviour of $C(q, k)$ by introducing a pair of scaled variables. It is noted that the joint photon-atom amplitude possesses a sharp resonance in the denominator of the Lorentzian. Because of the conditions $Mc^2 \gg \hbar\omega_0, \hbar\sigma_p c \gg \hbar\gamma$, Eq. (21) can be transformed into

$$C(q, k) = \frac{N'}{\delta k + \delta q + i} \times \exp\left[-\left(\frac{\gamma Mc\delta q}{\omega_0 \hbar \sigma_p}\right)^2\right], \tag{23}$$

where the scaled wavevectors are

$$\delta q \equiv \frac{\hbar\omega_0}{Mc^2\gamma}(qc - \omega_0), \quad \text{and} \quad \delta k \equiv \frac{1}{\gamma}\left(kc - \omega_0 + \frac{\hbar k_0^2}{2M}\right), \tag{24}$$

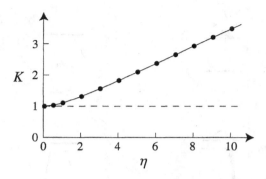

Figure 3.   A plot of the participation ratio K as a function of the observability index η.

with $N' = N/\gamma$. The combination of parameters in the Gaussian enables us to define an "observability" index

$$\eta \equiv \frac{\omega_0 \hbar \sigma_p}{\gamma M c} = \frac{\Delta \omega}{\gamma}, \tag{25}$$

in which $\Delta \omega = \omega_0 (\hbar \sigma_p / M)/c$ is interpreted as the thermal (motional) linewidth broadening. The observability index sets the effective range of δq giving an observable value of $C(q, k)$. It can now be also interpreted as the ratio of motional linewidth and natural linewidth. It is interesting that the amplitude expression is solely controlled by this single index. As a result, it is not surprising that the participation ratio K has to be related η. A plot of K as a function of η is shown in Fig. 3. It is found empirically that $K \approx 1 + 0.28(\eta - 1)$ for $\eta > 1$.

Experiments in which entanglement entanglement of an atom and a photon played a role have also been reported by Pfau, *et al.* [17], and Chapman, *et al.* [18]. They go beyond spontaneous emission in a suggestive way which we examine next.

5.   Beyond Spontaneous Emission

The atomic spontaneous emission considered in the previous section is an example of so-called "cross-modular" entanglement entanglement of two different quantum entities, and in this case one entangled partner is detectable with nearly 100% efficiency (the atom) and the other (the photon) is an ideally high-speed information carrier. This kind of entanglement is interesting as it may be used to build photonics devices that are useful to quantum information processing. However, it is found that the realistic values for K that can be reached via momentum entanglement in spontaneous emission are very small; they are estimated to be of the order of 1.

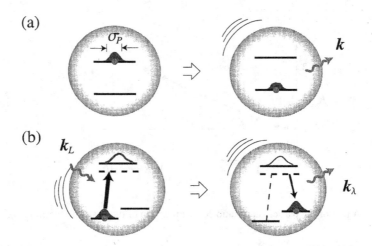

Figure 4.   Two generic atom-photon entanglement experiments: (a) spontaneous emision, (b) Raman scattering.

Since spontaneous emission is a fundamental process for an atom-photon interaction, it is not surprising that there are other systems that can simulate the same type of momentum entanglement. Some of these may have a greatly enhanced K value. First of all, we have shown that the control parameter η is inversely dependent on the radiative decay rate. Since $\eta \sim 1/\gamma$, we see that by modifying the decay rate, η and K could be controlled. There are at least two approaches. One is to utilize an atom with a very long lifetime, such as a Rydberg atom Rydberg atom. This method has the drawback that it lacks a true control of γ by externally tunable means. On the other hand, instead of modifying the actual decay rate one might control the population of the decaying level.

Spontaneous emission is essentially the same as Raman or Rayleigh scattering, except for the excitation process. The upper level is taken to be 100% occupied in the initial state of spontaneous emission, but this is reduced at steady state by the factor $(\Omega/2\Delta)^2$ in both Raman and Rayleigh cases, where Ω and Δ are the Rabi frequency and detuning of the incident field. In effect the radiative decay rate will be reduced by $(2\Delta/\Omega)^2$ which in turn gives an enhancement for η by the same factor. Here we will take this enhancement for granted, and propose 500 as an easily achievable value for the ratio $(2\Delta/\Omega)^2$ in optical excitation of a resonance transition. This leads to a hypothetical value of K greater than 100, which certainly is larger than any value yet reported.

An illustration of the similarity of spontaneous emission and Raman scattering leading to momentum entanglement entanglement is depicted in Fig. 4. Because of the much lower population in the excited state, or in other words, an effectively longer decay time, the atom undergoing the scattering process has to stay long enough in the pump laser laser beam so that the interaction is sufficient for yielding a photon. At the same time the longer interaction time means that either the drift velocity of the atom has to be small or the size of the interaction zone of the atom with the pump has to be large. Such practical issues have to be addressed when a real experiment is to be done. Optimistically, we suppose such a balance is not a great technical challenge.

The Raman and Rayleigh configurations also permit one to manipulate the polarization polarization and angular momentum Angular momentum degrees of freedom of both the atom and the photon. These are expected to be more robust to decoherence decoherence than the linear momentum entanglement entanglement invoked in the studies above.

6. Summary

A bi-partite quantum system involving entangled continuum variables has the potential for infinite information content. Conveniently, the continuous nature of the Hilbert space can be made discrete with the application of the Schmidt decomposition, which at the same time provides the useful mode-counting number K to give a measure of entanglement through the participation ratio (8).

Shared information is a nonlocal property of quantum pairs. The Schmidt decomposition provides the unique generalization of the Bell state Bell states used in many ways in studies of quantum information, for example in teleportation teleportation. It is possible that infinite and continuous Hilbert space (in contrast to the finite qubit space) can enable larger as well as more robust information flow in quantum channels. It is expected that experiments will be able to explore the high-K domain via photon-photon, atom-photon, and electron-photon interactions.

We have only sketched a very simple cross-modular entangled system which consists of an atom and a single photon. There are additional cross-platform, and possibly more exotic, instances of beyond-Bell pairing such as electron-photon Schmidt modes, quantum soliton pairs in optical fibers and waveguides, etc., including phased arrays with a large particle number. We are optimistic that retaining focus on high-dimensional contexts will be challenging and possibly most valuable.

Acknowledgments

This work was supported by the DoD Multidisciplinary University Research Initiative (MURI) program administered by the Army Research Office under Grant DAAD19-99-1-0215.

References

[1] For a review of quantum information devices, see H. E. Brandt, Prog. Quantum Electron. **22**, 257 (1998), and references therein.

[2] C. P. Williams and S. H. Clearwater, *Explorations in Quantum Computing* (TELOS, Santa Clara, 1998); and also *Ultimate Zero and One: Computing at the Quantum Frontier* (Copernicus, New York, 2000).

[3] M. A. Nielson and I. L. Chuang, *Quantum Computation and Quantum Information* (Cambridge University Press, New York, 2000).

[4] C. K. Hong and L. Mandel, Phys. Rev. Lett. **56**, 58 (1986), see also a similar idea in C. Adlard, E. R. Pike and S. Sarkar, Phys. Rev. lett. **79**, 1585 (1997).

[5] H. Huang and J. H. Eberly, J. Mod. Opt. **40**, 915 (1993).

[6] C. K. Law, I. A. Walmsley and J. H. Eberly, Phys. Rev. Lett. **84**, 5304 (2000).

[7] H. H. Arnaut and G. Barbosa, Phys. Rev. Lett. **85**, 286 (2000).

[8] See other Lectures in this Institute by L. A. Lugiato and A. Gatti.

[9] A. Ekert and P. L. Knight, Am. J. Phys. **63**, 415 (1995), and see references therein.

[10] L. Mandel and E. Wolf, *Optical Coherence and Quantum Optics* (Cambridge University Press, New York, 1995).

[11] S. Parker, S. Bose and M. B. Plenio, Phys. Rev. A **61**, 032305 (2000).

[12] R. Grobe, K. Rzążewski and J. H. Eberly, J. Phys. B **27**, L503 (1994).

[13] W.-C. Liu and J. H. Eberly, Phys. Rev. Lett. **83**, 523 (1999).

[14] K. W. Chan, C. K. Law and J. H. Eberly, Phys. Rev. Lett. **88**, 100402 (2002).

[15] C. Kurtsiefer, *et al.*, Phys. Rev. A **55**, R2539 (1997).

[16] K. Rzążewski and W. Zakowicz, J. Phys. B **25**, L319 (1992).

[17] T. Pfau *et al.*, Phys. Rev. Lett., **73**, 1223 (1994).

[18] M. S. Chapman *et al.*, Phys. Rev. Lett., **75**, 3783 (1995).

QUANTUM METROLOGY AND QUANTUM INFORMATION PROCESSING WITH HYPER-ENTANGLED QUANTUM STATES

A. V. Sergienko

Quantum Imaging Laboratory, Department of Electrical and Computer Engineering,
Boston Univ. 8 Saint Mary's St., Boston MA 02215
Department of Physics, Boston Univ., 590 Commonwealth Ave., Boston MA 02215
alexserg@bu.edu

G. S. Jaeger

Quantum Imaging Laboratory, Department of Electrical and Computer Engineering,
Boston Univ. 8 Saint Mary's St., Boston MA 02215
jaeger@bu.edu

G. Di Giuseppe

Quantum Imaging Laboratory, Department of Electrical and Computer Engineering,
Boston Univ. 8 Saint Mary's St., Boston MA 02215
gdg@bu.edu

Bahaa E. A. Saleh

Quantum Imaging Laboratory, Department of Electrical and Computer Engineering,
Boston Univ. 8 Saint Mary's St., Boston MA 02215
saleh@bu.edu

Malvin C. Teich

Quantum Imaging Laboratory, Department of Electrical and Computer Engineering,
Boston Univ. 8 Saint Mary's St., Boston MA 02215
Department of Physics, Boston Univ., 590 Commonwealth Ave., Boston MA 02215
teich@bu.edu

A.S. Shumovsky and V.I. Rupasov (eds.), Quantum Communication and Information Technologies, 13–45.
© 2003 *Kluwer Academic Publishers.*

Abstract   A pair of photons generated in the nonlinear process of spontaneous parametric down conversion is, in general, entangled so as to contain strong energy, time, polarization, and momentum quantum correlations. This entanglement involving more than one pair of quantum variable, known as hyper-entanglement entanglement, serves as a powerful tool in fundamental studies of foundations of the quantum theory, in the development of novel information processing techniques, and in the construction of new quantum measurement technologies, such as quantum ellipsometry.

Keywords: Quantum information, quantum optics, quantum communication, entanglement, quantum technology

1.   Introduction

Entangled-photon states produced by spontaneous parametric down- conversion provide a natural basis for quantum measurement and quantum information processing, because they are not prone to decoherence decoherence and are composed of photons that remain higher correlated even after propagating to widely separated locations in space. The strong quantum correlations naturally present between down-conversion photons allow for uniquely quantum mechanical forms of measurement to be performed, which offer advantages over their classical counterparts. These states also allow quantum information to be encoded, and their robust coherence allows information to be processed in uniquely quantum mechanical ways. In order to realize the full potential of entangled-photon states, it is vital to understand and exploit all those features present in their quantum states from the point of their creation, during their propagation and until their detection. Here we consider such states of multiphotons produced by spontaneous parametric down-conversion (SPDC). The unique properties of quantum systems captured by these states are strikingly manifested when quantum intensity correlation interferometry is performed.

Quantum-interference patterns generally arise in contexts where only a single parameter, such as polarization polarization, is actively manipulated. However, it has been shown [1] that by making use of all the parameters naturally relevant one can, in fact, modify the interference pattern associated with one parameter such as polarization, by manipulating others, such as the frequency and transverse wave vector. This interdependence of physical parameters has its origin in the non-factorizability of the quantum state produced in the process of spontaneous parametric down-conversion (SPDC) into a product of functions of single parameters, such as polarization, frequency and wave-vector. For example, inconsistencies between existing theoretical models and the results of femtosecond down-conversion experiments commonly arise as a result of failing to consider the full Hilbert space occupied by the entangled quantum state, with its dependency on multiple variables. In particular, femtosecond SPDC models have ignored transverse wave-vector components and, as a result, have not accounted for the previously demonstrated angular spread of such down-converted light [2]. However, a comprehensive approach to quantum states, such as has been recently pursued, permits intelligent engineering of quantum states. This approach will be discussed in the following section.

In many practical applications, technology can benefit from the fact that, though each individual subsystem may actually possess inherent uncertainties, the components of the entangled pair may exhibit no such uncertainty relative to one another. One can exploit this unique aspect of entanglement entanglement for the development of a new class of optical measurements, those of quantum optical metrology. Entangled states entangled states have been used with great effectiveness during the last twenty years for carrying out striking experiments, for example those demonstrating non-local dispersion cancellation [3], entangled-photon-induced transparency [4], and entangled-photon spectroscopy with monochromatic light [5]. The practical availability of entangled beams has made it possible to conduct such fundamental physics experiments without having to resort to costly instruments such as particle colliders and synchrotrons. A new generation of techniques for quantum metrology and quantum information processing is under development, which will be discussed in detail in following sections.

Entangled photons first became of interest in probing the foundations of quantum theory. The often deeply counterintuitive predictions of quantum mechanics have been the focus of intensive discussions and debates among physicists since the introduction of the formal theory in the 1930's. Since then, entangled states of increasing quality have been prepared in order to progressively better differentiate quantum behavior from classical behavior. Entangled quantum systems are composed of at least two component subsystems and are described by states that cannot be written as a product of independent subsystem states,

$$|\Psi\rangle \neq |\psi_1\rangle \otimes |\psi_2\rangle \,, \tag{1}$$

for *any* two quantum states $|\psi_n\rangle$ of the individual subsystems. Schrödinger [6], who first defined entanglement entanglement, stated, called entanglement "the characteristic trait of quantum mechanics." In 1935, Einstein, Podolsky and Rosen Einstein-Podolsky-Rosen (EPR) [7] presented an influential argument based on analyzing entangled states entangled states of two systems described by infinite superpositions of such as

$$|\Psi(x_1, x_2)\rangle = \sum_{i=1}^{\infty} c_n(x_1, x_2)|\psi_n(x_1)\rangle|\phi_n(x_2)\rangle \,, \tag{2}$$

where the ψ_i and ϕ_i are elements of orthogonal state bases, that quantum mechanics is an incomplete theory of physical objects, as judged from the perspective of metaphysical realism.

This theoretical argument was later followed by experiments by Wu and Shaknov [8] on electron-positron singlets that amounted to practical tests of nonclassical behavior. Simpler entangled states entangled states such as these,

$$|\Psi\rangle = \frac{1}{\sqrt{2}}(|\uparrow\rangle|\downarrow\rangle - |\downarrow\rangle|\uparrow\rangle), \tag{3}$$

were discussed by David Bohm [9] and analyzed by Bohm and Aharonov [10]. The systematic study of quantum-scale behavior that could not be explained by the class of local, deterministic hidden variables theories then began, with a focus on such states of strongly entangled particle pairs. In 1964, John Bell derived a general inequality that introduced a clear empirical boundary between local, classically explicable behavior and less intuitive forms of behavior, involving a notion that he called "nonlocality" [11].

By the early 1980s, advances in laser laser physics and optics had allowed for an entirely new generation of experiments by the group of Aspect at Orsay, France based on the use of photon pairs produced by nonlinear laser excitations of an atomic radiative cascade [12]. These experiments paved the way for future quantum-interferometric experiments involving entangled photons, finally producing an unambiguous violation of a Bell-type inequality by tens of standard deviations and strong agreement with quantum-mechanical predictions. By that time, a more powerful source of entangled photons, the optical parametric oscillator (OPO), had also already been developed, independently of tests of basic principles of quantum mechanics. Indeed, OPOs were operational in major nonlinear optics research groups around the world almost immediately following the development of the laser laser (for more details, see [13]). Shortly after the first OPOs were introduced, a number of experimental groups independently discovered the spontaneous emission of polarized photons in an optical parametric amplifier amplifier. OPO spontaneous noise, which was very weak, occupied a very broad spectral range, from near the blue pump frequency through to the infrared absorption band. A corresponding spatial distribution of different frequencies followed the well known and simple phase-matching phase conditions of nonlinear optical systems.

The statistics of photons appearing in such spontaneous conversion of one photon into a pair had been analyzed in the 1960s, demonstrating the very strong correlations between these photons in space, time and frequency. Burnham and Weinberg [14] first demonstrated the unique and explicitly nonclassical features of states of two-photons generated in the spontaneous regime from the parametric amplifier amplifier. Quantum correlations involving photon pairs were exploited again 10 years later in experimental work by Malygin, Penin and Sergienko [15]. The use of highly correlated pairs of photons for the explicit demonstration of Bell inequality Bell inequality violations has become popular and convenient since the mid-1980s and has been widely exploited since then. The process of generating these states is now known as "spontaneous parametric down-conversion" (SPDC), and has become widely utilized. New, high-intensity sources of SPDC have been developed over the last two decades (see, for example, [16]).

Spontaneous parametric down-conversion of one photon into a pair is said to be of one of two types, based on the satisfaction of "phase-matching" conditions of either type I or of type II, corresponding to whether the two photons of the down-conversion pair have the same polarization or orthogonal polarizations, respectively. The two photons of a pair can also leave the down-converting medium either in the same direction or in different directions, referred to as the collinear and non-collinear cases, respectively. A medium is required for down-conversion, as conservation laws exclude the decay of one photon into a pair in vacuum. The medium is usually some sort of birefringent crystal, such as potassium dihydrogen phosphate (KDP), having a $\chi^{(2)}$ optical nonlinearity. Upon striking such a nonlinear crystal, there is a small probability (on the order of 10^{-7}) that an incident pump photon will be down-converted into a two-photon (see Fig. 1). If down-conversion occurs, these conserved quantities

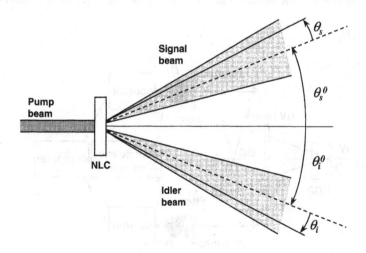

Figure 1.   Spontaneous parametric down-conversion [17].

are carried into that of the resulting photon pair under the constraints of their respective conservation laws, so that the phases phase of the corresponding wave-functions match, in accordance with the relations

$$\omega_s + \omega_i = \omega_p, \quad k_s + k_i = k_p, \tag{4}$$

referred to as the phase-matching conditions, where the k_i and ω_i are momenta and frequencies for the three waves involved. The individual photons resulting from down-conversion are often arbitrarily called "signal" (s) and "idler" (i), for historical reasons.

When the two photons of a pair have different momenta or energies, entanglement entanglement will arise in SPDC, provided that the alternatives are in principle experimentally indistinguishable. The two-photon state produced in type-I down-conversion can be written

$$|\Psi\rangle = \sum_{s,i} \delta(\omega_s + \omega_i - \omega_p)\delta(k_s + k_i - k_p)|k_s\rangle \otimes |k_i\rangle . \tag{5}$$

In this case, the two photons leave the nonlinear medium with the same polarization polarization, namely that orthogonal to the polarization of the pump beam. Down-conversion photons are thus produced in two thick spectral cones, one for each photon, within which two-photons appear each as a pair of photons on opposite sides of the pump-beam direction (see Fig. 1). In the mid-1980s,

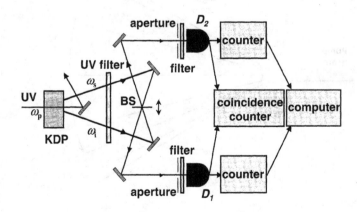

Figure 2.   Hong-Ou-Mandel interferometer [18].

Hong, Ou and Mandel [18] created noncollinear, type-I phase-matched SPDC

photon pairs in KDP crystal using an ultraviolet continuous-wave (cw) laser laser pump beam (see Fig. 2). This experiment empirically demonstrated the strong temporal correlation of the two-photons. Filters were placed in the apparatus, determining the frequency spread of the down-converted photons permitted to interfere. Since this experiment, the common approach to quantum interferometry has been to choose a single entangled parameter of interest and to eliminate the dependence of the quantum state on all other parameters. For example, when investigating polarization entanglement entanglement, strong spectral and spatial filtering filtering are typically imposed in an attempt to restrict attention to the polarization variable alone.

A more general approach to this problem is to exploit the multi-faceted nature of photon entanglement entanglement from the outset. In such an approach, the observed quantum-interference pattern in one parameter, such as polarization, can be modified by controlling the dependence of the state on the other parameters, such as frequency and transverse wave vector. A more complete theory of spontaneous parametric down-conversion allows us to understand the full character of fourth-order quantum interference in many valuable experiments. SPDC gives rise to a quantum state that is entangled in multiple parameters, such as three-dimensional wave-vector and polarization polarization . Many experiments designed to verify the non-factorizability of classes of quantum states, the mathematical essence of entanglement, are carried out in the context of models that fail to consider the overall relevant Hilbert space and are restricted to entanglement of only a single aspect of the quantum state, such as energy [19], momentum [20], or polarization [21]. Indeed, inconsistencies emerge in the analysis of quantum-interferometric experiments involving down-conversion under such circumstances, a fact that has been highlighted by the failure of the conventional theory of ultrafast parametric down-conversion to characterize quantum-interference experiments. This is because femtosecond SPDC models have ignored transverse wave-vector components and have thereby not accounted for the previously demonstrated angular spread of such down-converted light [22].

2.   Hyperentangled State Engineering

In order to better understand the quantum systems produced in spontaneous parametric down-conversion and the possibilities they provide, it is useful to analyze their behavior during each of the three distinct stages of their passage through any experimental apparatus, that is, the generation, propagation, and detection of the two-photon optical state. Begining with quantum state generation, one can see how to take advantage of the state by systematically controlling it during propagation and to extract useful information through the use of of a properly chosen detection scheme [23].

2.1   Generation

The relatively weak interaction between the three modes of the electromagnetic field participating in SPDC within the nonlinear crystal allows us to consider the two-photon state generated within the confines of first-order time-dependent perturbation theory. The two-photon state can be written

$$|\Psi^{(2)}\rangle \sim \frac{i}{\hbar} \int_{t_0}^{t} dt'\, \hat{H}_{\text{int}}(t')\, |0\rangle , \tag{6}$$

with the interaction Hamiltonian Hamiltonian

$$\hat{H}_{int}(t) \sim \chi^{(2)} \int_{V} d\mathbf{r}\, \hat{E}_p^{(+)}(\mathbf{r}, t)\hat{E}_o^{(-)}(\mathbf{r}, t)\hat{E}_e^{(-)}(\mathbf{r}, t) + H.c. , \tag{7}$$

where $\chi^{(2)}$ is the second-order susceptibility and V is the volume in which the interaction can take place. The field operator $\hat{E}_j^{(\pm)}(\mathbf{r}, t)$ represents the positive-(negative-) frequency portion of the electric-field operator E_j, with the subscript j representing the pump (p), ordinary (o), and extraordinary (e) waves at position $\mathbf{r}$ and time t, and H.c. stands for Hermitian conjugate. The high intensity of the pump field allows one to represent it by a classical c-number, with an arbitrary spatiotemporal profile given by

$$E_p(\mathbf{r}, t) = \int d\mathbf{k}_p\, \tilde{E}_p(\mathbf{k}_p)e^{i\mathbf{k}_p\cdot\mathbf{r}}e^{-i\omega_p(\mathbf{k}_p)t} , \tag{8}$$

where $\tilde{E}_p(\mathbf{k}_p)$ is the complex-amplitude of the field, as a function of the wave-vector $\mathbf{k}_p$. Decomposing the three-dimensional wavevector $\mathbf{k}_p$ into two-dimensional transverse wavevector $\mathbf{q}_p$ and a frequency part ω_p, this is

$$E_p(\mathbf{r}, t) = \int d\mathbf{q}_p\, d\omega_p\, \tilde{E}_p(\mathbf{q}_p; \omega_p)e^{i\kappa_p z}e^{i\mathbf{q}_p\cdot\mathbf{x}}e^{-i\omega_p t} , \tag{9}$$

where $\mathbf{x}$ spans the transverse plane perpendicular to the propagation direction z, as illustrated in Fig. 3. The ordinary and extraordinary modes of the field can similarly be expressed in terms of the creation operators $\hat{a}^\dagger(\mathbf{q}, \omega)$ as

$$\hat{E}_j^{(-)}(\mathbf{r}, t) = \int d\mathbf{q}_j\, d\omega_j\, e^{-i\kappa_j z}e^{-i\mathbf{q}_j\cdot\mathbf{x}}e^{i\omega_j t}\, \hat{a}_j^\dagger(\mathbf{q}_j, \omega_j) , \tag{10}$$

where $j = o, e$. The longitudinal component of momentum, $\mathbf{k}$, which we denote by κ, can be written in terms of the $(\mathbf{q}, \omega)$ pair [24] as $\kappa = \sqrt{[n(\omega, \theta)\,\omega/c]^2 - |\mathbf{q}|^2}$, where c is the speed of light in vacuum, θ is the angle between $\mathbf{k}$ and the optical

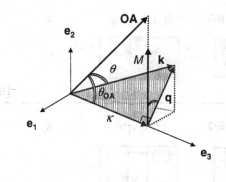

Figure 3. Decomposition of a three-dimensional wavevector (**k**) into longitudinal (κ) and transverse (**q**) components [23]. The angle between the optical axis of the nonlinear crystal (OA) and the wavevector **k** is θ. θ_{OA} is the angle between the optical axis and the longitudinal axis ($\mathbf{e}_3$). The spatial walk-off of the extraordinary polarization component of a field travelling through the nonlinear crystal is characterized by M.

axis of the nonlinear crystal (see Fig. 3), and $n(\omega, \theta)$ is the index of refraction in the nonlinear medium.

The quantum state after the nonlinear crystal is thus

$$|\Psi^{(2)}\rangle \sim \int d\mathbf{q}_o d\mathbf{q}_e \, d\omega_o d\omega_e \, \Phi(\mathbf{q}_o, \mathbf{q}_e; \omega_o, \omega_e) \hat{a}_o^\dagger(\mathbf{q}_o, \omega_o) \hat{a}_e^\dagger(\mathbf{q}_e, \omega_e)|0\rangle,$$

(11)

with

$$\Phi(\mathbf{q}_o, \mathbf{q}_e; \omega_o, \omega_e) = \tilde{E}_p(\mathbf{q}_o + \mathbf{q}_e; \omega_o + \omega_e) \, L \operatorname{sinc}\left(\frac{L\Delta}{2}\right) e^{-i\frac{L\Delta}{2}},$$

(12)

where L is the thickness of the crystal and $\Delta = \kappa_p - \kappa_o - \kappa_e$, κ_j ($j = p, o, e$) being related to the indices $(\mathbf{q}_j, \omega_j)$. Recall that the inextricable dependencies of the function $\Phi(\mathbf{q}_o, \mathbf{q}_e; \omega_o, \omega_e)$ on its several variables in Eqs. (11) and (12) is the essence of multi-parameter entanglement.

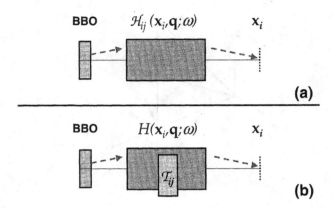

Figure 4.   (a) Illustration of the idealized setup for observation of quantum interference using SPDC [23]. BBO represents a beta-barium borate nonlinear optical crystal, $\mathcal{H}_{ij}(\mathbf{x}_i, \mathbf{q}; \omega)$ is the transfer function of the system, and the detection plane is represented by $\mathbf{x}_i$. (b) For most experimental configurations the transfer function can be factorized into diffraction-dependent $[H(\mathbf{x}_i, \mathbf{q}; \omega)]$ and diffraction-independent $(\mathcal{T}_{ij})$ components.

2.2   Propagation

Propagation of the down-converted light between the planes of generation and detection is described by the transfer function of the optical system. The photon-pair probability amplitude at the space-time points $(\mathbf{x}_A, t_A)$ and $(\mathbf{x}_B, t_B)$ of detection is defined by

$$A(\mathbf{x}_A, \mathbf{x}_B; t_A, t_B) = \langle 0|\hat{E}_A^{(+)}(\mathbf{x}_A, t_A)\hat{E}_B^{(+)}(\mathbf{x}_B, t_B)|\Psi^{(2)}\rangle \,, \qquad (13)$$

where the quantum operators at the detection locations are [25]

$$\hat{E}_A^{(+)}(\mathbf{x}_A, t_A) = \int d\mathbf{q}\, d\omega\, e^{-i\omega t_A}\, [\mathcal{H}_{Ae}(\mathbf{x}_A, \mathbf{q}; \omega)\hat{a}_e(\mathbf{q}, \omega)+$$
$$+ \mathcal{H}_{Ao}(\mathbf{x}_A, \mathbf{q}; \omega)\hat{a}_o(\mathbf{q}, \omega)] \,,$$

$$\hat{E}_B^{(+)}(\mathbf{x}_B, t_B) = \int d\mathbf{q}\, d\omega\, e^{-i\omega t_B}\, [\mathcal{H}_{Be}(\mathbf{x}_B, \mathbf{q}; \omega)\hat{a}_e(\mathbf{q}, \omega)+$$
$$+\mathcal{H}_{Bo}(\mathbf{x}_B, \mathbf{q}; \omega)\hat{a}_o(\mathbf{q}, \omega)] \,. \qquad (14)$$

The transfer function $\mathcal{H}_{ij}$ ($i = A, B$ and $j = e, o$) describes the propagation of a $(\mathbf{q}, \omega)$ mode from the nonlinear-crystal output plane to the detection plane.

The photon-pair probability amplitude is thus:

$$A(\mathbf{x}_A, \mathbf{x}_B; t_A, t_B) = \int d\mathbf{q}_o d\mathbf{q}_e \, d\omega_o d\omega_e \, \Phi(\mathbf{q}_o, \mathbf{q}_e; \omega_o, \omega_e)$$

$$\times \left[\mathcal{H}_{Ae}(\mathbf{x}_A, \mathbf{q}_e; \omega_e) \mathcal{H}_{Bo}(\mathbf{x}_B, \mathbf{q}_o; \omega_o) \, e^{-\mathrm{i}(\omega_e t_A + \omega_o t_B)} \right.$$

$$\left. + \mathcal{H}_{Ao}(\mathbf{x}_A, \mathbf{q}_o; \omega_o) \mathcal{H}_{Be}(\mathbf{x}_B, \mathbf{q}_e; \omega_e) \, e^{-\mathrm{i}(\omega_o t_A + \omega_e t_B)} \right] . \tag{15}$$

By appropriately choosing the optical system, the overall probability amplitude can be constructed as desired. The influence of the optical system on the photon-pair wave-function appears in the above expressions through the functions $\mathcal{H}_{Ae}$, $\mathcal{H}_{Ao}$, $\mathcal{H}_{Be}$, and $\mathcal{H}_{Bo}$.

2.3　Detection

The character of the detection process will depend on the nature of the apparatus. Slow detectors, for example, perform temporal integration while detectors of finite area perform spatial integration. One limit is that when the temporal response of a point detector is spread negligibly with respect to the characteristic time scale of SPDC, which is given by the inverse of down-conversion bandwidth. In this limit, the coincidence rate reduces to

$$R = |A(\mathbf{x}_A, \mathbf{x}_B; t_A, t_B)|^2 . \tag{16}$$

The other limit, typical of quantum-interference experiments, is reached as a result of the use of slow bucket detectors. In such a situation, the coincidence count rate R is given in terms of the photon-pair probability amplitude by

$$R = \int d\mathbf{x}_A d\mathbf{x}_B \, dt_A dt_B \, |A(\mathbf{x}_A, \mathbf{x}_B; t_A, t_B)|^2 . \tag{17}$$

2.4　Engineering basics

Having completed this formal analysis, one can begin to consider specific configurations of a quantum interferometer that might take practical advantage of the multiple parameter dependence of hyperentangled quantum states. This choice corresponds to a specific form of the the transfer function $\mathcal{H}_{ij}$. Almost all quantum-interference experiments performed to date have the common feature that $\mathcal{H}_{ij}$ in Eq. (14), with $i = A, B$ and $j = o, e$, can be separated into diffraction-dependent and polarization-dependent terms as

$$\mathcal{H}_{ij}(\mathbf{x}_i, \mathbf{q}; \omega) = \mathcal{T}_{ij} \, H(\mathbf{x}_i, \mathbf{q}; \omega), \tag{18}$$

where the diffraction-dependent terms are grouped in H and the diffraction-independent terms are grouped in $\mathcal{T}_{ij}$ (see Fig. 4). Free space, apertures, and lenses, for example, can be treated as diffraction-dependent elements while beam splitters and waveplates can be considered as diffraction-independent elements.

For example, in collinear SPDC configurations in the presence of a relative optical-path delay τ between the ordinary and the extraordinary polarized photons, $\mathcal{T}_{ij}$ is $\mathcal{T}_{ij} = (\mathbf{e}_i \cdot \mathbf{e}_j)\, e^{-i\omega\tau\delta_{ej}}$, where the δ_{ej} is the Kronecker delta, with $\delta_{ee} = 1$ and $\delta_{eo} = 0$, the unit vector $\mathbf{e}_i$ specifies the orientation of each polarization polarization analyzer in the experimental apparatus, and $\mathbf{e}_j$ is the unit vector that describes the polarization of each down-converted photon.

The general photon-pair probability amplitude given in Eq. (14), can be separated into diffraction-dependent and -independent elements:

$$A(\mathbf{x}_A, \mathbf{x}_B; t_A, t_B) = \int d\mathbf{q}_o d\mathbf{q}_e \, d\omega_o d\omega_e \, \Phi(\mathbf{q}_o, \mathbf{q}_e; \omega_o, \omega_e)$$

$$\times \left[\mathcal{T}_{Ae} H(\mathbf{x}_A, \mathbf{q}_e; \omega_e) \mathcal{T}_{Bo} H(\mathbf{x}_B, \mathbf{q}_o; \omega_o)\, e^{-i(\omega_e t_A + \omega_o t_B)} \right.$$

$$\left. + \mathcal{T}_{Ao} H(\mathbf{x}_A, \mathbf{q}_o; \omega_o) \mathcal{T}_{Be} H(\mathbf{x}_B, \mathbf{q}_e; \omega_e)\, e^{-i(\omega_o t_A + \omega_e t_B)} \right]. \tag{19}$$

Taking the angle between $\mathbf{e}_i$ and $\mathbf{e}_j$ to be $45°$, $\mathcal{T}_{ij}$ can be simplified by using $(\mathbf{e}_i \cdot \mathbf{e}_j) = \pm\frac{1}{\sqrt{2}}$ [21], and the photon-pair probability amplitude can be written

$$A(\mathbf{x}_A, \mathbf{x}_B; t_A, t_B) = \int d\mathbf{q}_o d\mathbf{q}_e \, d\omega_o d\omega_e \, \tilde{E}_p(\mathbf{q}_o + \mathbf{q}_e; \omega_o + \omega_e)\, L$$

$$\times \mathrm{sinc}\left(\frac{L\Delta}{2}\right) e^{-i\frac{L\Delta}{2}} e^{-i\omega_e\tau}$$

$$\times \left[H(\mathbf{x}_A, \mathbf{q}_e; \omega_e) H(\mathbf{x}_B, \mathbf{q}_o; \omega_o)\, e^{-i(\omega_e t_A + \omega_o t_B)} \right.$$

$$\left. - H(\mathbf{x}_A, \mathbf{q}_o; \omega_o) H(\mathbf{x}_B, \mathbf{q}_e; \omega_e)\, e^{-i(\omega_o t_A + \omega_e t_B)} \right]. \tag{20}$$

Substitution of Eq. (19) into Eq. (17) provides the coincidence-count rate, given an arbitrary pump profile and optical system.

The diffraction-dependent elements in most of these experimental arrangements are illustrated in Fig. 5(b). To describe this system via the function H, one needs to derive the impulse response function (the point-spread function) for these optical systems. A typical aperture diameter of $b = 1$ cm at a distance $d = 1$ m from the crystal output plane yields $b^4/4\lambda d^3 < 10^{-2}$ using $\lambda = 0.5$ μm, guaranteeing the validity of the Fresnel approximation. Without loss of generality, one can use a two-dimensional (one longitudinal, one transverse dimension) analysis of the impulse response function. The extension to three dimensions is then straightforward. The impulse response function of the full

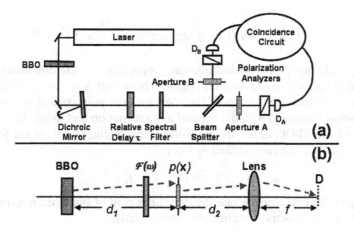

Figure 5.   (a) Schematic of the experimental setup for observation of quantum interference using type-II collinear SPDC [23]. (b) The path from the crystal output plane to the detector input plane. $\mathcal{F}(\omega)$ represents an (optional) filter transmission function, $p(\mathbf{x})$ is an aperture function, and f is the focal length of the lens.

optical system, from the crystal output plane to the detector input plane is

$$h(x_i, x; \omega) = \mathcal{F}(\omega)\, e^{i\frac{\omega}{c}(d_1+d_2+f)}\, e^{-i\frac{\omega x_i^2}{2cf}\left[\frac{d_2}{f}-1\right]} e^{i\frac{\omega x^2}{2cd_1}}$$
$$\int dx'\, p(x')e^{i\frac{\omega(x')^2}{2cd_1}}\, e^{-i\frac{\omega}{c}x'\left[\frac{x}{d_1}+\frac{x_i}{f}\right]}\,, \qquad (21)$$

which allows one to determine the transfer function of the system to be determined in terms of transverse wave-vectors via

$$H(\mathbf{x}_i, \mathbf{q}; \omega) = \int d\mathbf{x}\, h(\mathbf{x}_i, \mathbf{x}; \omega)\, e^{i\mathbf{q}\cdot\mathbf{x}}\,, \qquad (22)$$

so that the transfer function takes the explicit form

$$H(\mathbf{x}_i, \mathbf{q}; \omega) = \left[e^{i\frac{\omega}{c}(d_1+d_2+f)} e^{-i\frac{\omega|\mathbf{x}_i|^2}{2cf}\left[\frac{d_2}{f}-1\right]} e^{-i\frac{cd_1}{2\omega}|\mathbf{q}|^2}\, \tilde{P}\left(\frac{\omega}{cf}\mathbf{x}_i - \mathbf{q}\right) \right] \mathcal{F}(\omega),$$
$$(23)$$

where $\tilde{P}\left(\frac{\omega}{cf}\mathbf{x}_i - \mathbf{q}\right)$ is

$$\tilde{P}\left(\frac{\omega}{cf}\mathbf{x}_i - \mathbf{q}\right) = \int d\mathbf{x}'p(\mathbf{x}')e^{-i\frac{\omega\mathbf{x}'\cdot\mathbf{x}_i}{cf}}e^{i\mathbf{q}\cdot\mathbf{x}'} . \tag{24}$$

Using Eq. (23) one can now describe the propagation of the down-converted light from the crystal to the detection planes. Note that, since no birefringence is assumed, this transfer function is the same for both polarization modes (o,e).

Continuing the analysis in the Fresnel approximation and using the approximation that the SPDC fields are quasi-monochromatic, one finds the form of the coincidence-count rate defined in Eq. (17):

$$R(\tau) = R_0\left[1 - V(\tau)\right] , \tag{25}$$

where R_0 is the coincidence rate outside the region of quantum interference. In the absence of spectral filtering filtering, we have

$$V(\tau) = \Lambda\left(\frac{2\tau}{LD} - 1\right)\, \text{sinc}\left[\frac{\omega_p^0 L^2 M^2}{4cd_1}\frac{\tau}{LD}\Lambda\left(\frac{2\tau}{LD} - 1\right)\right]$$
$$\tilde{\mathcal{P}}_A\left(-\frac{\omega_p^0 LM}{4cd_1}\frac{2\tau}{LD}\mathbf{e}_2\right)\tilde{\mathcal{P}}_B\left(\frac{\omega_p^0 LM}{4cd_1}\frac{2\tau}{LD}\mathbf{e}_2\right) , \tag{26}$$

where $D = 1/u_o - 1/u_e$ with u_j denoting the group velocity for the j-polarized photon $(j = o, e)$, $M = \partial\ln n_e(\omega_p^0/2, \theta_{OA})/\partial\theta_e$, $\Lambda(x) = 1 - |x|$ for $-1 \leq x \leq 1$ and zero otherwise, and $\tilde{\mathcal{P}}_i$ (with $i = A, B$) is the normalized Fourier transform of the squared magnitude of the aperture function $p_i(\mathbf{x})$:

$$\tilde{\mathcal{P}}_i(\mathbf{q}) = \frac{\int\int d\mathbf{y}\, p_i(\mathbf{y})p_i^*(\mathbf{y})\, e^{-i\mathbf{y}\cdot\mathbf{q}}}{\int\int d\mathbf{y}\, p_i(\mathbf{y})p_i^*(\mathbf{y})} . \tag{27}$$

The profile of the function $\tilde{\mathcal{P}}_i$ within Eq. (25) plays a key role in our analysis.

Extremely small apertures are commonly used to reach the one-dimensional plane wave limit. However, the interest here is in just those effects that this excludes. As shown in Eq. (25), this gives $\tilde{\mathcal{P}}_i$ functions that are broad in comparison with Λ, so that Λ determines the shape of the quantum interference pattern, resulting in a symmetric triangular dip. The sinc function in Eq. (25) is approximately equal to unity for all practical experimental configurations, and therefore plays an insignificant role. On the other hand, this sinc function represents the difference between the familiar one-dimensional model (which predicts $R(\tau) = R_0\left[1 - \Lambda\left(\frac{2\tau}{LD} - 1\right)\right]$, a perfectly triangular interference dip) and a three-dimensional model in the presence of a very small on-axis aperture. It is clear from Eq. (25) that $V(\tau)$ can be altered dramatically by carefully selecting the aperture profile.

2.5   Realizations

These results allow one to go about quantum state engineering in a systematic fashion in various situations. For example, consider the effect on quantum interference of polarization of a circular aperture of diameter b. Both the aperture shape and size, via the function $\tilde{P}(\mathbf{q})$, will have significant effects on the polarization polarization quantum-interference patterns. Such an aperture can be described by the Bessel function J_1,

$$\tilde{\mathcal{P}}(\mathbf{q}) = 2\,\frac{J_1\,(b\,|\mathbf{q}|)}{b\,|\mathbf{q}|}\,. \tag{28}$$

Consider the experimental arrangement illustrated in Fig. 5(a). An experiment has been performed with SPDC with the pump being a single-mode cw argon-ion laser laser with a wavelength of 351.1 nm and a power of 200 mW delivered to a β-BaB$_2$O$_4$ (BBO) crystal with a thickness of 1.5 mm, aligned to produce collinear and degenerate Type-II spontaneous parametric down-conversion. Collinear down-conversion beams were sent through a delay line comprised of a z-cut crystalline quartz element (with its fast axis orthogonal to that of the BBO crystal) whose thickness was varied to control the relative optical-path delay between the photons of a down-converted pair. The characteristic thickness of the quartz element was much less than the distance between the crystal and the detection planes, yielding negligible effects on the spatial properties of the SPDC. The photon pairs were then sent to a non-polarizing beam splitter. Each arm of the polarization intensity interferometer following this beam splitter contained a Glan-Thompson polarization analyzer set to 45°, a convex lens to focus the incoming beam, and an actively quenched Peltier-cooled single-photon-counting avalanche photodiode detector [denoted D_A and D_B in Fig. 5(a)]. Note also that no spectral filtering filtering was used in the selection of the signal and idler photons for detection. The counts from the detectors were conveyed to a coincidence counting circuit with a 3-ns coincidence-time window.

For this situation involving a cw laser laser pump, the observed normalized coincidence rates correspond to the sort of quantum-interference pattern displayed in Fig. 6, as relative optical-path delay is varied, for various values of aperture diameter b placed 1 m from the crystal. The observed interference pattern is seen to be more strongly asymmetric for larger values of b. As the aperture becomes wider, the phase-matching condition between the pump and the generated down-conversion allows a greater range of $(\mathbf{q}, \omega)$ modes to be admitted. Those $(\mathbf{q}, \omega)$ modes that overlap less introduce more distinguishability and reduce the visibility of the quantum-interference pattern and introducing an asymmetric shape. The experimentally observed visibility for various aperture diameters, using the 1.5-mm thick BBO crystal employed in our experiments (symbols in Fig. 6), is consistent with this theory. When the pump field is

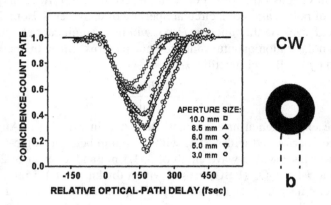

Figure 6.   Normalized coincidence-count rate $R(\tau)/R_0$, as a function of the relative optical-path delay τ, for different diameters of a circular aperture placed 1 m from the crystal [23]. The symbols represent experimental results and the solid curves are theoretical plots. The dashed curve represents the one-dimensional (1D) plane wave theory, which is seen clearly to be inadequate for large aperture diameters.

pulsed, there are additional limitations on the visibility that emerge as a result of the broad spectral bandwidth of the pump field [1, 26, 2, 27]. The asymmetry of the interference pattern for increasing crystal thickness is also more visible in the pulsed than in the cw regime.

Circular apertures are used after the down-conversion crystal in the majority of quantum-interference experiments involving relative optical-path delay. However, by again departing from convention, one can also use a vertical slit aperture to investigate the effect of and exploit *transverse* symmetry of the generated photon pairs. The solid curve of Fig. 7 shows the theoretical interference pattern expect for the vertical slit aperture. The data for the case of a horizontal slit shown by triangles is the normalized coincidence rate for a cw-pumped 1.5-mm BBO. Noting that the optical axis of the crystal falls along the vertical axis, these results verify that the dominant portion of distinguishability lies along the optical axis. The orthogonal axis (horizontal here) provides a negligible contribution to distinguishability, so that virtually full visibility is achievable, despite the wide aperture along the horizontal axis. The most dramatic effect observed is the symmetrization of the quantum-interference pattern and the recovery of the high visibility, despite the wide aperture along the horizontal axis.

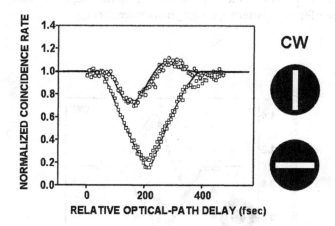

Figure 7. Normalized coincidence-count rate as a function of the relative optical-path delay for a 1 × 7-mm horizontal slit (circles) [23]. The data were obtained using a 351-nm cw pump and no spectral filters. Experimental results for a vertical slit are indicated by squares. Solid curves are the theoretical plots for the two orientations.

A practical benefit of such a slit aperture is that the count rate can be increased dramatically by limiting the range of transverse wave-vectors along the optical axis of the crystal, inducing indistinguishability and allowing a wider range along the orthogonal axis to increase the collection efficiency. A high count rate is required for many applications of entangled photon pairs. Researchers have generally suggested complex means of generating high-flux photon pairs [16], but we see that intelligent state engineering provides an elegant solution.

The optical elements in the situations above have been assumed to be placed concentrically about the longitudinal (z) axis. Under such circumstances, a single aperture before the beam splitter yields the same transfer function as two identical apertures placed in each arm after the beam splitter, as shown in Fig. 8. However, the quantum-interference pattern is also sensitive to a relative shift of the apertures in the transverse plane. To account for this, one must include an additional factor in Eq. (25):

$$\cos\left[\frac{\omega_p LM}{4cd_1}\frac{2\tau}{LD}\mathbf{e}_2\cdot(\mathbf{s}_A-\mathbf{s}_B)\right], \tag{29}$$

where s_i (with $i = A, B$) is the displacement of each aperture from the longitudinal (z) axis. This extra factor provides yet another degree of control on the quantum-interference pattern for a given aperture form.

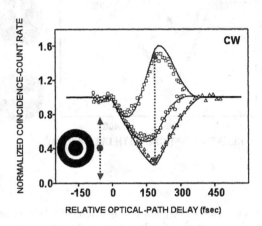

Figure 8.   Inversion of the central interference feature of polarization interference through the use of spatial parameters [23]. The normalized coincidence-count rate is shown as a function of the relative optical-path delay, for an annular aperture in one of the arms of the interferometer.

As in the case of the shifted slit, $V(\tau)$ becomes negative for certain values of the relative optical-path delay (τ), and the interference pattern displays a peak rather than a triangular dip, as shown in Fig. 8, something of significant practical value. One can thus use one physical parameters to effect a change in the interference pattern associated with another parameter. In this example, observations made at the center of the interference pattern show what would be a polarization polarization destructive interference minimum become an polarization interference maximum, when a spatial parameter (wave-vector) is altered.

The interference patterns generated in these experiments are seen to be influenced by the profiles of the apertures in the optical system, which admit wave vectors in specified directions. Including a finite bandwidth for the pump field strengthens this dependence on the aperture profiles, clarifying why the asymmetry was first observed in the ultrafast regime. The multi-parameter entangled nature of the two-photon state generated by SPDC allows transverse spatial effects to play a role in polarization-based quantum-interference experiments. In contrast to the usual single-direction polarization polarization entangled state, entangled states the wide-angle polarization-entangled state offers a richness that can be exploited in a variety of applications.

3.   Quantum Metrology and Quantum Information

There are several practical applications that become available when using entangled states of light. Such states have been used with great effectiveness during the last twenty years for innovative research out definitive experiments, such as realizing non-local dispersion cancellation, entangled-photon-induced transparency, and entangled-photon spectroscopy with monochromatic light. Though each individual subsystem of an entangled system exhibits inherent uncertainties, the elements of the entangled pair may exhibit no uncertainty relative to one another. For example, while the time of arrival of an individual particle may be totally random, an entangled pair following the same path always arrives simultaneously. One can exploit this unique aspect of entanglement entanglement for the development of a new class of optical measurements, those of quantum optical metrology. The availability of entangled beams has made it possible to conduct such experiments without having to resort to costly instruments such as particle colliders, and synchrotrons. Non-classical correlations between photons generated in SPDC are not diminished by separations, however large, between them during propagation, even when lying outside one another's the light cone. The interferometers developed for this purpose can exploit the persistence of entanglement over distances for the purposes of quantum communications, such as in quantum cryptography.

The non-local features of two-photon entangled states entangled states have opened up new realms of high-accuracy and absolute optical metrology. The research of the last two decades has produced several new technologies: 1) Quantum key distribution, more commonly known as quantum cryptography; 2) A method for absolute measurement of photodetector quantum efficiency that does not require the use of conventional standards of optical radiation, such as blackbody radiation; 3) A technique for determining polarization-mode-dispersion with attosecond precision; 4) A method of absolute ellipsometry which requires neither source nor detector calibration, nor a reference sample. While attention here is restricted to the above four techniques, a range of other new quantum techniques has also been pursued, including quantum imaging [28, 29], optical coherence tomography tomography [30] and quantum holography [31].

3.1   Quantum cryptography using entangled photon pairs

The currently most advanced form of quantum information experimentation is that taking place in quantum cryptography, in particular in quantum (cryptographic) key distribution (QKD). The security of QKD is not based on complexity, but on quantum mechanics, since it is generally not possible to measure an unknown quantum system without altering it and quantum states cannot be perfectly copied [32]. The basic QKD protocols are the BB84 scheme (Bennett and Brassard [33]) and the Ekert scheme [34]. BB84 uses single photons transmitted from sender (Alice) to receiver (Bob), which are prepared at random in four partly orthogonal polarization states: 0, 45, 90 and 135 degrees. When an eavesdropper, Eve, tries to obtain information about the polarization, she introduces observable bit errors, which Alice and Bob can detect by comparing a random subset of the generated keys. By contrast, the Ekert protocol uses entangled pairs and a Bell-type inequality to transmit quantum key bits. In the Ekert scheme, both Alice and Bob receive one particle of the entangled pair. They perform measurements along at least three different directions on each side, where measurements along parallel axes are used for key generation, and those along oblique angles are used for security verification.

Several innovative experiments have been made using entangled photon pairs to implement quantum cryptography, for example [35, 36, 37]. Quantum cryptography experiments have had two principal implementations: weak coherent state coherent state realizations of QKD and those using two-photons. The latter approach made use of the nonlocal character of polarization polarization Bell states Bell states generated by spontaneous parametric down-conversion. The strong correlation of photon pairs, entangled in both energy/time and momentum/space, eliminates the problem of excess photons faced by the coherent-state approach, where the exact number of photons actually injected is uncertain. In the entangled-photon technique, one of the pair of entangled photons is measured by the sender, confirming for the sender that the state is the appropriate one. It has thus become the favored experimental technique. The unique stability of a special two-photon polarization interferometer, initially designed to perform polarization mode dispersion measurements in optical materials, allows on to realization a method quantum key distribution that surpasses the performance of quantum cryptographic techniques using weak coherent states coherent state of light [35].

3.2   Absolute calibration of quantum efficiency of photon-counting detectors.

One of the important, and difficult, problems in optical measurement is the absolute calibration of optical radiation intensity and the measurement of the absolute value of quantum efficiency of photodetectors, especially when they operate at the single-photon level and in the infrared spectral range. A novel technique for the measurement of the absolute value of quantum efficiency of single-photon detectors in the $0.42\mu m$-wavelength region of the infrared spectrum with high precision has been developed [15]. This technique also does not exist in the realm of classical optics, since these required properties have their origin in vacuum fluctuations. They thus have a universal character that is everywhere present, allowing for a level of accuracy commensurate with that of a national metrology facility at every laboratory, astronomical observatory, and detector-manufacturing facility around the globe. This method also exploits the universal nature of the entangled super-correlation between entangled light quanta generated in spontaneous parametric down-conversion. As a result of the universal nature of vacuum fluctuations, this method does not require the use of an external optical standard.

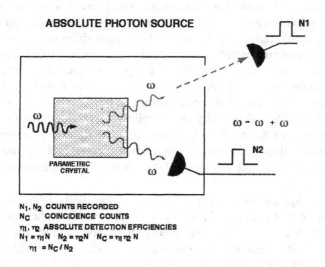

ABSOLUTE PHOTON SOURCE

N_1, N_2 COUNTS RECORDED
N_C COINCIDENCE COUNTS
η_1, η_2 ABSOLUTE DETECTION EFFICIENCIES
$N_1 = \eta_1 N$ $N_2 = \eta_2 N$ $N_C = \eta_1 \eta_2 N$
$\eta_1 = N_C / N_2$

Figure 9. Principle of absolute calibration of quantum efficiency without using standards.

Traditionally, two principal approaches for determining the absolute quantum efficiency have been used: 1) comparison with an optical signal which has well-known parameters (comprising different optical standards); and 2) measurement of an optical signal by using a preliminary calibrated photoelectric detector. The physical principle used for both the optical standard and photodetector calibration methods is the spectral distribution of the intensity of thermal optical radiation, which is characterized by the Planck blackbody radiation law. Unfortunately, these techniques are useful only for the measurement of intense optical signals. They cannot be used for the measurement of optical radiation at ultra-low levels, nor for the determination of the quantum efficiency of single-photon detectors such as those required in astronomy and spectroscopy. The principle of this technique is illustrated in Fig. 9. In addition to the number of pulses registered by detector 1 (N_1) and by detector 2 (N_2), one must detect the number of coincidence counts N_C. All down-conversion photons arrive only in pairs due to their temporal correlation. The number of single photons in the two arms will also be exactly the same due to momentum correlation between the photons of each pair, so that $N_1 = N_2 = N$, which is exactly the number of pairs $N_{pairs} = N$. As a result, the absolute value of quantum efficiency is simply determined by $\eta = N_C/N_2$. In case we have a photon-number resolving detector (such as has recently been developed at NIST-Boulder), the same calibration can be performed with a single detector.

The absolute value of quantum efficiency for the photon-counting photomultiplier is derived based on the distinction between its capability of distinguishing single-photon and double-photon events. This piece of information can be evaluated by measuring the pulse-height distribution. The photodetection process is usually characterized by the value of quantum efficiency, η, that can be used as a measure of successful conversion of optical quanta into macroscopic electric signal. If the average intensity of the photon flux, *i.e.* number of photons, arriving at the surface of photodetector in the unit of time is $< N >$, then the probability of successful photodetection will be $P_1 = \eta < N >$, while that of no detection will be defined by the complimentary value $P_0 = (1 - \eta) < N >$. The presence of SPDC radiation consisting of rigorously correlated photon pairs with continuous distribution in a broad spectral and angular range makes it possible to determine the spectral distribution of the measured quantities of photodetectors.

The average number of pairs $< N_{pairs} >$ per unit of time is clearly equal to the number of either signal $< N_s >$ or idler $< N_i >$ photons: $< N_s >=< N_i >=< N_{pairs} >$. From the theory of photodetection, the probability of having a double-photon event and a double-electron pulse will be $P_2 = \eta^2 < N_{pairs} >$. The probability of observing a single-photon detection event and a single-electron pulse will apparently involve the loss of one photon in the pair. Since this can happen in two different ways for every pair, the total probability of having a single-photon detection will be $P_1 = 2\eta(1 - \eta) < N_{pairs} >$. One can then conclude immediately that the value of quantum efficiency can be evaluated using the following formula:

$$\eta = (1 + P_1/2P_2) - 1. \tag{30}$$

However, the gain fluctuation and thermal noise in real photodetectors usually result in a very broad pulse-height distribution corresponding to single- and double-photon detection events. This has stimulated the development of a more realistic version of this technique that would be efficient, robust, and insensitive to such imperfections in real photon-counting detectors. In order to eliminate the influence of the broad pulse-height distribution, one can use a simple comparison between the numbers of registered detection events counted: when a photodetector is exposed to a pairs of entangled photons and when exposed to a signal (or idler) photon only. The total probability of detecting an electrical pulse when pairs of entangled photons arrive at the photocathode will consist of sum of probabilities P_1 and P_2:

$$P_{pair} = P_1 + P_2 = 2\eta(1 - \eta) < N_{pairs} > +\eta^2 < N_{pairs} > . \qquad (31)$$

The probability of detecting a photoelectric pulse in the case of exposure to signal (or idler) photons only will be

$$P_{single} = \eta < N_s >= \eta < N_i >= \eta < N_{pairs} > . \qquad (32)$$

The absolute value of quantum efficiency can thus be evaluated based on the results of these two measurements:

$$\eta = 2 - (P_{pair}/P_{single}) \qquad (33)$$

3.3   Quantum Metrology: measurement of polarization mode dispersion with attosecond resolution.

Conventional ellipsometry techniques were developed over the years to a very high degree of performance, and are used every day in many research and industrial applications. Traditional, non-polarization-based techniques for the measurement of optical delay usually make use of monochromatic light. The introduction of an optical sample in the one arm of the interferometer causes a sudden shift of interference pattern (sometimes over tens or hundreds of wavelengths) that is proportional to the absolute value of the optical delay. This approach requires one to keep track of the total number of shifted interference fringes in order to evaluate the absolute value of the optical delay. The accuracy of this approach is limited by the stability of the interferometer, the signal-to-noise level of the detector, and the wavelength of the monochromatic radiation used. Conventional polarization polarization interferometers used in ellipsometry measurements provide very high resolution, but have a similar problem of tracking the absolute number of 2π shifts of optical phase phase during the polarization-mode-dispersion measurement.

Optical engineers have come up with several ways to get around this problem, using additional complex measurement procedures. Since use of monochromatic classical polarized light does not allow one to measure the relative delay between two orthogonal waves in a single measurement, several measurements at different frequencies are used to reconstruct the polarization dispersion properties of materials. The use of highly monochromatic laser laser sources creates the additional problem of multiple reflections and strong irregular optical interference, especially in studying surface effects. Ellipsometry with low-coherence sources (white light) has received attention as a convenient method for the evaluation of dispersions in optical materials, particularly of communication fibers. While the technique provides the high timing resolution, along with the absolute nature of the optical delay measurement, its suffers from the problem of low visibility and instability of the interference pattern.

The unique double entanglement of the two-photon state in type-II phase-matched SPDC provides us with ultimate control of the relative position of photon pairs in space-time. The study of polarization entanglement entanglement and of the natural rectangular shape of the two-photon wave function in space-time in type-II phase-matched SPDC allows one to measure propagation time delay in optical materials with sub-femtosecond resolution [23]. This technique intrinsically provides an absolute value of polarization optical delay that is not limited by the usual value relative to one wave cycle of light. The probe light does not disturb the physical conditions of the sample under test, and can be used continuously during the growth and assembly processes to monitor major optical parameters of the device *in situ*. By manipulating the optical

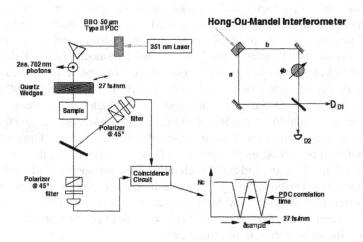

Figure 10.   Schematic of a two-photon polarization interferometer.

delay between the orthogonally polarized photons, a V-shaped correlation function feature is realized by a coincidence photon counting measurement. The general principle and a schematic experimental setup is illustrated in Fig. 10. The sharp V in the intensity correlation function can be made as narrow as 5-10 femtosecond wide. The introduction of any additional sample of optical material or photonic device with different group velocities for o-rays, (u_o,) than for e-rays, (u_e,) in the optical path before the beamsplitter will shift the V-shape distribution on a sub-femtosecond time scale. The shift is proportional to the optical delay in the sample of the length L:

$$d = (1/u_o - 1/u_e)L \approx (n_o - n_e)L/c. \tag{34}$$

In our realization, a 351nm Ar+ laser laser pumps the BBO crystal in a collinear and frequency-degenerate configuration. Orthogonally polarized photons generated in the BBO nonlinear crystal enter two spatially separated arms via a polarization-insensitive 50-50 beam-splitter (BS), so both ordinary and extraordinary polarized photons have equal probability of being reflected and transmitted. The two analyzers (oriented at 45 degrees) in front of each photon-counting detector D1 (D2) complete the creation of what are, in effect, two spatially separated polarization interferometers for the originally X (Y)-oriented signal and idler photons. Signal correlation is registered by coincidence events between detectors D1 and D2, as a function of a variable polarization delay (PD) in the interferometer. Spontaneous parametric down-conversion in a BBO nonlinear crystal with $L = 0.05$ mm to 1 mm generates signal and idler photons with coherence lengths of tens to hundreds of femtoseconds. Such an apparatus is illustrated in Fig. 11a.

A very useful new feature realized in such an apparatus is due to the non-symmetric manipulation of the relative optical delay t between ordinary and extraordinary photons in only one of the two spatially separated interferometers. As a result, the observed coincidence probability interferogram has its triangular envelope now filled with an almost 100 percent modulation, which is associated with the period of pump radiation. The additional introduction of a sample of optical or photonic material with different o-ray and e-ray group velocities in the optical path before the beam-splitter shifts the interference pattern proportional to $\tau_{sample} = d/c$, the difference in propagation times of the two polarizations.

This allows one to measure directly the absolute value of total optical delay between two orthogonally polarized waves in the sample on a very fine, sub-femtosecond time scale. The experimental result of the measurement of intensity correlations (coincidence probability) as a function of relative polarization delay d is illustrated in Fig. 11b. The SPDC signal is delivered to the detectors without the use of any limiting spectral filters. The full width at half-maximum (FWHM) of the correlation function envelope is defined by

$$\delta = (1/u_o - 1/u_e)L_{crystal}. \tag{35}$$

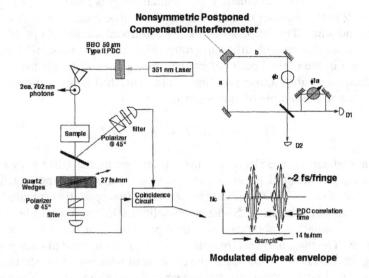

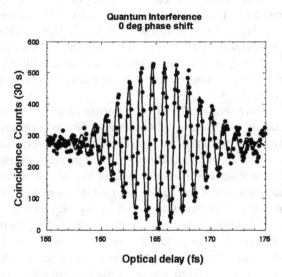

Figure 11.   a) Schematic of a two-photon polarization interferometer with a postponed optical delay. This non-symmetric delay is introduced only in one arm after the beam splitter. b) Measurement of intensity correlations as a function of relative polarization delay *d*.

The high visibility of the interference pattern and the extremely high stability of the polarization interferometer in such a collinear configuration allows one

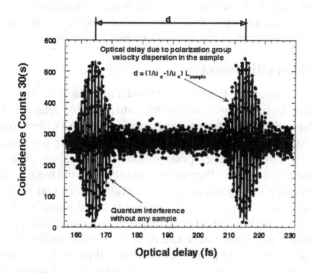

Figure 12. Measurement of the optical delay in the crystal quartz sample using a 50 μm nonlinear crystal. The horizontal scale is the time delay of the delay line located after the beamsplitter, BS.

to identify the absolute shift of the wide envelope with an accuracy defined by the fringe size of an internal modulation [39].

The observed quantum interference is very high contrast, approximately 90 percent. Resolution is further enhanced by reducing the total width of the envelope. This can be done by widening the phase matching spectrum by reducing the crystal length to 50 μm. This arrangement was used to measure the optical delay of a crystal quartz sample introduced into the optical path before the beam-splitter BS. The result of this measurement (performed with the 50μm nonlinear crystal) is illustrated in Fig. 12. The 25 fs width of the envelope enables one to clearly identify the central fringe position. Based on the signal-to-noise ratio, one can expect to resolve at least 1/100 of a fringe about (10^{-17} s).

This technique for the linear polarization dispersion measurement is easy convertible to the case of circular polarization. All advantages of using quantum correlation remain. This technique can be easily modified to study optical interactions at the surfaces of materials. One can use a reflection configuration, rather than the transmission mode, and to take advantage of the strong polarization dependence of evanescent waves. This approach is a uniquely sensitive tool for the analysis of the orientation, structure, morphology, and optical properties of single and multiple layers of atoms either grown or deposited on a substrate. Furthermore, if the technique can be demonstrated to be sensitive to the chemical identity of adsorbed molecules and atoms, then one can explore the application of the method to the field of chemical sensors.

3.4   Quantum ellipsometry

In an ideal ellipsometer, the light emitted from a reliable optical source is directed into an unknown optical system and then into a reliable detector. The user keeps track of the emitted and detected radiation in order to infer information about the optical system. This device can perform ellipsometry if the source can be made to emit light of any chosen polarization. The sample is characterized by the two parameters, ψ and Δ. Parameter ψ is related to the magnitude of the ratio of the polarization (complex) reflection coefficients, $\tilde{r}_1$ and $\tilde{r}_2$, for polarization eigenstates, with $\tan \psi = |\tilde{r}_1/\tilde{r}_2|$, while Δ is the phase shift between them. Because of the high accuracy required in measuring these parameters, an ideal ellipsometric measurement would require absolute calibration of both the source and the detector. In practice, however, ellipsometry makes use of an array of techniques for accomodating imperfect implementations, the most common techniques being null and interferometric ellipsometry.

In a null ellipsometer, the sample is illuminated with a beam of light that can be prepared in any state of polarization. The reflected light is then analyzed. The polarization of the incident beam is adjusted to compensate for the change in the relative amplitude and phase phase between the polarization eigen-states introduced by the sample. The resulting reflected beam is linearly polarized. If passed through an orthogonal linear polarizer, this beam will yield a null measurement at the optical detector. This method does not require a calibrated detector, since it does not measure intensity but instead records a null result. However, it does require a reference sample to calibrate the null, for example, to find the rotational axis of reference at which an initial null is obtained, to be compared with that after inserting the sample into the apparatus. The accuracy and reliability of all measurements thus depend on our knowledge of this reference sample. Another classical technique is to employ an interferometric configuration, usually created by beam splitters. The sample is placed in one of the two interferometric paths. One can then estimate the efficiency of the detector given the source intensity, by performing measurements when the sample is removed from the interferometer. This configuration thus alleviates the problem of an unreliable detector. However, this method depends on the reliability of the source and suffers from the drawback of requiring several optical components (beam splitters, mirrors, and so on), and so depends on the parameters of these as well.

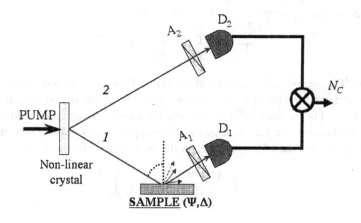

Figure 13.   Entangled twin-photon ellipsometer [40].

The quantum technique again takes advantage of the characteristics of quantum states produced in type-II-phase-matched SPDC [40]. Recall that type-II phase-matching requires the signal and idler photons to have orthogonal polarizations, one extraordinary and the other ordinary. These two photons emerge from the NLC with a relative time delay, due to the birefringence of the NLC. As has been discussed above, sending the pair through an appropriate birefringent element can compensate for this time delay. This temporal compensation allows one to extract ψ and Δ. The NLC of the apparatus is adjusted to produce SPDC in a type II non-collinear configuration, as illustrated in Fig. 13.

The fields at the detectors can be found as follows. Let $\hat{a}_s(\omega)$ and $\hat{a}_i(\omega')$ be bosonic annihilation operators are the annihilation operators for the signal-frequency mode ω and idler frequency mode ω'. The twin-photon Jones vector of the field following the beam splitter is then

$$\hat{\mathbf{J}}_1 = \begin{pmatrix} j[-\hat{A}_s(\omega) + \hat{A}_i(\omega')] \\ \hat{A}_s(\omega) + \hat{A}_i(\omega') \end{pmatrix} \tag{36}$$

where $\hat{A}_s(\omega) \begin{pmatrix} 1 \\ 0 \end{pmatrix} \hat{A}_i(\omega') \begin{pmatrix} 0 \\ 1 \end{pmatrix}$. The vectors $\begin{pmatrix} 1 \\ 0 \end{pmatrix}$ (horizontal) and $\begin{pmatrix} 0 \\ 1 \end{pmatrix}$ (vertical) are the Jones vectors representing orthogonal polarizations.

The first element in $\hat{\mathbf{J}}_1$ is the annihilation operator of the field in beam 1, while the second element is the annihilation operator of the field in beam 2.

One can define a *photon-pair* Jones matrix by

$$\mathbf{T} = \left(\begin{array}{cc} \mathbf{T}_{11} & \mathbf{T}_{12} \\ \mathbf{T}_{21} & \mathbf{T}_{22} \end{array} \right) \tag{37}$$

where the $\mathbf{T}_{kl}$ are the standard (single-photon) Jones matrices describing the action of a deterministic optical element, in terms of input and output beam pairings $\mathbf{k, l}$. The indices refer to the spatial modes of the input and output beams. The beams 1 and 2 impinge on the two polarization analyzers A_1 and A_2 directly, in absence of the sample, so the photon-pair Jones matrix is

$$\mathbf{T_p} = \left(\begin{array}{cc} \mathbf{P}(-\theta_1) & 0 \\ 0 & \mathbf{P}(\theta_2) \end{array} \right) , \tag{38}$$

where

$$\mathbf{P}(-\theta_1) = \left(\begin{array}{cc} \cos^2 \theta & \cos \theta \sin \theta \cos \theta \\ \cos \theta \sin \theta & \sin^2 \theta \end{array} \right) , \tag{39}$$

and θ_1 and θ_2 are the angles of the axes of the analyzers with respect to the horizontal direction.

The twin-photon Jones vector following the analyzers is therefore

$$\hat{\mathbf{J}}_2 = \mathbf{T_p}\hat{\mathbf{J}}_1 = \left(\begin{array}{c} j\mathbf{P}(-\theta_1)[-\hat{A}_s(\omega) + \hat{A}_i(\omega')] \\ \mathbf{P}(-\theta_2)\hat{A}_s(\omega) + \hat{A}_i(\omega') \end{array} \right) \tag{40}$$

which yields expressions for the fields at the detectors. The positive frequency components of the field at detectors D_1 and D_2, $\mathbf{E}_1^+$ and $\mathbf{E}_2^+$ respectively, are given by

$$\mathbf{E}_1^+ = j\left[-\cos \theta_1 \int d\omega \exp(-j\omega t)\hat{a}_s(\omega) + \sin \theta_1 \right.$$
$$\left. \int d\omega' \exp(-j\omega' t)\hat{a}_i(\omega') \right] \left(\begin{array}{c} \cos \theta_1 \\ \sin \theta_1 \end{array} \right) \tag{41}$$

$$\mathbf{E}_2^+ = \left[-\cos \theta_2 \int d\omega \exp(-j\omega t)\hat{a}_s(\omega) + \sin \theta_2 \right.$$
$$\left. \int d\omega' \exp(-j\omega' t)\hat{a}_i(\omega') \right] \left(\begin{array}{c} \sin \theta_2 \\ \cos \theta_2 \end{array} \right) , \tag{42}$$

while the negative frequency components are given by their Hermitian conjugates. Finally, with the sample present in beam one, there is an additional effect on the Jones vector before the polarization analyzer, described by the transformation $\mathbf{T_s}$, so that we have the Jones vector $\hat{\mathbf{J}}_3 = \mathbf{T_p}\mathbf{T_s}\hat{\mathbf{J}}_1$ where

$$\mathbf{T_s} = \begin{pmatrix} \mathbf{R} & 0 \\ 0 & \mathbf{I} \end{pmatrix}, \tag{43}$$

and

$$\mathbf{R} = \begin{pmatrix} \tilde{r}_1 & 0 \\ 0 & \tilde{r}_2 \end{pmatrix}. \tag{44}$$

In this case the field $\mathbf{E}_2^+$ is as before and $\mathbf{E}_1^+$ becomes

$$\mathbf{E}_1^+ = j\left[-\tilde{r}_1 \cos\theta_1 \int d\omega \exp(-j\omega t)\hat{a}_s(\omega) + \tilde{r}_2 \sin\theta_1 \right.$$
$$\left. \int d\omega' \exp(-j\omega't)\hat{a}_i(\omega') \right] \begin{pmatrix} \cos\theta_1 \\ \sin\theta_1 \end{pmatrix}. \tag{45}$$

The coincidence rate $N_c \propto \sin^2(\theta_1 - \theta_2)$ is therefore

$$R = C[\tan\psi \cos^2\theta_1 \sin^2\theta_2 + \sin^2\theta_1 \cos^2\theta_2$$
$$+2\sqrt{\tan\psi} \cos\Delta \cos\theta_1 \cos\theta_2 \sin\theta_1 \sin\theta_2]. \tag{46}$$

One can then obtain C, ψ and Δ with a minimum of three measurements with different analyzer settings $\theta_2 = 0°$, $\theta_2 = 90°$ and $\theta_2 = 45°$ with θ_1 at any fixed position other than $0°$ and $90°$.

An advantage of this setup over its idealized null ellipsometric competitor is that the two arms of the ellipsometer are separate, and the light beams traverse them independently in different directions. This allows various instrumentation errors of the classical setup to be avoided. In the present case, no optical components are placed between the source (NLC) and the sample; any desired polarization manipulation may be performed in the other arm of the entangled twin-photon ellipsometer. Entangled twin-photon ellipsometry is thus self-referencing, eliminating the necessity of constructing an interferometer altogether.

4. Conclusion

Entangled-photon states provide a natural basis for quantum measurement and quantum information processing due to their strong correlations, even when widely separated in space, in particular entangled-photon pairs created by spontaneous parametric down-conversion (SPDC). A comprehensive approach to these states has permits one to perform intelligent engineering of quantum states. The interdependence of physical parameters due to entanglement requires us to consider the full Hilbert space occupied by the entangled quantum state with its dependency on multiple variables. By exploiting this interdependence, one can modify the interference pattern associated with one parameter by manipulating other parameters.

Unique new forms of measurement can be performed using entangled quantum states. These states also allow quantum information to be encoded and this quantum information securely transmitted to a remote location, in a manner superior to other proposed quantum-mechanical methods. Practical use can thus be made of several of the correlation features present in entangled states. The unique properties of quantum systems were exploited in the practical quantum techniques that have been developed based on multiphoton interferometry.

References

[1] M., Atatüre, G. Di Giuseppe, M. D. Shaw, A. V. Sergienko, B. E. A. Saleh, and M. C. Teich, Phys. Rev. A **65**, 023808, 2002.

[2] G. Di Giuseppe, L. Haiberger, F. De Martini, and A. V. Sergienko, Phys. Rev. A **56**, R21, 1997; T. E. Keller and M. H. Rubin, Phys. Rev. A **56**, 1534, 1997; W. P. Grice, R. Erdmann, I. A. Walmsley, and D. Branning, Phys. Rev. A **57**, R2289, 1998.

[3] P. G. Kwiat, and R. Y. Chiao, Phys. Rev. A **45**, 6659 1992.

[4] H.-B. Fei, B. M. Jost, S. Popescu, B. E. A. Saleh, and M. C. Teich, Phys. Rev. Lett. **78**, 1679, 1997.

[5] B. E. A. Saleh, B. M. Jost, H.-B. Fei, and M. C. Teich, Phys. Rev. A **57**, 3972, 1998.

[6] E. Schrödinger, Naturwissenschaften **23**, 807, 1935; **23**, 823 (1935); **23**, 844, 1935 [Translation: J. D. Trimmer, Proc. Am. Phil. Soc. **124**, 323, 1980; reprinted in *Quantum Theory and Measurement*, edited by J. A. Wheeler and W. H. Zurek (Princeton University Press, Princeton, 1983)].

[7] A. Einstein, B. Podolsky and N. Rosen, Phys. Rev. **47**, 777, 1935.

[8] C.S. Wu, and I. Shaknov, The angular correlation of scattered annihilation radiation, Phys. Rev. Lett. **77**, 136, 1950 .

[9] D. Bohm, *Quantum Theory* (Prentice Hall, Englewood Cliffs, NJ, 1951) p. 614.

[10] D. Bohm, and Y. Aharonov, Phys. Rev. **108**, 1070, 1957.

[11] J. S. Bell, Physics **1**, 195, 1964.

[12] A. Aspect, J. Dalibard and G. Roger, Phys. Rev. Lett. **49**, 1804, 1982; A. Aspect, P. Grangier and G. Roger, Phys. Rev. Lett. **47**, 460, 1981.

[13] N. Bloembergen, Nonlinear optics and spectroscopy, Rev. Mod. Phys. **54**, 685, 1982.

[14] D. C. Burnham, and D. L. Weinberg, Phys. Rev. Lett. **25**, 84, 1970.

[15] A. A. Malygin, A. N. Penin and A. V. Sergienko, Sov. Phys. JETP Lett. **33**, 477, 1981; A. A. Malygin, A. N. Penin and A. V. Sergienko, Sov. Phys. Dokl. **30**, 227, 1981.

[16] P. G. Kwiat, E. Waks, A.J. White, I. Appelbaum and P.H. Eberhard, Phys. Rev. A **60**, R773, 1999.

[17] B. E. A. Saleh, A. Joobeur, and M. C. Teich, Phys. Rev. A **57**, 3991, 1998.

[18] C. K. Hong, Z. Y. Ou and L. Mandel, Phys. Rev. Lett. **59**, 2044, 1987.

[19] C. K. Hong, Z. Y. Ou, and L. Mandel, Phys. Rev. Lett. **59**, 2044, 1987); P. G. Kwiat, A. M. Steinberg, and R. Y. Chiao, Phys. Rev. A **47**, R2472, 1993.

[20] J. G. Rarity and P. R. Tapster, Phys. Rev. Lett. **64**, 2495, 1990.

[21] Z. Y. Ou and L. Mandel, Phys. Rev. Lett. **61**, 50, 1988; Y. H. Shih and C. O. Alley, Phys. Rev. Lett. **61**, 2921, 1988; Y. H. Shih and A. V. Sergienko, Phys. Lett. A **191**, 201, 1994; P. G. Kwiat, K. Mattle, H. Weinfurter, A. Zeilinger, A. V. Sergienko, and Y. H. Shih, Phys. Rev. Lett. **75**, 4337, 1995.

[22] A. Joobeur, B. E. A. Saleh, and M. C. Teich, Phys. Rev. A **50**, 3349, 1994; C. H. Monken, P. H. Souto Ribeiro, and S. Pádua, Phys. Rev. A, **57**, 3123, 1998.

[23] M. Atatüre, G. Di Giuseppe, M. D. Shaw, A. V. Sergienko, B. E. A. Saleh, and M. C. Teich, Phys. Rev. A **66**, 023822, 2002 .

[24] T. B. Pittman, D. V. Strekalov, D. N. Klyshko, M. H. Rubin, A. V. Sergienko, and Y. H. Shih, Phys. Rev. A **53**, 2804, 1996.

[25] B. E. A. Saleh, A. F. Abouraddy, A. V. Sergienko, and M. C. Teich, Phys. Rev. A **62**, 043816, 2000.

[26] M. Atatüre, A. V. Sergienko, B. M. Jost, B. E. A. Saleh, and M. C. Teich, Phys. Rev. Lett. **83**, 1323, 1999.

[27] J. Peřina Jr., A. V. Sergienko, B. M. Jost, B. E. A. Saleh, and M. C. Teich, Phys. Rev. A **59**, 2359, 1999.

[28] A. F. Abouraddy, B. E. A. Saleh, A. V. Sergienko, and M. C. Teich Phys. Rev. Lett. **87**, 123602, 2001.

[29] A. F. Abouraddy, B. E. A. Saleh, A. V. Sergienko, and M. C. Teich, J. Opt. Soc. Am. B, **19**, 1174, 2002.

[30] A. F. Abouraddy, M. B. Nasr, B. E. A. Saleh, A. V. Sergienko, and M. C. Teich, Phys. Rev. A **65**, 053817, 2002.

[31] A. F. Abouraddy, B. E. A. Saleh, A. V. Sergienko, and M. C. Teich, Optics Express, **9**, pp.498, 2001.

[32] W. K. Wootters and W. H. Zurek, Nature **299**, 802, 1982.

[33] C. H. Bennett and G. Brassard , Proceedings of the International Conference on Computer Systems and Signal Processing, Bangalore, 1984, p. 175.

[34] A. K. Ekert, Phys. Rev. Lett., **67**, 661, 1991.

[35] A.V. Sergienko, M. Atat'ure, Z. Walton, G. Jaeger, B.E.A. Saleh and M.C. Teich, Phys. Rev. A **60**, R2622, 1999.

[36] T. Jennewein, C. Simon, G. Weihs, H. Weinfurter and A. Zeilinger, Phys. Rev. Lett. **84**, 4729, 2000.

[37] W. Tittel, J. Brendel, H. Zbinden and N. Gisin, Phys. Rev. Lett. **84**, 4737, 2000.

[38] E. Dauler, G. Jaeger, A. Muller, A. Migdall and A. V. Sergienko, J. Res. NIST, **104**, 1, 1999.

[39] D. Branning, A. L. Migdall, A. V. Sergienko, Phys. Rev. A, **62**, 063808, 2000.

[40] A. F. Abouraddy, K. C. Toussaint, Jr., A. V. Sergienko, B. E. A. Saleh, and M. C. Teich, J. Opt. Sci. Am. B **19** 656, 2002.

OPTIMAL MANIPULATIONS WITH QUANTUM INFORMATION: UNIVERSAL QUANTUM MACHINES

Vladimír Bužek

Department of Mathematical Physics, National University of Ireland, Maynooth, Co. Kildare, Ireland

Research Center for Quantum Information, Slovak Academy of Sciences, Dúbravská cesta 9, Bratislava, Slovakia

1.   Introduction

One of the cornerstones of our present understanding of the Nature is quantum physics. This theory has "nonlocal" characteristics [1-3], due to quantum entanglement - let us just mention the EPR Einstein-Podolsky-Rosen paradox [4] and the violation of the Bell's inequalities [5, 6] which have been experimentally confirmed in several experiments, see e.g. Refs. [7, 8] (more on quantum-mechanical non-locality non-locality see also Refs.[9-18].

The most remarkable property of the *non-relativistic* quantum mechanics (which is inherently *nonlocal*) is that it peacefully coexists with the special theory of relativity in a sense that one cannot exploit quantum-mechanical entanglement between two space-like separated parties for communication of classical messages faster than light [19-28].

It turns out that quantum correlations first discussed in their seminal paper by Einstein, Podolsky, and Rosen [4] result in the measured probabilities which satisfy the *causal communication constraint* [18] [1]. This means that the probability of a particular measurement outcome on any one part of the system should be independent of measurement performed on the other parts. This requirement should guarantee the absence of faster-than-light signals [19] that is usually called as the *no-signaling*. [2].

This fundamental feature of quantum theory, that is that quantum-mechanical correlations (entanglement) entanglement cannot be used for superluminal communications has been challenged in 1982 by Nick Herbert [36] in his proposal

---

[1]This "signal locality" [25, 29], is also referred to in the literature as the "simple locality" [30, 31], the "parameter independence" [32], or "physical locality" [33].

[2]Probably I should note here that the signal locality can be formulated independently of quantum theory [25, 30, 31, 34, 35]

A.S. Shumovsky and V.I. Rupasov (eds.), Quantum Communication and Information Technologies, 47–84.
© 2003 *Kluwer Academic Publishers.*

of the FLASH - the superluminal communicator [3]. The key idea of the FLASH relies on the possibility to copy (clone) unknown states of quantum systems. Herbert has shown that if the quantum cloning cloning would be possible, then we would be able to enjoy a comfort of super-luminal communication mediated by entangled quantum systems [4].

Herbert's idea has been criticized instantly. Namely, Dieks [37] and Wootters and Zurek [38] have shown that perfect cloning cloning of unknown quantum states is impossible (see Section III). This seemingly closed the whole issue. Years later Mark Hillery and myself [39] have asked a question: "Even though the perfect copying of unknown states is impossible, how well we can clone quantum states?" This question then has triggered a whole series of papers devoted to investigation of universal quantum machines (UQM).

UQM's are quantum mechanical devices that take a certain number of quantum systems (e.g. qubits) as an input and produce an output that is as close as possible to some ideal target state (in general target state depends on input). A typical example is a process of universal cloning. The universal cloning machine produces copies of an input qubit in an *unknown* state, such that the fidelity fidelity of the copies does not depend on the input state [39-59]. In fact exactly this type of machine has been implicitly assumed by Herbert [36]. It has been recently shown [60-64] that for the simplest case (namely a cloner producing 2 copies from 1 qubit) a bound on the quality of the clones could be derived from the no-signaling condition is identical to the bounds derived from quantum mechanics. This means that if the clones were only a little bit better than allowed by quantum mechanics, one could immediately use them for superluminal communication.

So in my lectures, I am going to use the intriguing connection between cloning cloning of quantum states and faster-than-light signaling [65-71] to describe several interesting problems related to some aspects of optimal manipulations with quantum information. To make my presentation selfcontent in Section 2 I will discuss in some detail physical origin of Herbert's proposal. Section 3 will be devoted to the no-cloning theorem, while in Section 4 I will describe universal quantum cloners. In Section 5 I will show how bounds on cloning can be derived from the no-signaling condition and in Section 6 I will

---

[3]FLASH is an acronym for the *First Laser-Amplified Superluminal Hookup*.

[4]At the *IV Adriatico Conference on Quantum Interferometry* (12 – 15 March 2002, Trieste, Italy) in general discussion Asher Peres has admitted that he was one of the referees of Herbert's paper [36] and even though he found it wrong he had recommend it for publication. According to Peres, the paper was so fundamentally wrong that it deserved the publication. Interestingly enough, Giancarlo Ghirardy who was at the conference as well, has responded to the confession of Asher Peres, that he was the second referee of the same Herbert's paper. Unlike Peres, he has recommend the paper to be rejected and in his referee report he has proved the no-cloning theorem which has been proved independently and published by William Wootters and Wojciech Zurek [38] (for more details see Section III).

discuss the connection between cloning, signaling and generalized measurements. The rest of the notes are devoted to universal NOT gate that "flips" a qubit in an unknown state. t In Section 7 I will describe the U-NOT gate and in Section 8 I will present experimental realization of this gate. I conclude my notes in Section 9.

2.   Bell Telephone & FLASH

Let us start an overview of the signaling based on quantum non-locality non-locality with a description of the "Bell telephone" and then will continue with a more detailed description of Herbert's idea for superluminal communication based on amplification (cloning) of individual events in the measurement of entangled particles.

2.1   Bell telephone

The Bell telephone is supposed to be a physical device which "uses" for communication quantum non-locality. In order to operate such device first maximally entangled particles (let these be qubits) have to be distributed to two parties (Alice and Bob). This can be achieved by using polarization states of two-photons. The polarization-entangled pairs of photons are experimentally generated in a parametric down-conversion process in a nonlinear crystal [72]. These entangled photons are generated in a singlet state

$$|\psi\rangle_{AB} = \frac{1}{\sqrt{2}}(|\uparrow\rangle_A|\downarrow\rangle_B - |\downarrow\rangle_A|\uparrow\rangle_B)\,, \qquad (1)$$

where $|\uparrow\rangle$ and $|\downarrow\rangle$ describe two polarization states of the photon in a given basis (e.g. horizontal/vertical polarization). The singlet state (1) exhibits perfect quantum correlation (entanglement) for polarization measurements along parallel but *arbitrary* axes. This means that the state (1) is equivalent under *local* transformations to the state

$$|\psi\rangle_{AB} = \frac{1}{\sqrt{2}}(|\leftarrow\rangle_A|\rightarrow\rangle_B - |\rightarrow\rangle_A|\leftarrow\rangle_B)\,, \qquad (2)$$

where $|\substack{\leftarrow \\ \rightarrow}\rangle = (|\uparrow\rangle \pm |\downarrow\rangle)/\sqrt{2}$. Here I stress that even though results of measurements are perfectly correlated the actual outcome of an individual measurement on each of the particles is inherently random.

Alice and Bob receive their particles well before any communication via quantum channel is performed. It is also assumed that the singlet state does not decohere under the influence of the environment. Once the entangled particle are distributed Alice might like to send a message to Bob. To do so she decides to perform a measurement on her particle. She can rotate her measurement device arbitrarily, but in order to make our discussion simple let us assume that she is going to perform a measurement in one of the two bases ($\{|\uparrow\rangle, |\downarrow\rangle\}$ and $\{|\leftarrow\rangle, |\rightarrow\rangle\}$). After Alice performs her measurement in the given basis (let say, $\{|\uparrow\rangle, |\downarrow\rangle\}$) then she can predict with certainty what Bob's result would be if he performs the measurement in the *same* basis. Hence if Bob can attain somehow the information about the measurement basis he would then have a possibility of receiving signals at the superluminal speed.

Even though Alice can determine which basis she will use she definitely is not able to predetermine the outcome of her measurement to be $|\uparrow\rangle$ or $|\downarrow\rangle$ since both these results materialize with probability a $1/2$. Therefore the message can best be defined to be Alice's choice in which basis she is going to perform a measurement on her qubit. That is, which observable she is going to measure (we will not consider an option that she choose not to measure at all). In this situation Bob's task is to determine the state of a qubit on his side of the Bell telephone via a single measurement on his qubit.

Formally, the signal from Alice is encoded into a *binary* alphabet with each of the two letters corresponding to a specific choice of Alice's basis, i.e. the measurement procedures $\mathcal{A}^{(s)}$: $(s = 0, 1)$,

$$
\begin{aligned}
\mathcal{A}^{(0)} &: \quad \{|\uparrow\rangle_A\langle\uparrow| \otimes \mathbf{1}_B; |\downarrow\rangle_A\langle\downarrow| \otimes \mathbf{1}_B\}, \\
\mathcal{A}^{(1)} &: \quad \{|\leftarrow\rangle_A\langle\leftarrow| \otimes \mathbf{1}_B; |\rightarrow\rangle_A\langle\rightarrow| \otimes \mathbf{1}_B\},
\end{aligned}
\tag{3}
$$

respectively. This means that if Alice wants to send a logical "0" ("1") then she performs a measurement $\mathcal{A}^{(0)}$ ($\mathcal{A}^{(1)}$). There is no way to say *a priori* what the outcome of her measurement will be. So after measuring e.g. $\mathcal{A}^{(0)}$ Bob's particle will be either in the state $\rho_\uparrow^{(0)} = |\uparrow\rangle\langle\uparrow|$ or in the state $\rho_\downarrow^{(0)} = |\downarrow\rangle\langle\downarrow|$. These outcomes are realized with the same probability. Analogous situation takes place when Alice sends a logical "1", i.e. performs the measurement $\mathcal{A}^{(1)}$.

The main task of Bob is to determine which state his particle is in. To do so he performs a measurement on his particle. For simplicity, let us assume that he performs a projective measurement $\mathcal{B} = \{O_r\}$, such that $O_r = |\psi_r\rangle\langle\psi_r|$ and $\sum_r O_r = \mathbf{1}$. Ideally, Bob wants to perform a measurement such that $\mathrm{Tr}[O_r \rho_{j(r)}^{(s)}] = 1$ for just one of the "input" states $\rho_{j(r)}^{(s)}$ while for all others the trace is equal to zero. But this requires the signal (input) states corresponding to different outcomes of Alice measurements to be orthogonal, i.e. $\mathrm{Tr}[\rho_i^{(s)}, \rho_j^{(s')}] = 0$. Which obviously, is not the case in the present scheme. Therefore no projection measurement exists which would yield a *reliable* signal analysis from *individual* outcomes. If many entangled pairs of qubits have been used for communication using the above scheme, then depending which of the two measurements

$$\mathcal{B}^{(0)} : \{ \mathbf{1}_A \otimes |\uparrow\rangle_B \langle\uparrow|; \ \mathbf{1}_A \otimes |\downarrow\rangle_B \langle\downarrow| \}, \tag{4}$$
$$\mathcal{B}^{(1)} : \{ \mathbf{1}_A \otimes |\leftarrow\rangle_B \langle\leftarrow|; \ \mathbf{1}_A \otimes |\rightarrow\rangle_B \langle\rightarrow| \},$$

Bob performs his resulting ensembles of qubits are represented by two density operators

$$\rho_{\uparrow\downarrow}^{(B)} = \frac{1}{2} (|\uparrow\rangle\langle\uparrow| + |\downarrow\rangle\langle\downarrow|) = \frac{1}{2}\mathbf{1}; \tag{5}$$
$$\rho_{\rightarrow}^{(B)} = \frac{1}{2} (|\leftarrow\rangle\langle\leftarrow| + |\rightarrow\rangle\langle\rightarrow|) = \frac{1}{2}\mathbf{1},$$

which are identical. Consequently, within the linear quantum mechanics (for further discussion see Section 4) these two ensembles are indistinguishable and no information can be signalled this way. In other words, Bob is not able to find what decision has been made by Alice. Here I stress once again, that even though the two ensembles $\rho_{\uparrow\downarrow}^{(B)}$ and $\rho_{\rightarrow}^{(B)}$ are indistinguishable, the individual outcomes of particular measurements are in specific pure states.

2.2   Herbert's FLASH

In his proposal of the FLASH Herbert [36] has clearly indicated that his "hookup" is based on the "novel" type of the quantum measurement performed on Bob's side. According to Herbert: "Superluminal message (alternation of individual events) can be sent but not decoded". Therefore the FLASH was supposed to operate in such a way that via a measurement of an *individual* quantum system (polarization states of a photon) one can determine the state of the system. The "novel" aspect of the FLASH was an idea to use *active* detectors such that before they register the incoming state they first multiply (copy, clone) this state into a large collection of particles all in the same (incoming) state. This would mean that the FLASH realizes the copying (cloning) cloning of an unknown pure state $|\psi\rangle$ of the incoming particle with the Hilbert space H, on N other particles of the same physical origin, i.e. the cloner is described by a map $\mathcal{C}(H) \rightarrow \mathcal{C}(H^{\otimes N})$ such that

$$|\psi\rangle|0\rangle^{\otimes(N-1)} \rightarrow |\psi\rangle^{\otimes N}, \tag{6}$$

where $|0\rangle$ is some *known* state of the systems onto which the information is going to be copied (see below).

Once this collection of clones is generated then with the help of an optimal measurement better (perfect) determination of the incoming state can be performed. So to quote Herbert "FLASH does not really deal with statistical aggregates of photons but with the aid of perfect xeroxing provided by the laser laser effect is able to examine the polarization of each individual photon." Herbert calls this as the *third type* of measurement (as we will see later this measurement us nothing else but the well known generalized POVM measurement [73, 74]). So the question is whether this type of the active measurement can be physically realized and whether via the improvement of the detection scheme faster-than-light communication can be established.

The main Herbert's idea is to substitute the standard projection measurement by a generalized measurement which includes also an amplifier amplifier (cloner). In quantum theory any amplification is inevitable accompanied by quantum noise which irreversibly spoils the signal [75, 76]. Herbert was aware that "a serious objection to FLASH concerns the noise ... of the copying process". Nevertheless he has not performed a detailed analysis of the problem. He just presented a vague argument that the noise induced by the stimulated emission via which copying of the incoming photon is performed, is after all not a serious obstacle.

3.   No-cloning Theorem

So can it be that within a linear quantum mechanics the transformation (6) can be realized? In order to illuminate this question I will formulate the no-cloning cloning theorem. The issue of cloning, or (self) reproducing of quantum states has been first discussed by Wigner [77] (see below). The no-cloning theorem itself has been proved independently by Dieks [37], and Wootters and Zurek [38] and others [78-81]. I will also present Mandel's proposal [82] for cloning of polarization states of photons via down conversion. Mandel was the first to present an explicit calculation of the amount of noise which is inevitable for cloning of quantum states.

3.1   Pre-no-cloning history

Probably the first account on quantum cloning was done by E.P. Wigner [77] in his analysis of earlier work by W.M. Elsasser devoted to a discussion of the origin of life and the multiplication of organisms [83]. Wigner has presented a quantum-mechanical argument according to which "the probability is zero for existence of self-reproducing states". The argument is based on two assumptions: Firstly just systems with finite-dimensional Hilbert spaces are considered. Secondly, it is assumed that the Hamiltonian Hamiltonian which governs the behavior of a complicated system is a *random* symmetric matrix, with no particular properties except for its symmetric nature. Under this assumptions Wigner has shown that "it is infinitely unlikely that there be*any* state of the nutrient [5] which would permit the multiplication of any set of states which is much smaller than all possible states of the system." This conclusion is based on the fact that the input state of the system and the cloner and their ideal output states do not specify the S matrix describing the self-reproducing (cloning) process.

I will not go into further detail of Wigner's argument. The reason why I comment on this work is to show that the first idea of cloning cloning quantum states has not been related to the problem of signaling, but has been based on information aspects of quantum theory [6]

3.2   Wootters-Zurek theorem

Wootters and Zurek [38] have presented a very simple proof that the perfect cloning transformation (6) for unknown quantum states is impossible. Their proof goes as follows: in order to clone an unknown state $|\psi\rangle$ a device (quantum cloner) is needed. This cloner is initially prepared in a state $|S\rangle$ which does not depend on $|\psi\rangle$. In addition, a set of $(N-1)$ particles onto which the information is going to be copied is available. These particles are prepared in a known state denoted as $|0\rangle$. Then the perfect copying transformation $\mathcal{U}$ can be written as

$$|\psi\rangle|0\rangle^{\otimes(N-1)}|S\rangle \xrightarrow{\mathcal{U}} |\psi\rangle^{\otimes N}|S'\rangle,\qquad(7)$$

where $|S'\rangle$ is the state of the quantum copier after the cloning has been performed. Since the input is totally unknown, the transformation $\mathcal{U}$ has to work for an arbitrary input. So let us assume two input states $|\psi\rangle$ and $|\tilde{\psi}\rangle$ both transformed according to Eq. (7)

$$\mathcal{U}\left(|\psi\rangle|0\rangle^{\otimes(N-1)}|S\rangle\right) = |\psi\rangle^{\otimes N}|S'\rangle\,;$$
$$\mathcal{U}\left(|\tilde{\psi}\rangle|0\rangle^{\otimes(N-1)}|S\rangle\right) = |\tilde{\psi}\rangle^{\otimes N}|S'\rangle\,.\qquad(8)$$

Taking the inner product of the left-hand sides of above equations we find

$$\langle\psi|\tilde{\psi}\rangle = (\langle\psi|\tilde{\psi}\rangle)^2\,,\qquad(9)$$

which can only be fulfilled if the two states $|\psi\rangle$ and $|\tilde{\psi}\rangle$ are either identical or orthogonal. So the ideal (perfect) cloning device for *arbitrary* states does not

---

[5]The "nutrient" is equivalent to two quantum systems the clone and the ancilla (i.e. the cloner itself) via which the information is distributed.

[6]Without any further exploration of the idea I just note that the impossibility of self-reproducing units has a direct consequence in a theory of reproducing quantum cellular automata which has not been properly explored yet.

exist [7]. Perfect cloning machines would require nonlinear dynamics [38]. It is well known (see Refs. [65, 66, 67]) that non-linear dynamics might lead to violation of the signaling constraint. So it is not surprising that the nonlinear dynamics allows for perfect cloning of unknown states, since this process again leads to the violation of the signaling constraint.

Very similar argument as the one presented by Wootters and Zurek [38] has been presented independently by Dieks [37]. Yet another version of the no-cloning theorem is due to Yuen [78] who pointed out a relation between cloning cloning and generalized measurements.

The no-cloning theorem has been generalized (the so-called no-broadcasting theorem) for statistical mixtures by Barnum *et al.* [79]. The extension of the no-cloning theorem for entangled systems has been presented by Koashi and Imoto [80]. Lindblad [81] has recently presented the most general version of the no-cloning theorem.

3.3   Mandel's analysis

The no-cloning theorem says us that unknown states cannot be *perfectly* cloned. But what is the limit on cloning? How good the copies can be made? If the imperfection is small enough then Herbert's FLASH can still be used for superluminal communication. The Wootters-Zurek theorem itself does not rule out a possibility of signaling due to *imperfect* cloning. Interestingly enough, this crucial question for the FLASH has not attracted due attention. To be more specific, in his short note Mandel [82] was the first to study an unavoidable noise due to amplification (cloning) of a polarization state of a photon via interaction with a two-level atom two-level atom. In particular, Mandel has considered an incoming photon with two possible polarizations $\vec{\varepsilon}_1$ and $\vec{\varepsilon}_2$. The state of a single incoming photon is denoted as $|1_{\vec{\varepsilon}_s}\rangle$ (with $s = 1, 2$). The atomic amplifier amplifier in the ideal case generates out of the incoming photon two photons with the same polarization, i.e. $|1_{\vec{\varepsilon}_s}\rangle \rightarrow |2_{\vec{\varepsilon}_s}\rangle$. So, following Mandel [82], let us consider a two-level atom with the transition dipole moment $\vec{\mu}$. The atom is initially prepared in the excited state $|e\rangle$. The interaction Hamiltonian Hamiltonian governing the interaction between the atom and an electromagnetic field is considered in the electric-dipole and the rotating-wave approximation, i.e.

$$H_I = g \sum_{s=1}^{2} \left[\vec{\mu} \cdot \vec{\varepsilon}_s^* \, \sigma^- a_s^\dagger + h.c. \right] , \qquad (10)$$

---

[7]It has to be stressed that the linearity argument based on which the no-cloning theorem is proved does not forbid the amplification of *known* states. That is a cloning device designed specifically for a given input state can be constructed without any violation of the no-signaling constraint.

where $\sigma^{\pm}$ are Pauli spin-flip operators and $a_s^{\dagger}$ (a_s) is the creation (annihilation) operator of a photon with a polarization $\vec{\varepsilon}_s$. If we consider the initial state of the electromagnetic field to be $|1_{\vec{\varepsilon}_1}, 0_{\vec{\varepsilon}_2}\rangle$ then dynamics governed by the Hamiltonian (10) forces the field to evolve in a short-time limit into the state

$$|1_{\vec{\varepsilon}_1}\rangle|0_{\vec{\varepsilon}_2}\rangle \rightarrow \frac{\sqrt{2}\vec{\mu} \cdot \vec{\varepsilon}_1^* |2_{\vec{\varepsilon}_1}, 0_{\vec{\varepsilon}_2}\rangle + \vec{\mu} \cdot \vec{\varepsilon}_2^* |1_{\vec{\varepsilon}_1}, 1_{\vec{\varepsilon}_2}\rangle}{(2|\vec{\mu} \cdot \vec{\varepsilon}_1^*|^2 + |\vec{\mu} \cdot \vec{\varepsilon}_2^*|^2)^{1/2}}. \tag{11}$$

The state $|2_{\vec{\varepsilon}_1}, 0_{\vec{\varepsilon}_2}\rangle$ attributes to the stimulated emission while $|1_{\vec{\varepsilon}_1}, 1_{\vec{\varepsilon}_2}\rangle$ describes a spontaneous emission. So the perfect cloning cloning is accompanied with a noise due to spontaneous emission.

The amplifier amplifier (10) is designed for cloning of a specific input state. Mandel has also presented a generalization of the amplifier amplifier such that it clones an arbitrary polarization state with the same fidelity fidelity (i.e. this is a prototype of the universal cloning machine - see Section 4). This input-state independent amplifier amplifier consists of two resonant atoms with orthogonal transition dipole moments $\vec{\mu}_a = |\mu|\vec{\varepsilon}_a$; $\vec{\mu}_b = |\mu|\vec{\varepsilon}_b$, where $\vec{\varepsilon}_a$ and $\vec{\varepsilon}_b$ are two complex orthogonal unit polarization vectors. The interaction Hamiltonian Hamiltonian between the atoms and the electromagnetic field is taken in the form

$$H_I = g \sum_{s=1}^{2} \left[(\sigma_a^- \vec{\mu}_a + \sigma_b^- \vec{\mu}_b) \cdot \vec{\varepsilon}_s^* a_s^{\dagger} + h.c. \right]. \tag{12}$$

If the two atoms are initially excited then in a short-time limit the dynamics (12) generates the electromagnetic field in a state described by a density operator (the atomic degrees of freedom are traced-off):

$$\rho_{\vec{\varepsilon}_1, \vec{\varepsilon}_1} = \frac{2}{3}|2_{\vec{\varepsilon}_1}, 0_{\vec{\varepsilon}_2}\rangle\langle 2_{\vec{\varepsilon}_1}, 0_{\vec{\varepsilon}_2}| + \frac{1}{3}|1_{\vec{\varepsilon}_1}, 1_{\vec{\varepsilon}_2}\rangle\langle 1_{\vec{\varepsilon}_1}, 1_{\vec{\varepsilon}_2}| \tag{13}$$

As we will see later the fidelity fidelity of the cloning process (12) does not depend on the polarization of the incident photon and of two atomic transition dipole moments (providing these are orthogonal). So out of a state $|1_{\vec{\varepsilon}_1}, 0_{\vec{\varepsilon}_2}\rangle$ with an *arbitrary* polarization we obtain the two-photon state $|2_{\vec{\varepsilon}_1}, 0_{\vec{\varepsilon}_2}\rangle$ with a probability $2/3$ which does not depend on the polarization. In addition to this a noise, represented by the state $|1_{\vec{\varepsilon}_1}, 1_{\vec{\varepsilon}_2}\rangle$ is generated in this process. I will turn to this result of Mandel later [8]. Here, following Wootters and Zurek I note that to each "perfect clone" there is also one randomly polarized, spontaneously

---

[8]Mandel in his paper [82] refers to the Wootters-Zurek argument as to ingeniously simple. We might like to say that Mandel's ingenious insight into the physics of amplification lead him to a correct answer to the question raised many years later.

emitted photon. But still the question is not answered whether this spontaneous emission definitely rules out a possibility of superluminal signaling.

In order to have a clear answer to this question I present in Section 4 the optimal universal quantum cloner which introduces the smallest amount of noise during the cloning procedure and will show you that this cloner cannot be utilized for the FLASH.

Concluding this section I note that the recent proposal for experimental realization of the universal cloning machine presented by Simon *et al.* [54, 55], as well as an experimental demonstration of the quantum cloning machine due to Lamas-Linares *et al.* [56] and De Martini *et al.* [57, 58], share the basic features of the original Mandel's idea. I will describe in a more detail in Section 8 the experiment by De Martini *et al.* [59] in which universal NOT gate and quantum cloning has been realized.

4.   Universal Quantum Cloners

Within classical physics we can imagine "machines" which take as an input a classical physical object in a state which corresponds to a classical information encoded in the system and perform an arbitrary operation prescribed by a specific map (transformation). After the transformation the output is obtained which corresponds to a result of a classical information processing. The fidelity fidelity of such a process in classical physics for an *arbitrary* unknown input can in principle always be equal to unity.

Quantum mechanics offers new perspectives in information processing [84], which in part is due to the fact, that quantum information can be represented by *qubits* which are two-level quantum systems with one level labeled $|0\rangle$ and the other $|1\rangle$. Qubits can not only be in one of the two levels, but in any superposition of them as well. This fact makes the properties of quantum information quite different from those of its classical counterpart. A typical example is the quantum cloner.

The impossibility of copying (cloning) quantum information puts fundamental limits on amount of information extractable from finite ensembles of identically prepared quantum systems [42]. Or *vice versa*, since finite ensembles do not allow for a complete determination of states [73,74,85-89], unknown quantum states cannot be copied perfectly.

So let us assume quantum machines which take as an input a qubit in an unknown state and generate at the output qubits in a state according to a specific prescription, e.g. one would like to have copies of the original. Important feature of the quantum cloning cloning machine is that it generates an output with the fidelity fidelity which does not depend on the input. Obviously, if the input state is known (i.e. complete classical information about the preparation of the given state is available) then an arbitrary transformation can be performed with the fidelity fidelity equal to unity. If the state is unknown then the transformation cannot be performed perfectly. In this case it is desirable that all states are transformed equally well (with the fidelity independent on a particular input state). This covariance covariance property of quantum machines with respect to unitary transformations of inputs makes the machines *universal*.

4.1 Quantum cloning $1 \rightarrow 2$

Let $\mathcal{H} = \mathbf{C}^2$ denote a two-dimensional Hilbert space of a single qubit. The action of the cloning machine is equivalent to a a *completely positive trace preserving map* $\mathcal{T} : \mathcal{H} \rightarrow \mathcal{H}$ The machine is design so, that for any pure one-particle state ρ at the input, the output $\mathcal{T}(\rho)$ is as close as possible to an input state.

In order to derive the cloning transformation $\mathcal{T}$ we have to specify the constraints which have to be met:

(i)The state of the original system and its quantum copy at the output of the quantum copier, described by density operators $\rho_a^{(out)}$ and $\rho_b^{(out)}$, respectively, are identical, i.e.,

$$\rho_a^{(out)} = \rho_b^{(out)}. \tag{14}$$

(ii) If no *a priori* information about the *in*-state of the original system is available, then it is reasonable to require that *all* pure states should be copied equally well. One way to implement this assumption is to design a quantum copier such that the distances between density operators of each system at the output ($\rho_x^{(out)}$ where $x = a, b$) and the ideal density operator $\rho^{(id)}$ which describes the *in*-state of the original mode are input-state independent. Quantitatively this means that if we employ a fidelity fidelity

$$\mathcal{F} = \mathrm{Tr}\left[\rho^{(out)}\rho^{(id)}\right] \tag{15}$$

as a measure of distance between two operators, then the quantum copier should be such that $\mathcal{F} = const$ for all possible input states.

(iii) Finally, we would also like to require that the copies are as close as possible to the ideal output state, which is, of course, just the input state. This means that we want our quantum copying transformation to minimize the distance between the output state $\rho_x^{(out)}$ of the copied qubit and the ideal state $\rho_x^{(id)}$. The distance is minimized, that is the fidelity fidelity $\mathcal{F}$ is maximized, with respect to all possible unitary transformations $\mathcal{U}$ acting on the Hilbert space $\mathcal{H}$ of two qubits and the quantum copying machine (i.e., $\mathcal{H} = \mathcal{H}_a \otimes \mathcal{H}_b \otimes \mathcal{H}_c$)

$$\mathcal{F}(\rho_x^{(out)}; \rho_x^{(id)}) = \max\left\{\mathcal{F}^{(\mathcal{U})}(\rho_x^{(out)}; \rho_x^{(id)}); \forall \mathcal{U}\right\}, \tag{16}$$

where $x = a, b$.

It has been shown recently [39] that in the case of $1 \rightarrow 2$ cloning the machine can be represented as a third qubit c via which the information from a is transfered to the qubit b. The cloning cloning transformation itself can be expressed in a covariant form as

$$|\psi\rangle_a |\Xi\rangle_{bc} \longrightarrow \sqrt{\tfrac{2}{3}}|\psi\rangle_a|\psi\rangle_b|\psi^\perp\rangle_c - \tfrac{1}{\sqrt{3}}|\{\psi^\perp,\psi\}\rangle_{ab}|\psi\rangle_c, \qquad (17)$$

where $|\{\psi^\perp,\psi\}\rangle_{ab} = (|\psi\rangle_a|\psi^\perp\rangle_b + |\psi^\perp\rangle_a|\psi\rangle_b)/\sqrt{2}$ is a symmetrically entangled state of the two qubits which are in the states $|\psi\rangle$ and $|\psi^\perp\rangle$, respectively. The copy and the cloner qubits at the input are always prepared in the state $|\Xi\rangle_{bc}$ which I do not specify here. The two qubits a and b at the output of this cloner are both in the same state

$$\rho_x^{(out)} = \frac{2}{3}\rho + \frac{1}{6}\mathbb{1}; \qquad x = a, b. \qquad (18)$$

The scaling factor $s = 2/3$ corresponds to the fidelity fidelity of the cloning equal to $\mathcal{F} = 5/6$ which is much higher than the fidelity fidelity of estimation of the state which is equal to $\mathcal{F} = 2/3$ (for more details see Section 8).

There exists a simple logical network [41] which realizes the transformation (17). This network is composed of four control-NOT gates via which the information from the original qubit is transfered to the qubit b. It is interesting to note that depending on a specific preparation of the qubits b and c we can control the flow of information in the network: we can even swap the roles of the output qubits or perform an asymmetric cloning, etc. The proof has been presented by Gisin and Massar [42], Bruß *et al.* [43], and Werner [46], that the transformation (17) is indeed optimal, in a sense that it generates best clones under the constrains (14)-(16).

4.2   Quantum cloning $N \rightarrow N + M$

Let us now suppose that we have N qubits all prepared in the same state ψ at the input of the cloner, but we want to generate $N + M$ clones at the output under the same conditions as discussed above.

In order to perform $N \rightarrow N + M$ cloning we need M "blank" qubits (we label them with the subscript b) and M additional qubits which play the role of the cloner (c) and are used for the transfer of information. These $2M$ qubits are alway prepared in the same state $|\Xi\rangle_{bc}$. The universal optimal cloning transformation for every vector $|\psi\rangle \in \mathcal{H}$ can be expressed as [42, 45]

$$U_{NM}|N\psi\rangle_a \otimes |\Xi\rangle_{bc} = \sum_{j=0}^{M} \gamma_j^{(N,M)}|\Xi_j(\psi)\rangle_{ab} \otimes |\{(M-j)\psi^\perp; j\psi\}\rangle_c \qquad (19)$$

with

$$\gamma_j^{(N,M)} = (-1)^j \binom{N+M-j}{N}^{1/2} \binom{N+M+1}{M}^{-1/2}, \tag{20}$$

where $|N\psi\rangle_a = |\psi\rangle^{\otimes N}$ is the input state consisting of N qubits in the same state $|\psi\rangle$. On the right hand side of Eq.(19) $|\{(M-j)\psi^\perp; j\psi\}\rangle_c$ denotes symmetric and normalized states with $(M-j)$ qubits in the complemented (orthogonal) state $|\psi^\perp\rangle$ and j qubits in the original state $|\psi\rangle$. Similarly, the vectors $|\Xi_j(\psi)\rangle_{ab}$ consist of $N+M$ qubits, and are given explicitly by

$$|\Xi_j(\psi)\rangle_{ab} = |\{(N+M-j)\psi; j\psi^\perp\}\rangle_{ab}. \tag{21}$$

Let us note that with this choice of the coefficients $\gamma_j^{(M,N)}$, the scalar product of the right hand side with a similar vector, with ψ replaced by ϕ, becomes $\langle\psi,\phi\rangle^N$. This is consistent with the unitarity of the operator U_{NM}.

The $N+M$ qubits in the systems a and b at the output of the gate are individually in the state described by the density operator

$$\rho_j^{(out)} = s\rho + \frac{1-s}{2}\mathbb{1}, \qquad j = 1, \ldots, N+M, \tag{22}$$

with the scaling factor

$$s = \frac{N}{N+2} + \frac{2N}{(N+M)(N+2)}, \tag{23}$$

i.e. these qubits are the *clones* of the original state with a fidelity fidelity of cloning larger than the fidelity of estimation (see Sec. 6). This fidelity depends on the number, M, of clones produced out of the N originals, and in the limit $M \rightarrow \infty$ the fidelity of cloning becomes equal to the fidelity of estimation. These qubits represent the output of the *optimal* $N \rightarrow N+M$ cloner introduced by Gisin and Massar [42].

So this is the optimal universal quantum cloning process [42, 46] which is possible within the linear quantum mechanics. Now we have to answer the question whether the minimal noise represented by the term $\frac{1-s}{2}\mathbb{1}$ in the expression for a single-qubit density operators at the out-put of the cloner, is enough to preserve the no-signaling condition.

5.   No-signaling & Linearity of Quantum Mechanics

In this section I will show how the no-signaling condition determines bounds on possible dynamics of a physical system [65]. Before I proceeding I want to note that there is a long-standing discussion on the relation between the no-signaling constraint and the linearity of quantum mechanics [65-71]. In what follows I will not study subtle details of this relation (I refer those who are interested in the details of the problem to Ref. [71]). The only argument from all these discussions I am going to use for the present purpose is that I will associate the no-signaling condition with the linearity of quantum mechanics. I will adopt the usual quantum kinematics. The possible states of the system are described by one-dimensional projectors $P_\psi = |\psi\rangle\langle\psi|$ in a Hilbert space. The projectors in this Hilbert space are the measurable physical quantities (taking only values 0 and 1).

The time evolution of the system is described by a map $\mathcal{G}$ on the set of pure states:

$$\mathcal{G} : P_\psi \rightarrow \mathcal{G}(P_\psi).$$ (24)

Consequently, a mixture $\{P_{\psi_i}, x_i\}$ of states P_{ψ_i} with weights x_i evolves into a mixture of states $\mathcal{G}(P_{\psi_i})$ with the same weights x_i. Therefore the corresponding density operators evolve as follows:

$$\sum_i x_i P_{\psi_i} \rightarrow \sum_i x_i \mathcal{G}(P_{\psi_i})$$ (25)

Now consider two mixtures $\{P_{\psi_i}, x_i\}$ and $\{P_{\phi_j}, y_j\}$ such that

$$\sum_i x_i P_{\psi_i} = \sum_j y_j P_{\phi_j} = \rho_0,$$ (26)

i.e., they correspond to the same density operator. Assume that

$$\sum_i x_i \mathcal{G}(P_{\psi_i}) \neq \sum_j y_j \mathcal{G}(P_{\phi_j}),$$ (27)

then the two mixtures can be distinguished after a finite time.

But, as shown in Ref. [65] and as I will show below, any two mixtures corresponding to the same density operator can be (instantaneously) prepared at a distance making use of an appropriate entangled state. entangled states As stated above, I assume the usual quantum kinematics, which in particular implies existence of entangled states. The assumption (27) thus contradicts the requirement that there should be no superluminal communication.

Under the assumption that relativity is correct, this implies that the time development of the system can only depend on the initial density matrix density matrix ρ_0 (and not on the specific mixture):

$$\mathcal{G} : \rho_0 \rightarrow \mathcal{G}(\rho_0)$$ (28)

where

$$\mathcal{G}(\rho_0) = \mathcal{G}\left(\sum_i x_i P_{\psi_i}\right) = \sum_i x_i \mathcal{G}(P_{\psi_i})$$ (29)

for every mixture having a density operator ρ_0.

Let us notice that the assumption (27) would mean that the density operator formalism is not appropriate for determining the time evolution of mixtures, but it could remain useful for the computation of expectation values of observables. That any two mixtures corresponding to the same density operator can be prepared at a distance can be shown as follows: The equation (29) is based on two facts: (1) the time evolution depends only on the density operator; (2) the time evolution doesn't mix composite states of the mixture.

5.1   Bloch vectors and no-signaling

Let us consider a process in which a single particle input state is mapped into a two particle output state. The input state can be represented as

$$\rho^{(in)}(\vec{m}) = \frac{1}{2}(\mathbb{1} + \vec{m} \cdot \vec{\sigma}) \equiv |\vec{m}\rangle\langle\vec{m}|, \tag{30}$$

where $\vec{m}$ is a real vector whose length is less than or equal to unity. The most general two-particle output state, which is hermitian and has a trace equal to one, can be expressed in the basis of matrices $\{\mathbb{1} \otimes \mathbb{1}, \ \mathbb{1} \otimes \sigma_i, \ \sigma_i \otimes \mathbb{1}, \ \sigma_i \otimes \sigma_k\}$ as

$$\rho_{ab}^{(out)}(\vec{m}) = \frac{1}{4}[\mathbb{1} \times \mathbb{1} + \vec{\eta}_1 \cdot \vec{\sigma} \otimes \mathbb{1} + \mathbb{1} \otimes \vec{\eta}_2 \cdot \vec{\sigma}$$
$$+ \sum_{j,k=x,y,z} t_{jk}\sigma_j \otimes \sigma_k], \tag{31}$$

where $\vec{\eta}_1$, $\vec{\eta}_2$, and t_{jk} are functions of $\vec{m}$. The requirement that the reduced density matrices of the two output particles be the same, which I shall impose, implies that $\vec{\eta}_1 = \vec{\eta}_2$.

Let us now impose the requirement of covariance. covariance This means that if $\rho_a^{(in)}(\vec{m})$ is mapped onto $\rho_{ab}^{(out)}(\vec{m})$, and if U is a matrix in $SU(2)$, then the input state $U\rho_a^{(in)}(\vec{m})U^{-1}$ will be mapped onto the output state $U \otimes U\rho_{ab}^{(out)}(\vec{m})U^{-1} \otimes U^{-1}$. Another way of stating this condition is obtained by noting that if we express U as

$$U = \exp(-i\theta\vec{e} \cdot \vec{\sigma}/2), \tag{32}$$

where $\vec{e}$ is a unit vector corresponding to the rotation axis and θ is the rotation angle, then

$$U(\vec{m} \cdot \vec{\sigma})U^{-1} = \vec{m}' \cdot \vec{\sigma}, \tag{33}$$

where $\vec{m}' = R(\vec{e}, \theta)\vec{m}$. The rotation matrix, $R(\vec{e}, \theta)$, is the 3×3 matrix which rotates a vector about the axis $\vec{e}$ by an angle θ, and it is given explicitly by

$$R(\vec{e}, \theta) = \exp(\theta\vec{e} \cdot \vec{K}), \tag{34}$$

where

$$K_x = \begin{pmatrix} 0 & 0 & 0 \\ 0 & 0 & -1 \\ 0 & 1 & 0 \end{pmatrix},$$

$$K_y = \begin{pmatrix} 0 & 0 & 1 \\ 0 & 0 & 0 \\ -1 & 0 & 0 \end{pmatrix}, \tag{35}$$

$$K_z = \begin{pmatrix} 0 & -1 & 0 \\ 1 & 0 & 0 \\ 0 & 0 & 0 \end{pmatrix}.$$

We have that

$$U\rho_a^{(in)}(\vec{m})U^{-1} = \rho_a^{(in)}(R\vec{m}), \tag{36}$$

which will be mapped onto $\rho_{ab}^{(out)}(R\vec{m})$, so that the covariance covariance condition can now be expressed as

$$\rho_{ab}^{(out)}(R\vec{m}) = U \otimes U\rho_{ab}^{(out)}(\vec{m})\,U^{-1} \otimes U^{-1}. \tag{37}$$

Now let us examine the consequences of this relation: Let us first consider the terms linear in $\vec{\sigma}$ and let R be a rotation about $\vec{m}$ by a very small angle θ. We have that

$$\vec{\eta}_1(R\vec{m}) = R\,\vec{\eta}_1(\vec{m}), \tag{38}$$

which for our choice of rotation becomes

$$\vec{\eta}_1(\vec{m}) = (\mathbf{1} + \theta\,\hat{m} \cdot \vec{K})\vec{\eta}_1(\vec{m}), \tag{39}$$

or

$$\vec{e}_{\vec{m}} \cdot \vec{K}\vec{\eta}_1(\vec{m}) = 0, \tag{40}$$

where $\vec{e}_{\vec{m}}$ is a unit vector in the direction of $\vec{m}$. This implies that $\vec{e}_{\vec{m}} \times \vec{\eta}_1(\vec{m}) = 0$, so that $\vec{\eta}_1(\vec{m})$ is parallel to $\vec{m}$, and we can write $\vec{\eta}_1(\vec{m}) = \eta_1(\vec{m})\,\vec{m}$. If we now substitute this result back into Eq. (38) and consider a general rotation R, we have that

$$\eta_1(R\vec{m}) = \eta_1(\vec{m}). \tag{41}$$

This implies that $\eta_1(\vec{m})$ is a constant, which we shall denote by η_1. Analogous arguments lead to a conclusion that $\vec{\eta}_2 = \eta_2\,\vec{m}$.

Now let us see what covariance covariance implies about the terms quadratic in $\vec{\sigma}$. Application of the covariance condition, Eq. (37), to these terms gives

$$t_{jk}(R\vec{m}) = \sum_{j',k'} R_{jj'} R_{kk'} t_{j'k'}(\vec{m}). \tag{42}$$

If we again choose R to be a rotation about $\vec{m}$ by a small angle θ, we find the condition

$$0 = \sum_{j'} (\vec{e}_{\vec{m}} \cdot \vec{K})_{jj'} t_{j'k}(\vec{m}) + \sum_{k'} (\vec{e}_{\vec{m}} \cdot \vec{K})_{kk'} t_{jk'}(\vec{m}). \tag{43}$$

Since we consider universal machines we can assume without a loss of generality, that $\vec{m}$ be in the z direction, in particular, $\vec{m} = \vec{e}_z$. In this case we find that $t_{xx} = t_{yy}$, $t_{xy} = -t_{yx}$, and $t_{xz} = t_{zx} = t_{yz} = t_{zy} = 0$, where all of these are evaluated at $\vec{m} = \vec{e}_z$. We now want to impose the no signaling condition

$$\rho_{ab}^{(out)}(\vec{e}_z) + \rho_{ab}^{(out)}(-\vec{e}_z) = \rho_{ab}^{(out)}(\vec{e}_x) + \rho_{ab}^{(out)}(-\vec{e}_x), \tag{44}$$

and to do so we need to find all of the density matrixes in the above equation in terms of $t_{jk}(\vec{e}_z)$. This can be done by applying the covariance covariance condition, Eq. (37), to $\rho_{ab}^{(out)}(\vec{e}_z)$ and making the proper choice of R. When these results are substituted into Eq. (44) we find that $t_{xx}(\vec{e}_z) = t_{yy}(\vec{e}_z) = t_{zz}(\vec{e}_z)$, and I shall designate this common value by t. We then can rewrite the general expression for the two-qubit density operator satisfying the covariance covariance condition (37) as (here we again use the general orientation of the vector $\vec{m}$)

$$\rho_{ab}(\vec{m}) = \frac{1}{4} \left(\mathbf{1} \otimes \mathbf{1} + \eta_1 \vec{m}\vec{\sigma} \otimes \mathbf{1} + \eta_2 \mathbf{1} \otimes \vec{m}\vec{\sigma} + t\, \vec{\sigma} \otimes \vec{\sigma} + t_{xy}\vec{m}[\vec{\sigma} \times \vec{\sigma}] \right) \tag{45}$$

where $\eta_1, \eta_2, t, t_{xy}$ are real parameters. In order for $\rho(\vec{m})$ to be a physical density matrix, density matrix its eigenvalues have to be non-negative. A simple calculation shows that this implies constraints

$$1 + t \pm (\eta_1 + \eta_2)$$
$$0$$

$$1 - t \pm \sqrt{4t^2 + 4t_{xy}^2 + (\eta_1 - \eta_2)^2}$$
$$0 \tag{46}$$

In what follows I will apply the no-signaling constraint (46) to the universal quantum cloners of Section (4).

5.2   Bounds on cloning due to signaling

In the case of the $1 \rightarrow 2$ cloning, cloning the task is to optimize the fidelity fidelity $\mathcal{F} = \text{Tr}[\rho_{ab}^{(out)}(\vec{m})\rho_a^{(in)} \otimes \mathbb{1}] = \text{Tr}[\rho_{ab}^{(out)}(\vec{m}) \, P_{\vec{m}} \otimes \mathbb{1}]$, where $P_{\vec{m}} = |+\vec{m}\rangle\langle+\vec{m}|$, assuming $\eta_1 = \eta_2 \equiv \eta$. A simple solution of Eqs. (46) then leads to the optimal values $t_{xy} = 0, t = 1/3, \eta = 2/3$, for which $\mathcal{F} = \frac{5}{6}$, which corresponds to a single-qubit density operator at the output given by Eq. (18). Note that this also optimizes the fidelity describing generation of two clones, i.e. $\text{Tr}[\rho_{ab}^{(out)}(\vec{m}) \, P_{\vec{m}} \otimes P_{\vec{m}}] = \frac{2}{3}$ These are exactly the bounds that are valid in quantum mechanics as discussed in Section 4 and analyzed by Werner [46]. Therefore I can conclude that the *optimal* universal quantum cloning one cannot violate the no-signaling constraint. It is exactly the noise [represented by the term $1/6$ in Eq. (18)] induced by cloning procedure which prevents the FLASH to transmit signals superluminally.

One may wonder whether there still might be a possibility, that generating more than two clones out of an incoming qubit one would be able to operate the FLASH. In order to rule out this possibility one can check that the bound on $1 \rightarrow 1 + M$ cloning imposed by the no-signaling (linearity) constraint is satisfied by the optimal universal cloners.

6.   Cloning, Signaling & POVM

To understand the connection between the optimal universal cloning and the no-signaling constraint let us turn our attention to a problem of optimal estimation of states of quantum systems. Let us consider a finite ensemble of N qubits all prepared in the same pure state $|\psi\rangle$. If the state is totally unknown, i.e. we have no *a priori* information about its preparation, then we have to assume that all pure state are equally probable. This corresponds to a uniform probability distribution on a state space of a given system, i.e. in the case of qubits - the Bloch sphere (see Fig. 1).

It is well known [73, 74, 85, 86, 87, 88] (for a review see Ref. [89]) that there exists an optimal measurement of the finite set of N qubits via which the *best* possible estimation of the state $|\psi\rangle$ can be performed. Holevo [73] has shown that it is possible to realize the best estimation via the so-called *covariant* measurement, which is a continuous POVM measurement performed on the whole finite ensemble. Obviously, in this case the problem is, that physically it is difficult to perform experimentally a measurement with a continuous number of observables. Later it has been shown by Massar and Popescu [85] and Derka *et al.* [86], that the optimal measurement on a finite ensemble of qubits can be realized via a finite-dimensional POVM. Such POVM can be realized when we imagine projective measurements performed on the whole set of N qubits (that is the qubits are not measured sequentially, but simultaneously, in one "shot"). Once this optimal measurement is performed then the best possible estimation

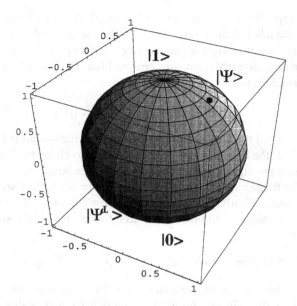

Figure 1. The state space of a qubit is a Bloch sphere. Pure states $|\psi\rangle$ are represented by points on the sphere, while statistical mixtures are points inside the sphere. The state $|\psi^{\perp}\rangle$ that is orthogonal to $\psi\rangle$ is its antipode.

of the measured state can be expressed in a form of the density operator

$$\rho^{(est)} = s_N \rho + \frac{1 - s_N}{2} \mathbf{1}, \tag{47}$$

where the "scaling" factor s_N is given by the expression

$$s_N = \frac{N}{N + 2}, \tag{48}$$

and is directly related to the mean fidelity

$$\overline{\mathcal{F}} = \int d\Omega_\rho \langle\psi|\rho^{(est)}|\psi\rangle, \tag{49}$$

where the integration is performed over all input states ρ and $d\Omega_\rho = \sin\vartheta d\vartheta\, d\varphi/4\pi$ is the integration measure associated with the state space , i.e. the Bloch sphere. When we insert $\rho^{(est)}$ given by Eq.(47) into Eq.(49) we find

$$\overline{\mathcal{F}} = s_N + \frac{1 - s_N}{2} = \frac{N + 1}{N + 2}. \tag{50}$$

There does not exist a measurement which would give us more information than the one POVM just considered (for more details see the review article [89]).

6.0.1   Single qubit case.   In the case of a single qubit a simple projective measurement is the most optimal one. Specifically, the *optimal* way to estimate the state, is to measure it along a *randomly* (we have no prior knowledge about the state) chosen direction in the two-dimensional Hilbert space[73, 85, 86]. So let us choose a random vector $|\eta\rangle$, where

$$|\eta\rangle = \cos(\vartheta'/2)|0\rangle + e^{i\varphi'}\sin(\vartheta'/2)|1\rangle, \qquad (51)$$

and measure $|\psi\rangle$ along it. If the result is positive, then the output is taken to be $|\eta\rangle$, and if negative, the output is $|\eta^\perp\rangle$. This would also correspond to our best estimation of the input state given the result of the measurement.

To evaluate the fidelity fidelity of the estimation I present a statistical picture of the measurement. Firstly, let us average over all possible orientations of the measurement apparatus. In order to do so let us write down a single-qubit density operator

$$\rho^{(meas)}(\eta) = |\langle\psi|\eta\rangle|^2\,|\eta\rangle\langle\eta| + |\langle\psi|\eta^\perp\rangle|^2\,|\eta^\perp\rangle\langle\eta^\perp|\,. \qquad (52)$$

which describes statistics of the measurements for a given orientation of the measurement apparatus. To get the final output density matrix density matrix we average (52) over all possible choices of the measurement (i.e. over all vectors $|\eta\rangle$)

$$\rho^{(est)} = \int d\,\Omega_\eta \rho^{(meas)}(\eta). \qquad (53)$$

where $d\,\Omega_\eta = \frac{1}{4\pi}\sin\vartheta' d\vartheta'\,d\varphi'$ is the integration measure on the state space of the "measurement" apparatus. After the integration is performed we find

$$\rho^{(est)} = s\rho + \frac{1-s}{2}\mathbf{1}, \qquad (54)$$

where for a single input qubit we have $s = 1/3$ and $\rho = |\psi\rangle\langle\psi|$.

In order to find the mean fidelity fidelity of the estimation itself we have to average the fidelity, i.e. $\langle\psi|\rho^{(est)}|\psi\rangle$ over all possible preparations, i.e.

$$\mathcal{F} = \int d\,\Omega_\rho \langle\psi|\rho^{(out)}|\psi\rangle = \frac{2}{3}. \qquad (55)$$

Obviously, instead of the projective measurement one can consider some other optimal generalized measurement to be performed on the input qubit. We can even consider a continuous POVM. Nevertheless, since in the given case the projective measurement, described above, is the optimal one, no other measurement can give us more information about the input state $|\psi\rangle$.

Now it is clear that the quantum cloning cloning can be represented as a specific generalized POVM measurement. It is a particular physical realization of the Naimark theorem [73] - the information contained in the original qubit (i.e. the state $|\psi\rangle$) is spread in between many clones. But when the optimal measurement on these clones is performed [44] the mean fidelity fidelity of the estimation is again equal to 2/3. In other words we cannot generate information via cloning and therefore we cannot violate the no-signaling constraint. This is the reason why the FLASH does not work. The argument can be generalized when the optimal $N \rightarrow N + M$ cloning cloning is considered [42, 46] Information about the input qubit(s) cannot be "generated". It only can be redistributed [94]. This is in perfect accordance with the no-signaling constraint.

7. Flipping Qubits: Universal NOT Gate

As follows from our previous discussion, quantum cloning can be considered as a form of a redistribution of quantum information from a set of incoming qubits to a large set of output qubits. In the cloning the main task has been to generate a copy (copies) of the original input qubit which are as close as possible to the input. But we can also assume other tasks, such as spin-flipping of unknown qubits.

One of the most striking difference between the classical and quantum information is as follows: it is not a problem to flip a classical bit, i.e. to change the value of a bit, a 0 to a 1 and vice versa. This is accomplished by a NOT gate. Flipping a qubit, however, is another matter: there exists the fundamental bound which prohibits to flip a qubit prepared in an *arbitrary* state $|\psi\rangle = \alpha|0\rangle + \beta|0\rangle$ and to obtain the state $|\psi^{\perp}\rangle = \beta^*|0\rangle - \alpha^*|1\rangle$ which is *orthogonal* to it, i.e. $\langle\psi^{\perp}|\psi\rangle = 0$.

Let us assume the Bloch sphere which represents a state space of a qubit. The points corresponding to $|\psi\rangle$ and $|\psi^{\perp}\rangle$ are antipodes of each other. The desired spin-flip operation is therefore the *inversion of the Bloch sphere* (see Fig. 1).

It is well known that this inversion preserves angles (which is related to the scalar product $|\langle\phi,\psi\rangle|$ of rays). Therefore, by the arguments of the Wigner theorem the ideal spin-flip operation must be implemented either by a unitary or by an anti-unitary operation. Unitary operations correspond to proper rotations of the Bloch sphere, whereas anti-unitary operations correspond to orthogonal transformations with determinant -1. The spin-flip is an anti-unitary operation, i.e. it is not completely positive.

Due to the fact that the tensor product of an anti-linear and a linear operator is not correctly defined the spin-flip operation cannot be applied to a qubit while the rest of the world is governed by unitary evolution. On the other hand if we consider a spin-flip operation we should have in mind a universal NOT gate flipping an input qubit to its orthogonal state. The gate itself is an operation applied to the qubit, that is just a subsystem of a "whole universe". Therefore it must be represented a completely positive operation. It is well known that any completely positive operation on a qubit can be realized by a unitary operation performed on the qubit and the ancillary system. Following this arguments we see that the ideal *universal* NOT gate which would flip a qubit in an *arbitrary* state does not exist.

Obviously, if the state of the qubit is known, then we can always perform a flip operation. In this situation the classical and quantum operations share many similar features, since the knowledge of the state is a classical information, which can be manipulated according to the rules of classical information processing (e.g. known states can be copied, flipped etc). But, the universality of the operation is lost. That is, the gate which would flip the state $|0\rangle \rightarrow |1\rangle$, is not able to perform a flip $|(0) + |1\rangle)/\sqrt{2} \rightarrow (|0\rangle - |1\rangle)/\sqrt{2}$.

Since it is not possible realize a perfect Universal-NOT gate which would flip an arbitrary qubit state, it is of interest to study, what is the best approximation to the perfect Universal-NOT gate. Here one can consider two possible scenarios. The first one is based on the measurement of input qubit(s) – using the results of an optimal measurement one can manufacture an orthogonal qubit, or any desired number of them. Obviously, the fidelity fidelity of the NOT operation in this case is equal to the fidelity of estimation of the state of the input qubit(s). The second scenario would be to approximate an anti-unitary transformation on a Hilbert space of the input qubit(s) by a unitary transformation on a larger Hilbert space which describes the input qubit(s) and ancillas.

It has been shown recently, that the best achievable fidelity fidelity of both flipping scenarios is the same [90, 91, 92, 93]. That is, the fidelity of the optimal Universal NOT gate is equal to the fidelity of the best state-estimation performed on input qubits [73, 85, 86] (one might say, that in order to flip a qubit we have to transform it into a bit). In what follows I briefly describe the unitary transformation realizing the quantum scenario for the spin-flip operation, that is, I present the optimal Universal NOT gate. Then I describe the recent experiment by De Martini *et al.* (see Ref. [59]) in which qubits encoded in polarization states of photons have been flipped.

7.1   Theoretical description of spin flipping

Let $\mathcal{H} = \mathbf{C}^2$ denote the two-dimensional Hilbert space of a single qubit Then the input state of N systems prepared in the pure state $|\psi\rangle$ is the N-fold tensor product $|\psi\rangle^{\otimes N} \in \mathcal{H}^{\otimes N}$. The corresponding density matrix density matrix is $\sigma \equiv \rho^{\otimes N}$, where $\rho = |\psi\rangle\langle\psi|$ is the one-particle density matrix. An important observation is that the vectors $|\psi\rangle^{\otimes N}$ are invariant under permutations of all N sites, i.e., they belong to the symmetric, or "Bose"-subspace $\mathcal{H}_+^{\otimes N} \subset \mathcal{H}^{\otimes N}$. Thus as long as we consider only pure input states we can assume all the input states of the device under consideration to be density operators on $\mathcal{H}_+^{\otimes N}$. I will denote by $S(\mathcal{H})$ the density operators over a Hilbert space $\mathcal{H}$. Then the U-NOT gate must be a completely positive trace preserving map $T \; : \; S\left(\mathcal{H}_+^{\otimes N}\right) \rightarrow S(\mathcal{H})$. My aim is to design T in such a way that for any pure one-particle state $\rho \in S(\mathcal{H})$ the output $T(\rho^{\otimes N})$ is as close as possible to the orthogonal qubit state $\rho^\perp = \mathbb{1} - \rho$. In other words, I am trying to make the fidelity fidelity $\mathcal{F} := \mathrm{Tr}[\rho^\perp T(\rho^{\otimes N})] = 1 - \Delta$ of the optimal complement with the result of the transformation T as close as possible to unity for an arbitrary input state. This corresponds to the problem of finding the minimal value of the error measure $\Delta(T)$ defined as

$$\Delta(T) = \max_{\rho, \text{pure}} \mathrm{Tr}\left[\rho \, T(\rho^{\otimes N})\right]. \tag{56}$$

Note that this functional Δ is completely unbiased with respect to the choice of input state. More formally, it is invariant with respect to unitary rotations (basis changes) in $\mathcal{H}$: When T is any admissible map, and U is a unitary on $\mathcal{H}$, the map $T_U(\sigma) = U^* T(U^{\otimes N} \sigma U^{* \otimes N}) U$ is also admissible, and satisfies $\Delta(T_U) = \Delta(T)$. I will show you later on that one may look for optimal gates T, minimizing $\Delta(T)$, among the *universal* ones, i.e., the gates satisfying $T_U = T$ for all U. For such U-NOT gates, the maximization can be omitted from the definition (56), because the fidelity $\mathrm{Tr}\left[\rho \, T(\rho^{\otimes N})\right]$ is independent of ρ.

7.2   Measurement-based scenario

An estimation device (see also previous section) by definition takes an input state $\sigma \in S(\mathcal{H}_+^{\otimes N})$ and produces, on every single experiment, an "estimated pure state" $\rho \in S(\mathcal{H})$. As in any quantum measurement this will not always be the same ρ, even with the same input state ρ, but a random quantity. The estimation device is therefore described completely by the probability distribution of pure states it produces for every given input. Still simpler, I will characterize it by the corresponding probability density with respect to the unique normalized measure on the pure states (denoted "$d\phi$" in integrals), which is also invariant under unitary rotations. For an input state $\sigma \in S(\mathcal{H}_+^{\otimes N})$, the value of this probability density at the pure state $|\phi\rangle$ is

$$p(\phi, \sigma) = (N+1)\langle \phi^{\otimes N}, \sigma \phi^{\otimes N}\rangle. \tag{57}$$

To check the normalization, note that $\int d\phi \, p(\phi, \sigma) = \mathrm{Tr}[X\sigma]$ for a suitable operator X, because the integral depends linearly on σ. By unitary invariance of the measure "$d\phi$" this operator commutes with all unitaries of the form $U^{\otimes N}$, and since these operators, restricted to $\mathcal{H}_+^{\otimes N}$ form an irreducible representation of the unitary group of $\mathcal{H}$ [for $d = 2$, it is just the spin spin $N/2$ irreducible representation of SU(2)], the operator X is a multiple of the identity. To determine the factor, one inserts $\sigma = 1$, and uses the normalization of "$d\phi$" to verify that $X = 1$.

Note that the density (57) is proportional to $|\langle \phi, \psi \rangle|^{2N}$, when $\sigma = |\psi^{\otimes N}\rangle \langle \psi^{\otimes N}|$ is the typical input to such a device: N systems prepared in the same pure state $|\psi\rangle$. In that case the probability density is clearly peaked sharply at states $|\phi\rangle$ which are equal to $|\psi\rangle$ up to a phase.

Suppose now that we combine the state estimation with the preparation of a new state, which is some function of the estimated state. The overall result will then be the integral of the state valued function with respect to the probability distribution just determined. In the case at hand the desired function is $f(\phi) = (1 - |\phi\rangle\langle\phi|)$. So the result of the whole measurement-based ("classical") scheme is

$$\rho^{(est)} = T(\sigma) = \int d\phi \, p(\phi, \sigma) \, (1 - |\phi\rangle\langle\phi|). \tag{58}$$

The fidelity fidelity required for the computation of Δ from Eq.(56) is then equal to (see also [85, 86])

$$\Delta = (N + 1) \int d\phi \, |\langle\phi, \psi\rangle|^{2N} (1 - |\langle\phi, \psi\rangle|^2) = \frac{1}{N + 2}, \tag{59}$$

where I have used that the two integrals have exactly the same form (differing only in the choice of N), and that the first integral is just the normalization integral. Since this expression does not depend on ρ, we can drop the maximization in the definition (56) of Δ, and find $\Delta(T) = 1/(N + 2)$, from which we find that the fidelity fidelity of creation of a complement to the original state ρ is

$$\mathcal{F} = \frac{N + 1}{N + 2}. \tag{60}$$

Finally I note, that the result of the operation (58) can be expressed in the form

$$\rho^{(out)} = s_N \rho^\perp + \frac{1 - s_N}{2} \mathbf{1}, \tag{61}$$

with the "scaling" parameter $s_N = \frac{N}{N+2}$. From here it is seen that in the limit $N \to \infty$, perfect estimation of the input state can be performed, and, consequently, the perfect complement can be generated. For finite N the mean fidelity fidelity is always smaller than unity. The advantage of the measurement-based scenario is that once the input qubit(s) is measured and its state is estimated an arbitrary number M of identical (approximately) complemented qubits can be produced with the same fidelity, simply by replacing the output function $f(\phi) = (1 - |\phi\rangle\langle\phi|)$ by $f_M(\phi) = (1 - |\phi\rangle\langle\phi|)^{\otimes M}$.

7.3   Quantum scenario

In follows I will present a transformation which produces complements whose fidelity is the same as those produced by the measurement-based method. Assume we have N input qubits in an unknown state $|\psi\rangle$ and we are looking for a transformation which generates M qubits at the output in a state as close as possible to the orthogonal state $|\psi^\perp\rangle$. The universality of the proposed transformation has to guarantee that all input states are complemented with the same fidelity. If we want to generate M approximately complemented qubits at the output, the U-NOT gate has to be represented by $2M$ qubits (irrespective of the number, N, of input qubits), M of which serve as ancilla, and M of which become the output complements. Let us denote these subsystems by subscripts "a"=input, "b"=ancilla, and "c"=(prospective) output. The U-NOT gate transformation, U_{NM}, acts on the tensor product of all three systems. The gate is always prepared in some state $|\Xi\rangle_{bc}$, independently of the input state $|\psi\rangle^{\otimes N}$. Interestingly enough the optimal U-NOT gate is realized by the same transformation as the quantum cloning. cloning That is, the U-NOT is described by Eq. (19).

Each of the M qubits under consideration at the output of the U-NOT gate is described by the density operator (61) with $s_N = \frac{N}{N+2}$, *irrespective* of the number of complements produced. The fidelity fidelity of the U-NOT gate depends only on the number of inputs. This means that this U-NOT gate can be thought of as producing an approximate complement and then cloning it, with the quality of the cloning independent of the number of clones produced. The universality of the transformation is directly seen from the "scaled" form of the output operator (61).

Let us stress that the fidelity of the U-NOT gate (19) is exactly the same as in the measurement-based scenario. Moreover, it also behaves as a classical (measurement-based) gate in a sense that it can generate an arbitrary number of complements with the same fidelity. In fact, the transformation (19) represents the *optimal* U-NOT gate via quantum scenario. That is, the measurement-based and the quantum scenarios realize the U-NOT gate with the same fidelity. In Appendix A I will present a proof of the following theorem:

Theorem I. *Let $\mathcal{H}$ be a Hilbert space of dimension $d = 2$. Then among all completely positive trace preserving maps $T : \mathcal{S}\left(\mathcal{H}_+^{\otimes N}\right) \to \mathcal{S}(\mathcal{H})$, the measurement-based U-NOT scenario (58) attains the smallest possible value of the error measure defined by Eq.(56), namely $\Delta(T) = 1/(N+2)$.*

I conclude this section by saying that in the quantum world governed by unitary operations anti-unitary operations can be performed with the fidelity which is bounded by the amount of classical information potentially available about states of quantum systems.

8.   Experimental Realization of U-NOT Gate

In what follows I will describe the experiment by DeMartini *et al.* [59] in which a flipping of a single qubit has been realized. In this case of just single input qubit, the flipping transformation reads exactly the same as the cloning cloning transformation (17) To be specific this transformation describes a process when the original qubit is encoded in the system a while the flipped qubit is in the system c. The density operator describing the state of the system c at the output is

$$\rho_c^{(out)} = \frac{1}{3}|\psi^\perp\rangle\langle\psi^\perp| + \frac{1}{3}\mathbf{1}, \tag{62}$$

and the fidelity of the spin flipping is $\mathcal{F} = 2/3$.

A natural way to encode a qubit into a physical system is to utilize polarization polarization states of a single photon. In this case the Universal NOT gate can be realized via the stimulated emission. The key idea of DeMartini's experiment is based on the proposal that universal quantum machines [40] such as quantum cloner can be realized with the help of stimulated emission in parametric down conversion [82, 54, 56]. Specifically let us consider a qubit to be encoded in a polarization state of a photon. This photon is injected as the input state into an optical parametric amplifier amplifier (OPA) physically consisting of a nonlinear (NL) BBO (β-barium-borate) crystal cut for Type II phase matching and excited by a pulsed mode-locked ultraviolet laser laser UV having pulse duration $\tau \approx 140f$ sec and wavelength (wl) $\lambda_p = 397.5nm$ associated to pulse duration. The relevant modes of the NL 3-wave interaction were the spatial modes with wave-vector (wv) k_1 and k_2 each supporting the two horizontal (H) and vertical (V) linear-polarizations $((\Pi)$ of the interacting photons, e.g. Π_{1H} is the horizontal polarization unit vector associated with k_1. The OPA was frequency degenerate, i.e. the interacting photons had the same wl's $\lambda = 795nm$. The action of OPA under suitable conditions can be described by a simplified Hamiltonian

$$\hat{H}_{int} = \kappa\left(\hat{a}_\psi^\dagger\hat{b}_{\psi^\perp}^\dagger - \hat{a}_{\psi^\perp}^\dagger\hat{b}_\psi^\dagger\right) + h.c. \tag{63}$$

A property of the device, of key importance in the context of the present work, is its amplifying behavior with respect to the polarization Π of the interacting photons. It has been shown theoretically in Ref. [54] and in a recent experiment on universal quantum cloning [56, 59] that the amplification efficiency of this type of OPA under injection by *any* externally injected quantum field, e.g. consisting of a single photon or of a classical "coherent" field, can be made *independent* of the polarization state of the field. In other word the OPA "gain" is independent of any (*unknown)* polarization state of the injected field: This precisely represents the necessary *universality (U)* property of the U-NOT gate. For this reason in Eq. (63) we have denoted the creation $\hat{a}^\dagger_\psi$ $(\hat{b}^\dagger_\psi)$ and annihilation $\hat{a}_\psi$ $(\hat{b}_\psi)$ operators of a photon in mode k_1 (k_2) with subscripts ψ (or $\psi^\perp$) indication the invariance of the process with respect to polarization states of the input photon.

Let us consider the input photon in the mode k_1 to have a polarization ψ. I will describe this polarization state as $\hat{a}^\dagger_\psi|0,0\rangle_{k_1} = |1,0\rangle_{k_1}$, where I have used notation introduced by Simon et al. [54], i.e. the state $|m,n\rangle_{k_1}$ represents a state with m photons of the mode k_1 having the polarization ψ, while n photons have the polarization $\psi^\perp$. Initially there are no excitations in the mode k_2. The initial polarization state of these two modes reads $|1,0\rangle_{k_1} \otimes |0,0\rangle_{k_2}$ and it evolves according the Hamiltonian Hamiltonian (63):

$$
\begin{aligned}
\exp(-i\hat{H}_{int}t)&|1,0\rangle_{k_1} \otimes |0,0\rangle_{k_2} \simeq |1,0\rangle_{k_1} \otimes |0,0\rangle_{k_2} \\
&- i\kappa t \left(\sqrt{2}|2,0\rangle_{k_1} \otimes |0,1\rangle_{k_2} - |1,1\rangle_{k_1} \otimes |1,0\rangle_{k_2} \right)
\end{aligned}
\tag{64}
$$

This approximation for the state vector describing the two modes at times $t > 0$ is sufficient since the values κt are usually very small (see below). The zero order term corresponds to the process when the input photon in the mode k_1 did not interact in the nonlinear medium, while the second term describes the first order process in the OPA. This second term is formally equal (up to a normalization factor) to the right-hand side of Eq. (17). Here the state $|2,0\rangle_{k_1}$ describing two photons of the mode k_1 in the polarization state ψ corresponds to the state $|\psi\psi\rangle$. This state-vector describes the cloning cloning of the original photon [54, 56]. The vector $|0,1\rangle_{k-2}$ describes the state of the mode k_2 with a single photon in with the polarization $\psi^\perp$. That is, this state vector represents the flipped version of the input.

To see that the stimulated emission is indeed responsible for creation of the flipped qubit, let us compare the state (64) with the output of the OPA when the vacuum is injected into the nonlinear crystal. In this case, to the same order of approximation as above we obtain

$$
\exp(-i\hat{H}_{int}t)|0,0\rangle_{k_1} \otimes |0,0\rangle_{k_2} \simeq |0,0\rangle_{k_1} \otimes |0,0\rangle_{k_2}
$$

$$
i\kappa t \left(|1,0\rangle_{k_1} \otimes |0,1\rangle_{k_2} - |0,1\rangle_{k_1} \otimes |1,0\rangle_{k_2} \right)
\tag{65}
$$

We see that the flipped qubit described by the state vector $|0,1\rangle_{k_2}$ in the right-hand sides of Eqs.(64) and (65) does appear with different amplitudes corresponding to the ratio of probabilities to be equal to $1 : 2$. This ratio has been measured in the experiment [59].

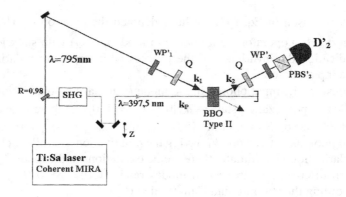

Figure 2.   Schematic description of the experimental verification of the universality of the flip operation. A coherent state coherent state of attenuated laser laser field with wl $\lambda = 795nm$ has been used in the experiment. The source is Ti:Sa Coherent MIRA pulsed laser providing by Second Harmonic Generation Second harmonic generation(SHG) the OPA "pump" field associated with the spatial mode with wv k_p and wl λ_p. A small portion of the laser radiation at wl λ was directed along the OPA injection mode k_1. The parametric amplification, with calculated "gain" $g = 0.31$, was detected at the OPA output mode k_1 by D_1', a Si linear photodiode SGD100. The time superposition in the NL crystal of the "pump" and of the "injection" pulses was assured by micrometric displacements (Z) of a 2-mirror optical "trombone". Various Π−states of the injected pulse were prepared by the set $(WP\prime_1 + Q)$ consisting of a Wave-plate (either $\lambda/2$ or $\lambda/4$) and of a 4.3mm X-cut Quartz plate. These states were then analyzed after amplification and before detection on mode k_2 by an analogous optical set $(WP\prime_2 + Q + \Pi-analyzer)$, the last device being provided by the *Polarizing Beam Splitter PBS_2'*. In the experiment all the 4.5mm thick X-cut quartz plates (Q) provided the compensation of the unwanted beam walk-off effects due to the birefringence of the NL crystal.

8.1   Universality

On the "microscopic" quantum level the justification of this U-property of the OPA amplifier amplifier resides in the SU(2) rotational invariance of the NL interaction Hamiltonian Hamiltonian when the spatial orientation of the OPA NL Type II crystal makes it available for the generation of 2-photon entangled "singlet" states by Spontaneous Parametric Down Conversion (SPDC), i.e. by injection of the "vacuum field" [54]. However I should note that in the present context the *universality* property, i.e. the $\Pi-insensitivity$ of the parametric amplification "gain" g, is a "macroscopic" classical feature of the OPA device. As a consequence, it can be tested equally well either by injection of "classical", e.g. coherent (Glauber) fields or of a "quantum" states of radiation, e.g. a single-photon Fock-state. De Martini *et al.* [59] carried out successfully both tests, leading to identical results. Below I describe the experiment

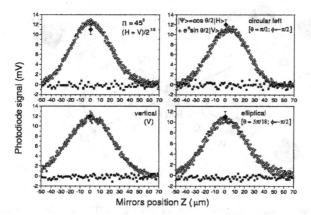

Figure 3. Experimental verification of the universality of OPA. The plots show the amplification pulses detected by D'_2 on the OPA output mode, k_2. Each plot corresponds to a definite $\Pi-$state, $|\psi\rangle = [\cos(\vartheta/2)\,|H\rangle + \exp(i\phi)\,\sin(\vartheta/2)\,|V\rangle]$, either *linear* $-\,\Pi$, i.e. $\vartheta = 0, \pi/2$, π; $\phi = 0$, or *circular* $-\Pi$, i.e. $\vartheta = \pi/2$; $\phi = \pm\pi/2$, or *elliptical* $-\,\Pi$, in the very general case: $\vartheta = 5\pi/18$; $\phi = -\pi/2$.

corresponding to the injection by attenuated "coherent" laser laser field – see Fig. 2

The *universality* condition is demonstrated by the plots in Fig. 3 showing the amplification pulses detected by D'_2 on the OPA output mode, k_2. Each plot corresponds to a definite $\Pi-$state, $|\psi\rangle = [\cos(\vartheta/2)\,|H\rangle + \exp(i\phi)\,\sin(\vartheta/2)\,|V\rangle]$, either *linear* $-\,\Pi$, i.e. $\vartheta = 0$, $\pi/2$, π; $\phi = 0$, or *circular* $-\Pi$, i.e. $\vartheta = \pi/2$; $\phi = \pm\pi/2$, or *elliptical* $-\,\Pi$, in the very general case: $\vartheta = 5\pi/18$; $\phi = -\pi/2$. We may check that the corresponding amplification curves, each corresponding to a standard injection pulse$_1$ with an average photon number $N \approx 5 \times 10^3$, are almost identical. For more generality, the *universality* condition as well as the insensitivity of this condition to the value of N is also demonstrated by the *single* experimental data reported, with different scales, at the top of each amplification plot and corresponding to injection pulses with: $N \approx 5 \times 10^2$. Single-photon tests of the same conditions were also carried out with a different experimental setup, as said.

8.2   Optimality

Let us move to the main subject of the experiment under consideration, i.e. the quantum U-NOT gate. In virtue of the tested *universality* of the OPA amplification, it is of course sufficient to consider here the OPA injection by a *single-photon* in *just one* $\Pi-$state, for instance in the *vertical* $-\,\Pi$ state.

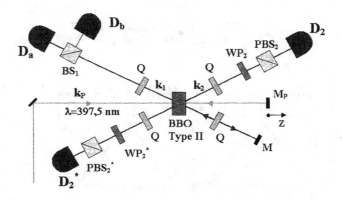

Figure 4.   Experimental realization of the quantum U-NOT gate. Consider the k_p pump mode, i.e., the "towards R" excitation. A SPDC process created single photon-pairs with wl $\lambda = 795nm$ in entangled *singlet* $\Pi -$ *states*, i.e. rotationally invariant, as said. One photon of each pair, emitted over k_1 was reflected by a spherical mirror M onto the NL crystal where it provided the $N = 1$ *quantum injection* into the OPA amplifier amplifier excited by the UV "pump" beam associated to the back reflected mode $-k_p$. In the experiment the flipping of a single photon in a state $|\psi\rangle = |V\rangle$ has been considered. In the experiment, owing to a spherical mirror M_p with 100% reflectivity and micrometrically adjustable position Z, the UV pump beam excited the same NL OPA crystal amplifier amplifier in both directions k_p and $-k_p$, i.e. correspondingly oriented towards the Right (R) and Left (L) sides of the figure. Because of the low intensity of the UV beam, the 2 photon injection probability, $N = 2$ has been evaluated to be $\approx 3.5 \times 10^{-4}$ smaller that for the $N = 1$ condition. The twin photon emitted over k_2 was $\Pi -$ *selected* by the devices $(WP_2 + PBS_2)$ and then detected by D_2, thus providing the "trigger" of the overall *conditional experiment*. All detectors in the experiment were equal active SPCM-AQR14 with quantum efficiency: $QE \approx 55\%$. Because of the EPR non-locality non-locality implied by the singlet state, the $\Pi -$ *selection* on channel k_2 provided the realization on k_1 of the state $|\psi\rangle = |V\rangle$ of the injected photon. All the X-cut quartz plates Q provided the compensation of the unwanted beam walk-off effects due to the birefringence of the NL crystal. Consider the "towards L" amplification, i.e. the amplification process excited by the mode $-k_p$, and do account in particular for the OPA output mode k_2. The $\Pi-$state of the field on that mode was analyzed by the device combination $(WP_2^* + PBS_2^*)$ and measured by the detector D_2^*. The detectors D_a, D_b were coupled to the field associated with the mode k_1. The experiment was carried out by detecting the rate of the 4-coincidences involving all detectors $[D_2^* D_2 D_a D_b]$.

Accordingly, Fig. 4 shows a layout of the single-photon, $N = 1$, quantum-injection experiment with input state $|\psi\rangle = |V\rangle$.

From the analysis presented by Simon *et al.* [54, 56] it follows that the state of the field emitted by the OPA indeed realizes the U-NOT gate operation, i.e. the "optimal" realization of the "anti-cloning" of the original qubit originally encoded in the mode k_1. The flipped qubit at the output is in the mode k_2. As it has been shown earlier the state created by the U-NOT gate is not pure. The is a minimal amount of noise induced by the process of flipping which is inevitable

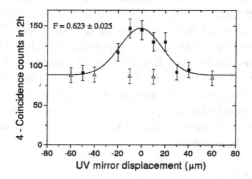

Figure 5. Experimental verification of the optimality of the U-NOT gate. The plots show experimental 4-coincidence data as function of the time superposition of the UV pump and of the injected single-photon pulses. That superposition was expressed as function of the micrometric displacement Z of the back-reflecting mirror M_p. The height of the central peak expresses the rate measured with the $\Pi - analyzer$ of mode k_2 set to measure the "correct" horizontal (H) polarization, i.e. the one orthogonal to the (V) polarization of the $\Pi - state$, $|\psi\rangle = |V\rangle$ of the injected, input single photon, $N = 1$. By turning by 90^0 the $\Pi - analyzer$, the amount of the "noise" contribution is represented by a "flat" curve. In the experiment the "noise" was provided by the OPA amplification of the unavoidable "vacuum" state associated with the mode k_1.

in order to preserve complete-positiveness of the Universal NOT gate. This mixed state is described by the density operator (62). The polarization state of the output photon in the mode k_2 in the experiment performed by De Martini *et al.* [59] is indeed described by this density operator.

The plots of Fig. 5 report the experimental 4-coincidence data as function of the time superposition of the UV pump and of the injected single-photon pulses.

The main result of the experiment [59] consists of the determination of the ratio R^* between the height of the central peak and the one of the flat "noise" contribution. To understand this ratio let us firstly note that the most efficient stimulation process in the OPA is achieved when a perfect match (overlap) between the input photon and the photon produced by the source is achieved. This situation corresponds the value of the mirror position Z equal to zero [see Eq.(64)]. As soon as the mirror is displaced from the position the two photons do not overlap properly and the stimulation is less efficient. Correspondingly, the spin flip operation is more noisy. In the limit of large displacements Z the spin flipping is totally random due to the fact that the process corresponds to injecting the vacuum into the crystal [see Eq.(65)]. The theoretical ration between the corresponding probabilities is 2. In the experiment [59] the ratio has been found to be $R^* = (1.66 \pm 0.20)$. This corresponds to a measured value of the fidelity fidelity of the U-NOT apparatus: $\mathcal{F}^* = (0.623 \pm 0.025)$ to be compared with the theoretical value: $F = 2/3 = 0.666$. Note that the height of the central peak does not decrease towards zero for large Z's. This effect is due to the finite time-resolution of the 4-coincidence electronic apparatus, which is in the nanosecond range. It would totally disappear if the resolution could be pushed into the sub-picosecond range, i.e. of the order of the time duration of the OPA pump and injection pulses. It can be easily found that the spurious out of resonance plateau of the central peak should indeed reproduce the size of the "noise" condition measured on the mode k_2. As we can see, this is indeed verified by the experiment.

9.   Conclusion

In these notes I have overviewed two types of universal quantum machines - universal quantum cloners and the universal NOT gate. In order to motivate my discussion I have briefly discussed the role of no-signaling condition in quantum mechanics and I have shown that quantum cloning cannot be used for super-luminal signaling as originally suggested by Nick Herbert. The main message of my notes can be summarized as follows: Even though quantum mechanics does open new perspectives in information processing it also imposses new bounds on how well we can manipulate with information encoded in quantum systems. In quantum processes we can redistribute information according to specific rules, but we cannot generate it (e.g. in a sense that we would be able to perform better state estimation).

Acknowledgments

I would like to thank many of my friends for long standing collaboration and discussions on issues presented in these notes. First of all I thank Mark Hillery for his contribution to my understanding of universal quantum machines. I want to thank Reinhard Werner, Radoslav Derka, Christoph Simon, Artur Ekert, Peter Knight, Francesco De Martini, Nicolas Gisin, and Sam Braunstein. This work was supported by the IST project EQUIP under the contract IST-1999-11053. I also acknowledges a support from the Science Foundation Ireland offered to me via the E.T.S. Walton award.

Appendix: Proof of Theorem I

I have already shown in Section 7 [see Eq.(59)] that for the measurement-based strategy the error Δ attains the value $1/(N + 2)$. The more difficult part, however, is to show that no other scheme [i.e., quantum scenario] can do better. In what follows I will largely follow the arguments in Ref. [50].

Note first that the functional Δ is invariant with respect to unitary rotations (basis changes) in $\mathcal{H}$: When T is any admissible map, and U is a unitary on $\mathcal{H}$, the map $T_U(\sigma) = U^* T(U^{\otimes N} \sigma U^{*\otimes N})U$ is also admissible, and satisfies $\Delta(T_U) = \Delta(T)$. Moreover, the functional Δ is defined as the maximum of a collection of linear functions in T, and is therefore convex. Putting these observations together we get

$$\Delta(\overline{T}) \leq \int dU \, \Delta(T_U) = \Delta(T), \tag{A.1}$$

where $\overline{T} = \int dU \, T_U$ is the average of the rotated operators T_U with respect to the Haar measure on the unitary group. Thus $\overline{T}$ is at least as good as T, and has the additional "covariance property" $\overline{T}_U = \overline{T}$. Without loss we can therefore assume from now on that $T_U = T$ for all U.

An advantage of this assumption is that a very explicit general form for such covariant operations is known by a variant of the Stinespring Dilation Theorem (see [50] for a version adapted to our needs).

The form of T is further simplified in our case by the fact that both representations involved are irreducible: the defining representation of SU(2) on $\mathcal{H}$, and the representation by the operators $U^{\otimes N}$ restricted to the symmetric subspace $\mathcal{H}_+^{\otimes N}$. Then T can be represented as a discrete convex combination $T = \sum_j \lambda_j T_j$, with $\lambda_j \geq 0, \sum_j \lambda_j = 1$, and T_j admissible and covariant maps in their own right, but of an even simpler form. Covariance covariance of T already implies that the maximum can be omitted from the definition (56) of Δ, because the fidelity no longer depends on the pure state chosen. In a convex combination of covariant operators we therefore get

$$\Delta(T) = \sum_j \lambda_j \Delta(T_j). \tag{A.2}$$

Minimizing this expression is obviously equivalent to minimizing with respect to the discrete parameter j.

Let us write the general form of the extremal instruments T_j in terms of expectation values of the output state for an observable X on $\mathcal{H}$:

$$\text{Tr}\big(T(\sigma)X\big) = \text{Tr}\big[\sigma V^*(X \otimes \mathbb{1})V\big], \tag{A.3}$$

where $V : \mathcal{H}_+^{\otimes N} \rightarrow \mathcal{H} \otimes \mathbb{C}^{2j+1}$ is an isometry intertwining the respective representations of SU(2), namely the restriction of the operators $U^{\otimes N}$ to $\mathcal{H}_+^{\otimes N}$ (which has spin $N/2$) on the one hand, and the tensor product of the defining representation (spin-$1/2$) with the irreducible spin-j representation. By the triangle inequality for Clebsch-Gordan reduction, this implies $j = (N/2) \pm (1/2)$, so only two terms appear in the decomposition of T. It remains to compute $\Delta(T_j)$ for these two values.

The basic idea is to use the intertwining property of the isometry V for the generators S_α, J_α, and $L_\alpha, \alpha = 1, 2, 3$ of the SU(2)-representations on $\mathcal{H}, \mathbb{C}^{2j+1}$ and $\mathcal{H}_+^{\otimes N}$, respectively. We will show that

$$V^*(S_\alpha \otimes 1_j)V = \mu_j \, L_\alpha, \tag{A.4}$$

where μ_j is some constant depending on the choice of j. That such a constant exists is clear from the fact that the left hand side of this equation is a vector operator (with components labeled by $\alpha = 1, 2, 3$), and the only vector operators in an an irreducible representation of SU(2) are multiples of angular momentum Angular momentum (in this case L_α). The constant μ_j can be expressed in terms of a $6j$-symbol, but can also be calculated in an elementary way by using the intertwining property, $V L_\alpha = (S_\alpha \otimes 1 + 1 \otimes J_\alpha)V$ and the fact that the angular momentum

Angular momentum squares $\mathbf{J}^2 = \sum_\alpha J_\alpha^2 = j(j+1)$, $\mathbf{S}^2 = 3/4$, and $\mathbf{L}^2 = N/2(N/2+1)$ are multiples of the identity in the irreducible representations involved, and can be treated as scalars:

$$
\begin{aligned}
\mu_j \mathbf{L}^2 &= \sum_\alpha V^*(S_\alpha \otimes \mathbb{1}_j) V L_\alpha \\
&= \mathbf{S}^2 + \sum_\alpha V^*(S_\alpha \otimes J_\alpha) V.
\end{aligned}
\tag{A.5}
$$

The sum on the right hand side can be obtained as the mixed term of a square, namely as

$$
\frac{1}{2}\left(\sum_\alpha V^*(S_\alpha \otimes \mathbb{1} + \mathbb{1} \otimes J_\alpha)^2 V - \mathbf{S}^2 - \mathbf{J}^2 \right)
\tag{A.6}
$$

$$
= (\mathbf{L}^2 - \mathbf{S}^2 - \mathbf{J}^2).
\tag{A.7}
$$

Combining these equations we find

$$
\mu_j = \frac{1}{2} + \frac{\mathbf{S}^2 - \mathbf{J}^2}{2\mathbf{L}^2} =
\begin{cases}
\frac{1}{N} & for\, j = N\frac{1}{2+\frac{1}{2}} \\
\frac{-1}{N+2} & for\, j = N\frac{1}{2-\frac{1}{2}}. \\
\\
\end{cases}
\tag{A.8}
$$

Let us combine equations (A.3) and (A.4) to get the error quantity Δ from equation (56), with the pure one-particle density matrix density matrix $\rho = \frac{1}{2}\mathbb{1} + S_3$:

$$
\Delta(T) = \mathrm{Tr}(V^*(\rho \otimes \mathbb{1}) V \rho^{\otimes N}) = \frac{1}{2}(1 + N\mu_j).
\tag{A.9}
$$

With equation (A.8) we find

$$
\Delta(T) =
\begin{cases}
1 & for\, j = N\frac{1}{2+\frac{1}{2}} \\
\frac{1}{N+2} & for\, j = N\frac{1}{2-\frac{1}{2}}. \\
\\
\end{cases}
\tag{A.10}
$$

The first value is the largest possible fidelity fidelity for getting the state ρ from a set of N copies of ρ. The fidelity 1 is expected for this trivial task, because taking any one of the copies will do perfectly. On the other hand, the second value is the minimal fidelity, which we were looking for. This clearly coincides with the value (59), so the Theorem is proved.

The Theorem as it stands concerns the task of producing just one particle in the U-NOT state of the input. From previous results we see that it is valid also in the case of many outputs. We see that the maximum fidelity is achieved by the classical process via estimation: in equation (58) we just have to replace the output state $(\mathbb{1} - |\phi\rangle\langle\phi|)$ by the desired tensor power. Hence once again the optimum is achieved by the scheme based on classical estimation. Incidentally, this shows that the multiple outputs from such a device are completely unentangled, although they may be correlated.

References

[1] D. Bohm, *Quantum Theory*, (Prentice-Hall, Englewood Cliffs, NJ, 1951), pp. 611–623, reprinted in [3], pp. 356–368.

[2] A. Peres *Quantum Theory: Concepts and Methods* (Kluwer, Dordrecht, 1993).

[3] J. A. Wheeler and W. H. Zurek, *Quantum Theory and Measurement*, (Princeton University Press, Princeton, NJ, 1983).

[4] A. Einstein, B. Podolsky, and N. Rosen, "Can quantum-mechanical description of physical reality be considered complete?" *Phys. Rev.* **47**, 777 (1935), reprinted in [3], p. 138.

[5] J. S. Bell, "On the Einstein-Podolsky-Rosen paradox," *Physics* (NY) **1**, 195 (1964), reprinted in [3], p. 403.

[6] J. S. Bell, *Speakable and Unspeakable in Quantum Mechanics: Collected papers on quantum philosophy* (Cambridge University Press, Cambridge, 1987).

[7] A. Aspect, J. Dalibar, and G. Roger, "Experimental test of Bell inequalities using time-varying analyzers" *Phys. Rev. Lett.* **49**, 1804 (1982).

[8] G. Weihs, T. Jennewein, C. Simon, H. Weifurter, and A. Zeilinger, "Violation of Bell's inequality under strict Einstein Locality conditions" *Phys. Rev. Lett.* **81**, 5039 (1998).

[9] J. F. Clauser, M. A. Horne, A. Shimony, and R. A. Holt, "Proposed experiment to test local hidden-variable theories," *Phys. Rev. Lett.* **23**, 880 (1969), reprinted in J. A. Wheeler and W. H. Zurek, in Ref. 3, pp. 409–413.

[10] J. F. Clauser and M. A. Horne, "Experimental consequences of objective local theories," *Phys. Rev. D* **10**, 526 (1974).

[11] J. F. Clauser and A. Shimony, "Bell's theorem: Experimental tests and implications," *Rep. Prog. Phys.* **41**, 1881 (1978).

[12] B. S. Cirel'son, "Quantum generalizations of Bell's inequality," *Lett. Math. Phys.* **4**, 93 (1980).

[13] D. M. Greenberger, M. A. Horne, and A. Zeilinger, "Going beyond Bell's theorem," in *Bell's Theorem, Quantum Theory, and Conceptions of the Universe*, edited by M. Kafatos (Kluwer Academic, Dordrecht, The Netherlands, 1989), pp. 69–72;

[14] D. M. Greenberger, M. A. Horne, A. Shimony, and A. Zeilinger, "Bell's theorem without inequalities," *Am. J. Phys.* **58**, 1131 (1990).

[15] S. Popescu and D. Rohrlich, "Quantum nonlocality as an axiom," *Found. Phys.* **24**, 379 (1994).

[16] L. Hardy, "Nonlocality for two particles without inequalities for almost all entangled states," *Phys. Rev. Lett.* **71**, 1665 (1993).

[17] P. G. Kwiat and L. Hardy, "The mystery of the quantum cakes," *Am. J. Phys.* **68**, 33 (2000).

[18] M. Hillery and B. Yurke, "Bell's theorem and beyond," *Quantum Opt.* **7**, 215 (1995).

[19] G. C. Ghirardi, A. Rimini, and T. Weber, "A general argument against superluminal transmission through the quantum mechanical measurement process," *Lett. Nuovo Cimento* **27**, 293 (1980).

[20] P. J. Bussey, " "Super-luminal communication" in Einstein-Podolsky-Rosen experiments," *Phys. Lett. A* **90**, 9 (1982).

[21] T. F. Jordan, "Quantum correlations do not transmit signals," *Phys. Lett. A* **94**, 264 (1983).

[22] P. J. Bussey, "Communication and non-communication in Einstein-Rosen experiments," *Phys. Lett. A* **123**, 1 (1987).

[23] K. Kraus, "Quantum theory does not require action-at-a-distance," *Found. Phys. Lett.* **2**, 1 (1989).

[24] H. Scherer and P. Busch, "Problem of signal transmission via quantum correlations and Einstein incompleteness of quantum mechanics," *Phys. Rev. A* **47**, 1647 (1993)

[25] S. M. Roy and V. Singh, "Hidden variable theories without non-local signaling and their experimental tests," *Phys. Lett. A* **139**, 437 (1989).

[26] J. Sarfatti "Design for a superluminal signal device" *Phys. Essays* **4**, 315 (1991).

[27] P. R. Holland and J. P. Vigier, "The quantum potential and signaling in the EPR experiment." *Found. Phys.* **18**, 741 (1988).

[28] Y. Aharonov and D. Z. Albert, "Can we make a sense out of the measurement process in relativistic quantum mechanics" *Phys. Rev. A* **24**, 359 (1981).

[29] S. M. Roy and V. Singh, "Tests of signal locality and Einstein-Bell locality for multiparticle systems," *Phys. Rev. Lett.* **67**, 2761 (1991).

[30] J. P. Jarrett, "On the physical significance of the locality conditions in the Bell arguments," *Noûs* **18**, 569 (1984).

[31] L. E. Ballentine and J. P. Jarrett, "Bell's theorem: Does quantum mechanics contradict relativity?" *Am. J. Phys.* **55**, 696 (1987).

[32] A. Shimony, "An exposition of Bell's theorem," in *Sixty-Two Years of Uncertainty: Historical, Philosophical, and Physical Inquiries into the Foundations of Quantum Mechanics*, edited by A. I. Miller (Plenum, New York, 1990), pp. 33–43, reprinted in A. Shimony, *Search for a Naturalistic World View. Vol. II: Natural Science and Metaphysics*, (Cambridge University Press, New York, 1993), pp. 90–103.

[33] N. D. Mermin, "The best version of Bell's theorem," in *Fundamental Problems in Quantum Theory: A Conference held in Honor of Professor John A. Wheeler* (edited by D. M. Greenberger and A. Zeilinger), *Ann. (NY) Acad. Sci.* **755**, pp. 616–623 (1995).

[34] A. Shimony, in *Proceedings of the International Symposium on Foundations of Quantum Mechanics, Tokyo, 1983*, ed. S. Kamefuchi *et al.* (Physical Society of Japan, Tokyo, 1984), p. 225.

[35] M. Redhead, *Incompleteness, Nonlocality and Realism* (Clarendon, Oxford, 1987).

[36] N. Herbert, "FLASH - a superluminal communicator based upon a new kind of quantum measurement" *Found. Phys.* **12**, 1171 (1982).

[37] D. Dieks, "Communication by EPR devices," *Phys. Lett. A* **92**, 271 (1982).

[38] W. K. Wootters and W. H. Zurek, "A single quantum cannot be cloned," *Nature* **299**, 802 (1982).

[39] V. Bužek and M.Hillery, "Quantum copying: Beyond the nonclonning theorem." *Phys. Rev. A* **54**, 1844 (1996).

[40] G.Alber, et al., 'Quantum Information: An Introduction to Basic Theoretical Concepts and Experiments." Springer Tracts in Modern Physics vol. 173 (Springer Verlag, Berlin, 2001).

[41] V.Bužek, S. Braunstein, M. Hillery, and D.Bruß, "Quantum copying: A network" *Phys. Rev. A* **56**, 3446 (1997).

[42] N. Gisin and S. Massar, "Optimal quantum cloning machines" *Phys. Rev. Lett.* **79**, 2153 (1997).

[43] D. Bruß, D. P. Di Vincenzo, A. Ekert, C. Machiavello, C. Fuchs, and J. A.Smolin, "Optimal universal and state-dependent quantum cloning." *Phys. Rev. A* **57**, 2368 (1998).

[44] D.Bruß, A. Ekert, and C. Machiavello: "Optimal universal quantum cloning and state estimation." *Phys. Rev. Lett.* **81**, 2598 (1998).

[45] V. Bužek, M. Hillery, and P.L.Knight, "Flocks of quantum clones: Multiple copying of qubits." *Fortschr. Phys.* **48**, 521 (1998).

[46] R.F. Werner, "Optimal cloning of pure states." *Phys. Rev. A* **58**, 1827 (1998).

[47] P. Zanardi, "Quantum cloning in d dimensions." *Phys. Rev. A* **58**, 3484 (1998).

[48] V.Bužek and M. Hillery, "Universal optimal cloning of qubits and quantum registers." *Phys. Rev. Lett.* **58**, 5003 (1998).

[49] L. M. Duan and G.G.Guo, "Probabilistic cloning and identification of linearly independent quantum states." *Phys. Rev. Lett.* **80**, 4999 (1998).

[50] M. Keyl and R. F. Werner, "Optimal cloning of pure states, judging single clones." *J. Phys. A* **40**, 3283 (1999).

[51] G. Lindblad, "Cloning the quantum oscillator." *J. Phys. A* **33**, 5059 (2000).

[52] N. J. Cerf, "Asymmetric cloning in any dimension" *J. Mod. Opt.* **47**, 187 (2000).

[53] N. J. Cerf, "Pauli cloning of a quantum bit." *Phys. Rev. Lett.* **84**, 4497 (2000).

[54] C. Simon, G. Weihs, and A. Zeilinger, "Optimal quantum cloning and universal NOT without quantum gates." *J. Mod. Opt.* **47**, 233 (2000).

[55] C. Simon, G.Weihs, and A. Zeilinger, "Optimal quantum cloning via stimulated emission." *Phys. Rev. Lett.* **84**, 2993 (2000).

[56] A.Lamas-Linares, Ch.Simon, J.C.Howell, and D.Bouwmeester "Experimental quantum cloning of single photons." *Science* **296**, 712 (2002).

[57] F. De Martini and V.Mussi: "Universal quantum cloning and macroscopic superposition in parametric amplification." *Fort. der Physik* **48**, 413 (2000).

[58] F. De Martini, V. Mussi, and F. Bovino, "Schrödinger cat states and optimum universal quantum cloning by entangled parametric amplification." *Opt. Commun.* **179**, 581 (2000).

[59] F. De Martini, V.Bužek, F.Sciarrino, and C.Sias, "Experimental realization of the quantum universal NOT gate." *Nature* **419** No. 6909, 815 (2002).

[60] N. Gisin, "Quantum cloning without signaling" *Phys. Lett. A* **242**, 1 (1998).

[61] L. Hardy and D. D. Song, "No signalling and probabilistic quantum cloning." *Phys. Lett. A* **259**, 331 (1999).

[62] C. Simon, G. Weihs, and A. Zeilinger, "Quantum cloning and signaling." *Acta Phys. Slov.* **49**, 755 (1999).

[63] S. Ghosh, G. Kar, and A. Roy, "Optimal cloning and no signalling." *Phys. Lett. A* **261**, 17 (1999).

[64] A. K. Pati, "Probabilistic exact cloning and probabilistic no-signaling." *Phys. Lett. A* **270**, 103 (2000).

[65] N. Gisin, "Stochastic quantum dynamics and relativity" *Helv. Phys. Acta* **62**, 363 (1989).

[66] N. Gisin, "Weinberg non-linear quantum-mechanics and superluminal communications" *Phys. Lett. A* **143**, 1 (1990).

[67] N. Gisin and M. Rigo, "Relevant and irrelevant nonlinear Schrodinger equations," *J. Phys. A* **28**, 7375 (1995).

[68] G. Svetlichny, "Quantum formalism with state-collapse and superluminal communication" *Found. Phys.* **28**, 131 (1998).

[69] G. Svetlichny, "The space-time origin of quantum mechanics: Covering law" *Found. Phys.* **30**, 1819 (1998).

[70] C. Simon, V. Bužek, and N. Gisin, "The no-signaling condition and quantum dynamics." *Phys. Rev. Lett.* **87**, 170405 (2001).

[71] M. Ziman and P. Štelmachovič, "Quantum theory: Kinematics, linearity & no-signaling condition" Los Alamost e-print arXiv quant-ph/0211149 (2002).

[72] P. G. Kwiat, K. Mattle, H. Weifurter, A. Zeilinger, A. V. Sergeinko, and A. V. Shih, "New high-intensity source of polarization-entangled photon pairs" *Phys. Rev. Lett.* **75**, 433 (1995).

[73] A. Holevo, *Probabilistic and Statistical Aspects of Quantum Theory* (Amsterdam, North Holland, 1982).

[74] C. W. Helstrom, *Quantum Detection and Estimation Theory* (Academic Press, New York, 1976).

[75] C. W. Gardiner, *Quantum noise* (Springer, Berlin, 1991).

[76] C. W. Gardiner and P. Zoller, *Quantum noise : a handbook of Markovian and non-Markovian quantum stochastic methods with applications to quantum optics* (Springer, Berlin, 2000).

[77] E. Wigner, "The probability of the existence of a self-reproducing unit," in *The Logic of Personal Knowledge* (The Free Press, 1961), p.231.

[78] H. P. Yuen, "Amplification of quantum states and noiseless photon amplifier" *Phys. Lett. A* **113**, 405 (1986).

[79] H. Barnum, C. M. Caves, C. A. Fuchs, R. Jozsa, and B. Schumacher, "Non-commuting mixed states cannot be broadcast" *Phys. Rev. Lett.* **76**, 2818 (1996).

[80] M. Koashi and N. Imoto, "No-cloning theorem of entangled states" *Phys. Rev. Lett.* **81**, 4264 (1998).

[81] G. Lindblad, "A general no-cloning theorem" *Lett. Math. Phys.* **47**, 189 (1999).

[82] L. Mandel, "Is a photon amplifier always polarization dependent?" *Nature* **304**, 188 (1983).

[83] W. M. Elsasser, *The Physical Foundations of Biology* (Pergamon Press, London, 1958). Without going into detail it is interesting to turn attention on one of Elsasser's conclusion according to which a computing machine which could store all the information contained in a germ-cell would be inconceivably large. Today, half a century later, we believe that Hilbert spaces of quantum computers [84] are large enough to accommodate comfortably all the relevant information.

[84] M. A. Nielsen and I. L. Chuang, *Quantum Computation and Quantum Information* (Cambridge University Press, Cambridge, 2000).

[85] S. Massar and S. Popescu, "Optimal extraction of information from finite ensembles." *Phys. Rev. Lett.* **74**, 1259 (1995).

[86] R. Derka, V. Bužek, and A. Ekert, "Universal algorithm for optimal state estimation from finite ensembles." *Phys. Rev. Lett.* **80**, 1571 (1998).

[87] S. Massar and S. Popescu, "Amount of information obtained by a quantum measurement." *Phys. Rev. A* **62**, art.no. 062303 (2000).

[88] S. Massar and R. D. Gill, "State estimation for large ensembles." *Phys. Rev. A* **61**, art. no. 042312 (2000)

[89] V. Bužek and R. Derka, "Quantum observations" in *Coherence and Statistics of Photons and Atoms*, ed. J.Peřina (John Wiley & Sons, New York, 2001) pp. 198–261.

[90] V. Bužek, M. Hilery, and R. Werner, "Optimal manipulations with qubits: Universal NOT gate." *Phys. Rev. A* **60**, R2626 (1999).

[91] V. Bužek, M. Hilery, and R. Werner, "Universal NOT gate." *J. Mod. Opt.* **47**, 211 (2000).

[92] P. Rungta, V. Bužek, C.M. Caves, M. Hillery, and G.J. Milburn, 'Universal state inversion and concurrence in arbitrary dimensions." *Phys. Rev. A* **64**, 042315 (2001).

[93] N. Gisin and S. Popescu, "Spin flips and quantum information for anti-parallel spins." *Phys. Rev. Lett.* **83**, 432 (1999).

[94] S. L. Braunstein, V. Bužek, and M. Hillery, "Quantum information distributors: Quantum network for symmetric and asymmetric cloning in arbitrary dimension and continuous limit." *Phys. Rev. A* **63**, 052313 (2001).

AN INTRODUCTION TO QUANTUM IMAGING

L.A. Lugiato

INFM, Dipartimento di Scienze Chimiche Fisiche e Matematiche, Università dell'Insubria, Via Valleggio 11, 22100 Como, Italy

A. Gatti

INFM, Dipartimento di Scienze Chimiche Fisiche e Matematiche, Università dell'Insubria, Via Valleggio 11, 22100 Como, Italy

E. Brambilla

INFM, Dipartimento di Scienze Chimiche Fisiche e Matematiche, Università dell'Insubria, Via Valleggio 11, 22100 Como, Italy

Abstract   In the field of quantum imaging one takes advantage of the quantum aspects of light and of the intrinsic parallelism of optical signals to develops new techniques in image processing and parallel processing at the quantum level.

Keywords: Quantum physics, optical imaging, quantum entanglement, parametric down-conversion and amplification, quantum.

1.   Introduction

In this article we provide a brief overview of the newly born field of quantum imaging. The starting point is provided by the general topic of the spatial aspects of quantum optical fluctuations. This has been the object of several studies in the past, but only recently there has been a constant focus of attention, which is mainly due to the fact that the spatial features may open new possibilities, e.g. in the direction of parallel processing and multichannel operation by quantum optical procedures. Once realized that these studies may have interesting perspectives, it is natural to coin a new name as quantum imaging to designate this field.

The field of quantum imaging encompasses several topics, which include,

- Detection of weak phase and amplitude objects beyond the standard quantum limit,

A.S. Shumovsky and V.I. Rupasov (eds.), Quantum Communication and Information Technologies, 85–99.
© 2003 *Kluwer Academic Publishers.*

- Amplification of weak optical images preserving the signal-to-noise ratio (noiseless amplification),

- Entangled two-photon microscopy, limits in the detection of small displacements and in image reconstruction,

- Quantum lithography,

- Quantum teleportation teleportation of optical images.

In this paper we will discuss some of these topics. Several of the results and approaches that we will illustrate have been pursued by the participants in the European Project QUANTIM (Quantum Imaging), that started in January 2001.

2.   The Optical Parametric Amplifier and Its Spatial Quantum Properties

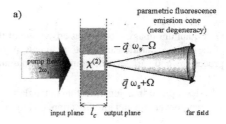

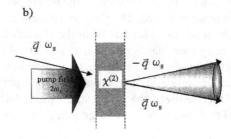

Figure 1.   (a) Scheme for parametric down-conversion. (b) Parametric amplification of a plane wave. $\vec{q}$ is the component of the wave-vector in the planes orthogonal to the pump propagation direction.

A key system which is considered in quantum imaging is the optical para-metric amplifier amplifier (OPA), in which one has a slab of $\chi^{(2)}$ material which is pumped by a coherent plane wave of frequency $2\omega_s$ (see Fig. 1 a). A fraction of the pump photons are down-converted into signal-idler photon pairs, which are distributed over a broad band of temporal frequencies around the degen-erate frequency ω_s. For each fixed temporal frequency, the photon pairs are distributed over a band of spatial frequencies labelled by the transverse com-ponent of the wave vector.

If one considers a single temporal and spatial mode, the state of the signal-idler system at the exit of the OPA has the form

$$|\psi\rangle = \sum_{n=0}^{\infty} c_n |n\rangle_1 |n\rangle_2 , \tag{1}$$

where $|n\rangle_1$ and $|n\rangle_2$ are the Fock states with n photons for the signal and idler mode, respectively, and the coefficients c_n are given by

$$c_n = \frac{[\tanh \alpha]^n}{\cosh \alpha} , \tag{2}$$

where α is the gain parameter. The state (1) is clearly entangled à la Einstein-Podolsky-Rosen Einstein-Podolsky-Rosen[1].

If, in addition to the pump field, we inject a coherent plane wave with fre-quency ω_s and transverse wave vector (Fig.1 b), in the output we have a signal wave which corresponds to an amplified version of the input wave and for this reason the system is called optical parametric amplifier amplifier. Because of the pairwise emission of photons, there is also an idler wave which, close to degeneracy, is symmetrical with respect to the signal wave.

Referring to the case in which only the pump is injected, two regimes can be distinguished. One is that of pure spontaneous parametric down-conversion, as in the case of a very thin crystal. In this case coincidences between partners of single photon pairs are detected. The other is that of dominant stimulated parametric down-conversion, in which a large number of photon pairs at a time is detected. In the following we will consider both cases alternatively.

Next, we want to illustrate the key spatial quantum properties of the field emitted by an OPA, in the linear regime of negligible pump depletion and in the configuration of a large photon number.

In the near field (see Fig. 2) one has the phenomenon of spatially multimode squeezing or local squeezing (squeezing in small regions) discussed in [2-5]. A good level of squeezing is found, provided the region which is detected has a linear size not smaller than the inverse of the spatial bandwidth of emission in the Fourier plane. If, on the other hand, one looks at the far field (which can be reached, typically, by using a lens as shown in Fig.2) one finds the phenomenon

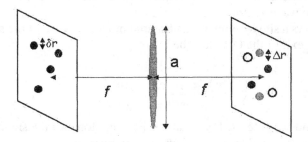

Figure 2.   Illustration of the near field/far field duality. f is the focal length of the lens.

of *spatial entanglement* between small regions located symmetrically with respect to the center. Precisely, if one considers two symmetrical pixels 1 and

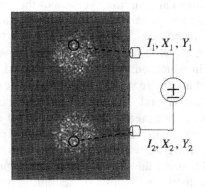

Figure 3.   Illustration of the concept of spatial entanglement. I_i, X_i, Y_i ($i = 1, 2$) denote the intensities and the quadrature components measured in the two pixels 1 and 2, respectively.

2 (Fig. 3), the intensity fluctuations in the two pixels are very well correlated or, equivalently, the fluctuations in the intensity difference between the two pixels are very much below the shot noise [6, 7]. Because this phenomenon arises for any pair of symmetrical pixels, we call it spatial entanglement en-

tanglement. The same effect occurs also for quadrature components, because in the two pixels the fluctuations of the quadrature component X are almost exactly correlated, and those of the quadrature component Y are almost exactly anticorrelated [8]. The minimum size of the symmetrical small regions, among which one finds spatial entanglement, is determined by the final aperture of optical elements, and is given, in the paraxial approximation, by $\lambda f/a$, where λ is the wavelength, f is the focal length of the lens and a is the aperture of optical elements (e.g. the lens aperture, see Fig. 2). In a more realistic model of the OPA, the finite waist of the pump field should be taken into account. In this case the minimum size of the regions where entanglement is detectable in the far field is mostly determined by the pump waist.

The spatial entanglement of intensity fluctuations in the far field is quite evi-

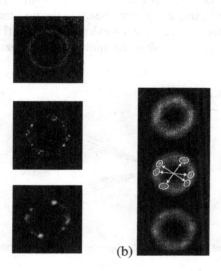

(b)

Figure 4. (a) Intensity distribution in the far field for a single shot of the pulsed pump field. a) Numerical simulations. The waist of the pump beam is 1000 μm, 300 μm, 150 μm in the three frames from top to bottom, respectively. (b) Experimental observations by Devaux and Lantz at University of Besancon (see [9]).

dent even in single shots (the pump field is typically pulsed). Fig. 4 a shows a numerical simulation in a case of non collinear phase matching at degenerated frequency. One observes the presence of symmetrical intensity peaks, which become broader and broader as one reduces the waist of the pump field. A similar situation is observed in an experiment performed using a LBO crystal [9].

3.   Detection of Weak Amplitude or Phase Objects Beyond the Standard Quantum Limit

Let us consider first the case of a weak amplitude object which is located, say, in the signal part of the field emitted by an OPA (Fig. 5). Both signal and idler are very noisy and therefore, in the case of large photon number, if the object is weak and we detect only the signal field, the signal-to-noise ratio for the object is low. But, because of the spatial entanglement entanglement, the fluctuations in the intensity difference between signal and idler are small. Hence if we detect the intensity difference, the signal-to-noise ratio for the object becomes much better.

Next, let us pass to the case of a weak phase phase object in which one can exploit, instead, the property of spatially multimode squeezing. The configuration is the standard one of a Mach-Zender interferometer in which, as it is well known, one can detect a small phase shift with a sensitivity beyond the standard quantum limit by injecting a squeezed beam in the port through which usually normal vacuum enters. If we have a weak phase image (Fig. 6) we can obtain the same result by injecting a spatially multimode squeezed light [10].

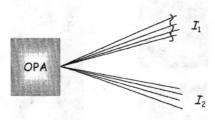

Figure 5.   Detection of a weak amplitude object by measuring the intensity difference $i_1 - i_2$.

4.   Quantum Imaging with Entangled Photon Pairs

The theory of entangled two-photon imaging was pioneered by Klyshko [11, 12] who formulated a heuristic approach. This research stimulated a number of key experiments, especially in the laboratory of Shih [13, 14, 15], similar experiments in the group of Zeilinger [16] were addressed to the discussion of fundamental issues of quantum physics. Recently the Boston team formulated a systematic theory [17, 18]. Let us now dwell on this last approach.

In a classical imaging configuration one has a source, an illumination system, an object, an imaging system and a detector. The field at the detection plane is related to that at the object plane by a linear integral transformation with a

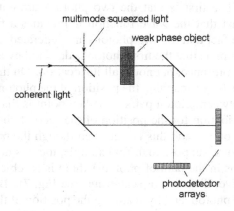

Figure 6. Detection of a weak phase object.

kernel h. Let us focus on the interesting case illustrated in Fig. 7, in which

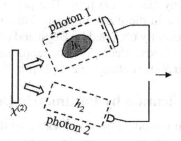

Figure 7. Quantum imaging with entangled photon pairs. Photon 1 is revealed by a bucket detector which does not reveal its transverse position; photon 2 is observed by a detector that scans its position in coincidence with photon 1 detection.

the imaging is performed using photon pairs, in which case the topography of the system allows for two branches with different kernels to be simultaneously illuminated, a system that was heuristically considered by Belinskii and Klyshko [12]. The two-photon wave function is denoted $\psi(\mathbf{x}, \mathbf{x}')$ ($\mathbf{x}$ and $\mathbf{x}'$ are source position vectors in the transverse plane) and the two kernels are denoted h_1 and h_2 for the signal and idler beams 1 and 2, respectively. Next, let us introduce two assumptions. The first is that the two photons are entangled. This is expressed by the fact that the wave function ψ includes a factor $\delta(\mathbf{x} - \mathbf{x}')$, which expresses the fact that the two photons are generated at the same point. The second assumption is that the first photon is detected by a bucket detector, which can reveal its presence, but not at all its location. On the other hand, the other detector is capable of scanning the position of the photon, and one detects the coincidences between photon pairs. The key point is that in this situation the probability distribution for the position of the second photon depends not only on h_2 but also on h_1, and this is true even though the position of photon 1 is not detected. Thus, suppose that, for example, there is nothing in the path of photon 2, whereas in the path of photon 1 there is an object which may be transmissive/reflective, diffusive or scattering (see Fig. 7). By using a bucket detector for the first photon, and by scanning the position of the second photon and detecting coincidences one can obtain an image of the object which is in the path of the first photon despite the fact that the position of this photon is not scanned. The essential point is that, when the two photons are entangled, such distributed quantum imaging is in general partially coherent, and can possibly be fully coherent in which case phase phase information about the object is preserved , whereas for classically correlated but unentangled photons the imaging would be incoherent. These authors have also applied the principle of entangled-photon imaging to quantum holography [19].

All works [11-19] refer to the regime of single photon pair detection. A theory which encompasses alone the regime of large signal-idler photon number has been recently developed by the authors [20]. This paper also clarifies which is the really quantum feature in these phenomena. This is necessary especially after the relevant result recently obtained by Boyd and collaborators who show that the experiment using bucket detectors can be reproduced also using classical correlations instead of quantum entanglement [21].

5.   Image Amplification by Parametric Down Conversion

Detection of a weak amplitude object by measuring the intensity difference $i_1 - i_2$. Let us come back to the configuration of Fig.1a in the case of a large number of photon pairs. Let us assume that now, instead of a plane wave at frequency close to ω_s, we inject a coherent monochromatic image (Fig.8) of frequency ω_s. Parametric image amplification has been extensively studied from a classical viewpoint (see e.g.[9]). A basic point in Fig.8 is that, if the image is injected off axis, one obtains in the output a signal image, that represents

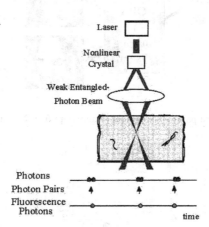

Figure 8.   Off-axis injection of an image and generation of twin entangled images.

an amplified version of the input image, and also a symmetrical idler image. An interesting situation arises if one has, in addition to the amplifier amplifier, a pair of lenses located at focal distances with respect to the object plane, to the amplifier amplifier and to the image plane (Fig.9). As it was shown by our group [6, 8, 22], in the limit of large amplification the two output images can be considered twin of each other even from a quantum mechanical viewpoint. As a matter of fact, they do not only display the same intensity distribution but also the same local quantum fluctuations. Precisely, let us consider two symmetrical pixels in the two images (Fig.10). It turns out that the intensity fluctuations in the two pixels are identical, i.e. exactly correlated/synchronized. On the other hand, the phase fluctuations are exactly anticorrelated. So in this way, from one image one obtains twin images in a state of spatial entanglement which involves also the quadrature components X and Y, as it was already described in the case of pure parametric fluorescence without any signal injection in [2]. There is, however, a negative point that concerns the signal-to-noise ratio. When the input image is injected off axis, this mechanism of amplification is phase insensitive and therefore, as it is well known, it adds 3 dB of quantum noise in the output [23].   In order to have noiseless amplification, i.e. amplification which preserves the signal-to-noise ratio, one must inject two coherent images symmetrically (see Fig.11 and [22]). In this case one has in the input two identical, but uncorrelated images and in the output two amplified images

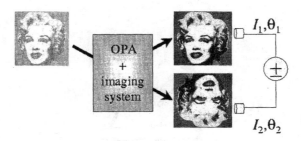

Figure 9. Scheme of the parametric optical image amplifier. Not shown in the figure is the pump field of frequency $2\omega_s$.

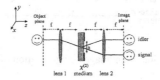

Figure 10. The spatial entanglement between the two output images concerns intensity and phase fluctuations, and also the fluctuations of quadrature components.

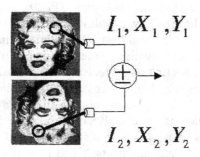

Figure 11. Symmetrical injection of an image.

in a state of spatial quantum entanglement entanglement. One can prove that this symmetrical configuration is phase phase sensitive and, in fact, as it was shown in [24, 25] the amplification can become noiseless. A couple of years ago there was a landmark experiment by Kumar and collaborators [26] which demonstrated the noiseless amplification of a simple test pattern.

6. Measurement of Small Displacements

This topic has been recently studied from a quantum mechanical standpoint by Fabre and his group [27]. They analyze the displacement of a light beam. According to the Rayleigh diffraction limit, the position of the beam can be measured with an error on the order of the beam section. However, one can use, rather, a split detector which measures the intensities i_1 and i_2 from the two halves of the beam cross section (Fig.12).

If one displaces gradually the beam with respect to the detector and plots

Figure 12.   Measurement of very small beam displacements in the transverse plane.

the intensity difference, one obtains a curve like that shown in Fig. 12 and the precision in the measurement of the displacement is limited only by noise. The standard quantum limit is given [28] by the ratio of the Rayleigh limit to the square root of the photon number. In this way one can measure shifts in the sub-nanometer range. In the case of the microscopic observation of biological samples, one cannot raise the photon number and therefore it is important to have the possibility of going beyond the standard quantum limit by reducing the noise in the intensity difference $(i_1 - i_2)$ below the shot noise level. To achieve this, a possibility is to use a beam in a state of spatial entanglement between its upper and its lower part, because in this way the number of photons in the two parts is basically the same and the fluctuations in the intensity difference are very small. Another possibility, which was proposed in [27], is to synthesize a special two-mode state which arises from the superposition of a Gaussian mode in a squeezed vacuum state and a spatially antisymmetrical mode which lies in a coherent state. coherent state The antisymmetrical mode is obtained from a Gaussian mode by flipping upside down one of the two halves. One shows that in this way one can measure displacements beyond the standard quantum limit. Experiments with this configuration are presently conducted in the laboratory of Bachor at the University of Canberra in collaboration with the group of Fabre [29].

7.   Image Reconstruction

Kolobov and Fabre have recently studied the quantum limits in the process of image reconstruction [30]. The very well known scheme they analyzed is shown in Fig. 13. The object is contained in a finite region of size X. The imaging system is composed by two lenses and an aperture. Because of the transverse finiteness of the imaging system, diffraction comes into play, with a consequence that some details of the object are blurred. In order to better reconstruct the object, one can proceed as follows. One considers the linear operator H which transforms the object into the image, its eigenvalues and its eigenfunctions, which are called prolate spheroidal functions:

$$H f_n(x) = \mu_n f_n(x) \tag{3}$$

If the imaging were perfect, H would be the identity operator and all eigenvalues would be equal to unity. An imperfect imaging introduces a sort of loss so that $\mu_n \leq 1$. Now one expands the image $e(x)$ on the basis of the eigenfunctions:

$$e(x) = \sum_n c_n f_n(x) \tag{4}$$

and obtains the coefficients c_n. The object $a(x)$ is reconstructed by the following expression

$$e(x) = \sum_n \frac{c_n}{\sqrt{\mu_n}} f_n(x) \tag{5}$$

where the coefficients c_n have been replaced by $c_n \sqrt{\mu_n}$.
In principle the reconstruction is perfect, but a problem is introduced by the

circumstance that when the index n is increased the eigenvalues μ_n quickly approach zero. As it is shown in [30], this feature makes the reconstruction of fine details very sensitive to noise, so that again quantum noise represents the ultimate limitation. The idea proposed in [30] is to illuminate the object by bright squeezed light instead of coherent light, and to replace the vacuum state in the part of the object plane outside the object itself by squeezed vacuum radiation. In this way one can obtain a definite improvement of the resolution in the reconstruction, beyond the standard quantum limit.

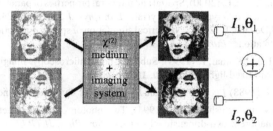

Figure 13.   Scheme of the imaging system.

Acknowledgments

We thank Claude Fabre, Mikhail Kolobov, Eric Lantz, Bahaa Saleh, Alexander Sergienko, Malvin Teich for giving us permission of including in this paper material provided by them. This work is supported by the European FET Project QUANTIM (Quantum Imaging).

References

[1] Einstein, A., Podolsky, B. and Rosen, N. (1935) Can Quantum-Mechanical Description of Physical Reality be considered Complete ? *Phys.Rev.* **47**, 777-780.

[2] Kolobov, M.I. and Sokolov, I.V. (1989) Spatial behavior of squeezed states of light and quantum noise in optical images. *Sov. Phys JETP* **96**, 1945-1957;
Kolobov, M.I. and Sokolov, I.V. (1989) Squeezed states of light and noise-free optical images. *Phys.Lett. A* **140**, 101-104;
Kolobov, M.I., Sokolov, I.V. (1991) Multimode squeezing, antibunching in space and noise-free optical images. *Europhys.Lett.* **15**, 271-276.

[3] Kolobov, M.I. (1999), The spatial behavior of nonclassical light. *Rev. Mod. Phys.* **71**, 1539-1589.

[4] Lugiato, L.A. and Gatti, A. (1993), Spatial structure of a squeezed vacuum. *Phys. Rev. Lett.* **70**, 3868-3871;
Gatti, A. and Lugiato, L.A. (1995), Quantum images and critical fluctuations in the optical parametric oscillator below threshold. *Phys. Rev. A* **52**, 1675-1690;
Lugiato, L.A., Brambilla, M. and Gatti, A. (1999), Optical Pattern Formation, in *Advances in Atomic, Molecular and Optical Physics*, vol. 40, Academic Press, Boston pp. 229-306

[5] Lugiato, L.A. and Grangier, Ph. (1997), Improving quantum-noise reduction with spatially multimode squeezed light. *J. Opt. Soc. Am.* B **14**, 225-231.

[6] Gatti, A., Brambilla, E., Lugiato, L.A. and Kolobov, M.I. (1999), Quantum Entangled Images. *Phys. Rev. Lett.* **83**, 1763-1766.

[7] Brambilla, E., Gatti, A., Lugiato, L.A. and Kolobov, M.I. (2001), Quantum structures in traveling-wave spontaneous down conversion *Eur. Phys. J. D* **15**, 127-135.

[8] Navez, P., Brambilla, E., Gatti, A. and Lugiato, L.A. (2001), Spatial entanglement of twin quantum images. *Phys. Rev. A.* **65**, 13813-13823.

[9] Devaud, F. and Lantz, E. (2000), Spatial and temporal properties of parametric fluorescence around degeneracy in a type I LBO crystal. *Eur. Phys. J. D* **8**, 117-124;
Lantz, E. and Devaud, F. (2001), Numerical simulation of spatial fluctuations in parametric image amplification. *Eur. Phys. J. D* **17**, 93-98.

[10] Kolobov, M.I. and Kumar, P. (1993), Sub-shot-noise microscopy: imaging of faint phase objects with squeezed light. *Opt. Lett.* **18**, 849-851.

[11] Klyshko, D.N. (1988) *Photons and Non-Linear Optics* Taylor and Francis.

[12] Belinskii, A.V. and Klyshko, D.N. (1994), Two-photon optics: diffraction, holography, and transformation of two-dimensional signals. *Sov. Phys. JETP bf 78*, 259-262.

[13] Ribeiro, P.H.S., Padua, S., Machado da Silva, J.C., Barbosa, G.A. (1994), Controlling the degree of visibility of Young's fringes with photon coincidence measurements. *Phys. Rev. A* **49**, 4176-4179.

[14] Strekalov, D.V, Sergienko, A.V., Klyshko, D.N. and Shih, Y.H. (1995), Observation of Two-Photon "Ghost" Interference and Diffraction *Phys. Rev. Lett* **74**, 3600-3603.

[15] Pittman, T.B. , Sergienko, A.V., Klyshko, D.N. and Shih, Y.H. (1995), Optical imaging by means of two-photon quantum entanglement. *Phys. Rev. A* **52**, R3429-R3432.

[16] Zeilinger, A. (1999) Experiment and the foundations of quantum physics *Rev. Mod. Phys.* **71**, S288-S297.

[17] Abouraddy, A.F., Saleh, B.E.A., Sergienko, A.V. and Teich, M.C. (2001), Role of Entanglement in Two-Photon Imaging. *Phys. Rev. Lett.* **87**, 123602-123605;

[18] Abouraddy, A.F., Saleh, B.E.A., Sergienko, A.V. and Teich, M.C. (2002), Entangled-photon Fourier optics. *J. Opt. Soc. Am.* B **19**, 1174-1184.

[19] Abouraddy, A.F., Saleh, B.E.A., Sergienko, A.V. and Teich, M.C. (2001), Quantum holography. *Opt. Express* **9**, 498-503.

[20] Gatti, A., Brambilla, E and Lugiato, L.A., Entangled Imaging and wave-particle duality in the macroscopic realm, submitted for publication.

[21] Bennink, R.S., Bentley, S.J. and Boyd, R.W. (2002), "Two-Photon" Coincidence Imaging with a Classical Source. *Phys. Rev. Lett.* **89**, 113601-113605.

[22] Gatti, A., Brambilla, E., Lugiato, L.A. and Kolobov, M.I. (2000), Quantum aspects of imaging. *J.Opt. B: Quantum Semiclass. Opt.* **2**, 196-203.

[23] Caves, C.M. (1982), Quantum limits on noise in linear amplifiers. *Phys. Rev. D* **26**, 1817-1839.

[24] Kolobov, M.I. and Lugiato, L.A. (1995), Noiseless amplification of optical images. *Phys Rev. A* **52**, 4930-4940.

[25] Sokolov, I.V., Kolobov, M.I. and Lugiato, L.A. (1998), Quantum fluctuations in traveling-wave amplification of optical images. *Phys. Rev. A* **60**, 2420-2430.

[26] Choi, S.-K., Vasilyev, M. and Kumar, P. (1999), Noiseless Optical Amplification of Images. *Phys.Rev. Lett.* 83, 1938-1941.

[27] Fabre, C.,Fouet, J.B., and Maitre, A. (2000), Quantum limits in the measurement of very small displacements in optical images. *Opt. Lett.* **25**, 76-78.

[28] Treps, N. (2001), Effets quantiques dans les images optiques, doctoral thesis at Universite' Paris VI.

[29] Fabre C. private communication.

[30] Kolobov, M.I. and Fabre, C. (2000), Quantum Limits on Optical Resolution. *Phys. Rev.Lett.* **85**, 3789-3792.

LECTURE NOTES ON QUANTUM-NONDEMOLITION MEASUREMENTS IN OPTICS

Victor V. Kozlov

Abteilung für Quantenphysik, Universität Ulm, Ulm, Germany, 89081;

Fock Institute of Physics, St.-Petersburg University, Ulyanovskaya 1, Petrodvoretz, St.-Petersburg, Russia, 198904

Abstract   A general overview of quantum nondemolition measurements is provided and illustrated with a few examples from quantum optics. Also given are basic principles and theoretical fundament.

1.   Introduction

Interpretation of quantum measurements [1] has long been the subject of controversial discussions. Typically, experiments with quantum objects involve *ensemble* measurements, i.e. series of measurements on a large number of identical quantum systems. This ensemble approach gives rise to a usual for standard textbooks, probabilistic interpretation of quantum mechanics that is of course correct. Sometimes it is however not necessary or even possible to prepare an ensemble. Instead, one measurement or series of measurements on a *single* object is preferable. Theory of ensemble measurements and therefore their probabilistic interpretation is of little use when only one object is involved. Different theoretical tools are needed to get an adequate explanation of the output data.

In these lecture notes we shall address the problem of measurements and more specifically repeated measurements, of single quantum objects. The closer definition excludes such straightforward examples like direct photodetection, when the photon is absorbed by an atom and therefore irretrievably lost for a further use. As we are interested mainly in optical experiments the exclusion of direct photodetection means the refusal of the major detection method. Instead, for repeated measurements on single objects (in an optical experiment, it is always a pulse of an e.m. field) it is necessary to use an *indirect* way of measuring the value of the observable of interest.

First, let us "distinguish possible from impossible" and use for that a referral to ensemble measurements. The last are aimed to accumulate statistics on a

101

A.S. Shumovsky and V.I. Rupasov (eds.), Quantum Communication and Information Technologies, 101–123.

physical variable or a set of variables and in some cases even reconstruct a phase-space distribution of the object that is doable only to a certain, though in principle any desirable, accuracy. Such knowledge is not possible to obtain without involving a vast number of identical objects, i.e. ensemble. Statistical properties of a single object is quite unnatural notion. It is therefore not surprising that precise knowledge of the quantum state of a single object is not permitted by quantum mechanics.

Another way to look at the impossibility of getting accurate knowledge about quantum state of the single object is to refer to the Heisenberg uncertainty relation. According to this rule, a pair of non-commuting variables can be measured simultaneously only to a limited precision. The phase-space distribution which carries full information about the quantum state can be characterized in terms of at least one pair of noncommuting variables, generalized position and momentum. Thereby it inherits the property of being known only to a strictly limited, and not desirable as with ensemble measurements, accuracy. This limitation is so severe that it makes a little sense to reconstruct a full set of quantum properties of the object from such poor knowledge. Needless to say that repeated measurements is of no help because a preceeding measurement already disturbed the system so that the following measurement will already address a new object, in the sense that the object is now characterized by a new quantum state [2].

So, from the very beginning we leave the hope to precisely determine the quantum state of the single object. However, frequently the experimenter is interested in a limited information about the system. This can be value of a physical variable, for instance the energy of e.m. field stored in a cavity. Moreover, the experimenter may wish to find out not only this value but also keep it for later manipulations, in other words for repeated measurements. For that he (she) would like to be sure that the value did not change during the first measurement and then next measurement would reveal this same value. If two successive measurements show same result, the same property can be held throughout arbitrary number of repeated measurements.

The measurement strategy that allows us to

(i) perform *repeated* quantum measurements on observable of the single quantum object;

(ii) do this *without disturbing* its value,

got the name quantum-nondemolition (QND) measurements nondemolition measurements, following Braginsky and co-workers [3]. For our example with the field in the cavity the QND strategy means measurement of the e.m. energy in a non-destructive manner, i.e. no photon is lost during the measurement. The possibility to know the value of one variable as accurate as desirable is by no means in violation of Heisenberg's uncertainty relation. A corresponding increase in uncertainty in the variable non-commuting with the variable of interest preserves validity of the principle. The interplay between uncertainties of two non-commuting variables is illustrated in some detail in next section.

2. Back-action in repeated quantum measurements

We consider a free quantum particle of mass m characterized by its position x, momentum p, and the Hamiltonian Hamiltonian

$$H = \frac{p^2}{2m}, \tag{1}$$

and set the goal to perform two successive measurements of position. With this example we demonstrate how the first measurement spoils accuracy of the following one and thereby makes it impossible to meet the QND requirements in measurements of position.

Let the particle be prepared at time instant $t = 0$ in a state characterized by root-mean-square (r.m.s.) position and momentum deviation Δx and Δp, correspondingly. Without loss of generality we assume that x and p are uncorrelated and the product of two deviations takes the minimal possible value allowed by the Heisenberg uncertainty relation

$$\Delta x \Delta p \geq \hbar/2, \tag{2}$$

i.e. we set $\Delta x \Delta p = \hbar/2$.

Position of the free particle evolves with time according to the Heisenberg equation of motion

$$\dot{x} = (i\hbar)^{-1}[x, H] \tag{3}$$
$$= p/m, \tag{4}$$

which has immediate solution

$$x(t) = x(0) + \frac{p(0)}{m}t. \tag{5}$$

We perform first measurement at time instant $t = 0_+$ immediately after the preparation so that $x(0_+) \approx x(0)$. The close time separation is assumed solely for simplicity. Resolution Δx_{meas} of the detection apparatus is assumed to be relatively high such that $\Delta x_{meas} < \Delta x$. So, after the measurement we say that the outcome is $x_0 \pm \Delta x_{meas}$, where $x_0 \equiv \langle x(0) \rangle$ and angular brackets stand for averaging over initial state. According to the basic postulate of quantum mechanics Eq. (2) the better knowledge of position must have an immediate impact on the momentum uncertainty Δp_{meas}. The last cannot be determined better than with the error

$$\Delta p_{meas} = \frac{\hbar}{2\Delta x_{meas}}, \tag{6}$$

which is clearly larger than initial value Δp. Realization of the equality Eq. (6) (and not inequality) is already an experimental challenge. In general, keeping minimal value of the uncertainty requires a clever choice of *realizable* inter-action Hamiltonian Hamiltonian between the quantum system and detection apparatus so that extra noise does not penetrate the signal system.

Second measurement is performed at a later time $t = \tau$. Between the two measurements the mean value of position evolves in a classical-like manner and yields at $t = \tau$:

$$x_1 \equiv \langle x(\tau) \rangle = x_0 + \frac{\langle p(0) \rangle}{m} \tau. \tag{7}$$

Here we assume for simplicity that the mean value of momentum was not shifted during the first measurement, i.e. $\langle p(0) \rangle = \langle p(0_+) \rangle$. QND strategy implies that the subsequent measurement on the system finds the observable of interest within interval Δx_{meas} around x_1. When $\Delta x_{meas} \to 0$ (note that $\Delta x_{meas} = 0$ is not physical) one ideally gets $x(\tau) \to x_1$. Formally speaking, the r.m.s. of position at time $t = \tau$ should not exceed that at time $t = 0_+$. To find whether it is the case we derive equation of motion for position error variance $(\Delta x)^2 \equiv \langle x^2 \rangle - \langle x \rangle^2$, and get for the r.m.s. deviation at $t = \tau$:

$$
\begin{aligned}
\Delta x(\tau) &= \sqrt{(\Delta x_{meas})^2 + \left(\frac{\Delta p_{meas}}{m} \tau \right)^2} \\
&= \sqrt{(\Delta x_{meas})^2 + \left(\frac{\hbar}{2m\Delta x_{meas}} \tau \right)^2},
\end{aligned}
\tag{8}
$$

where Eq. (6) is used to obtain second equality. Clearly, $\Delta x(\tau) > \Delta x_{meas}$ for all later times, and the later the next measurement the poorer its accuracy. Also, the better the resolution of the first measurement, the worse the accuracy of the second one, see Fig. 1 for pictorial representation of the uncertainties on different stages of evolution.

This example demonstrates the effect of back action. That is the accuracy of a subsequent measurement is deteriorated by the fact that the system was disturbed by a preceeding measurement. Therefore, position measurements cannot be arranged in the QND fashion. Absence of the back action is so crucial for the QND strategy that sometimes the term quantum nondemolition is changed to back-action-evading.

Though the QND strategy does not work for position measurements, it is nev-ertheless possible to find the situation when the back action can be avoided and the uncertainty between successive measurements is kept unchanged. Immedi-ate example is measurements of momentum. Momentum is conserved in free evolution and though the disturbance of position while the first measurement

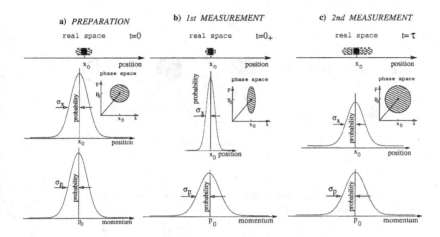

Figure 1. Position and momentum uncertainties for a free particle: a) preparation stage; b) immediately after the first measurement; c) immediately after the second measurement. Above: pictorial illustration of the particle and degree of randomness of its location in real space. Below, probability distributions of the position and momentum, both are assumed to be Gaussian. Also shown are phase-space distributions. Initially, the circle corresponds to the state which minimizes Heisenberg's uncertainty relation.

can be very large this does not produce a cumulative effect (like that demonstrated in Eq. (8)) on the r.m.s. deviation of momentum. It is equally important that not only evolution between the measurements is free of the back action but the disturbance is also avoided during the very process of measurement when it be caused by the detection apparatus (formally, via interaction Hamiltonian). Note that the part of the detection apparatus which has immediate contact to the object is frequently called meter system, see section 4 for details. Only when both back action evading conditions are met simultaneously the measurement becomes of truly the QND type. Formalization of these two requirements in terms of conditions imposed on the Hamiltonian Hamiltonian of the combined "signal + meter" system will be given in section 4.

3. QND measurements of photon number in a cavity

Many optical experiments involving QND strategy are reported up to date. A large part of them is summarized in review articles [4] and books [5, 6]. It is possible and also reasonable for us to touch only a few examples. The choice does not pretend to be ideal and is certainly biased by author's personal preference and taste.

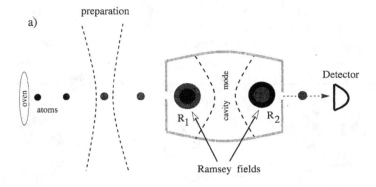

Figure 2.   Setup of Ref. [7] for QND measurements of a single photon: a) arrangement of principal parts; b) a sequence of pictorial illustrations of atomic excitations at main stages. The degree of excitation is shown by the size of small circles attached to the corresponding level. Coupling fields are shown by ellipses connecting a corresponding pair of levels.

3.1   experiment

We start with the QND experiment in cavity quantum electrodynamics performed in the group of S. Haroche in Paris, following [7] (see Ref. [10] for comparison). The idea is conceptually simple and involve only no more than one photon stored in a high-Q microwave cavity and one atom passing through this cavity, see Fig. 2. The experiment is arranged in such a way that the atom can detect the presence of a single photon inside the cavity without destroying this photon. In other words, the experimenters have managed to perform the measurement of the photon number by allowing the cavity field to interact with the single atom in such a way that the intra-cavity e.m. energy remained unperturbed.

Since the interest is in manipulation with the e.m. field stored inside the cavity the obvious requirement is that the field does not leak from the cavity during the experiment. The cavity used in [7] consists of two spherical, superconducting niobium mirrors surrounded by a cylindrical ring with 3-mm holes for atom access. The cavity is cooled to 1 K and has a Q-factor of about 3×10^8, which corresponds to a photon lifetime of 1 ms (compare to ~ 1 ns, characteristic to a conventional resonator). So, the experiment should be completed in a time period less than 1 ms, fairly long for major applications in quantum optics.

Second restriction is imposed on atom properties. It is quite natural to require that the atom clearly "sees" the field inside the cavity. As energy of the field is equal to at most one quantum of microwave radiation the atom-field coupling constant should be large. The greatest value is realized when mode of the cavity of frequency ω_c is tuned to exact resonance with the atomic transition, here $e \leftrightarrow g$ with frequency $\omega_{eg} = 51.1$ GHz. On the other hand, the larger the coupling strength the faster the decay rate from the upper level. In its turn, the decay causes the erasure of information gained by the atom while traversing the cavity, and therefore should be suppressed. The trade-off between desirable increase of the coupling constant and unwanted speed-up of the decay rate is in use of rubidium atoms excited to the Rydberg circular states with principal quantum numbers 49 (level i), 50 (level g), and 51 (level e). These states on one hand have exceptionally large coupling to microwave radiation (single-photon Rabi frequency at center of the cavity is $\Omega/2\pi = 47$ kHz) and on the other hand, long lifetimes ~ 30 ms.

After leaving the oven the atomic beam is velocity selected. Selected atoms, one at a time, move with velocity $v \approx 503$ m/s crossing cavity's transverse gaussian mode with waist $w = 6$ mm in time $\tau_{int} = \sqrt{\pi}w/v \approx 0.021$ ms which is considerably less than both atomic and cavity photon lifetimes. Such careful arrangement of time scales allows experimenters to perform repeated measurements with reliable accuracy.

Three Rydberg levels i, g, and e play a role in the experiment. The other levels are far off resonance and give negligible contribution. Levels i and g are of the same parity, opposite to the parity of level e. Often, such configuration is termed as of a cascade type. A rubidium atom is initially excited on g level at a controllable time. It is then subjected to a sequence of interactions with microwave fields known as Ramsey spectroscopy.

In the beginning each atom passes through first Ramsey zone R_1 interacting with an intense field of Rabi frequency Ω_R during time τ_R. The field is tuned to resonance with $g \leftrightarrow i$ transition at 54.3 GHz and does not couple e and g levels separated by 51.1 GHz. The Ramsey field induces Rabi oscillations between g and i levels. The goal is to prepare the atom in a coherent superposition of $|g\rangle$ and $|i\rangle$ states. Following the Ramsey zone the atom enters the cavity where the microwave field (signal) is stored. The interaction with the cavity mode lasts during transit time τ_{int}.

The experiment is arranged in such a way that the cavity can be loaded with one quantum or left empty. So, there are only two possible outcomes of the measurement, either to find one photon or conclude that the cavity was empty. Second Ramsey zone R_2 is next on the atom's path. Here $|g\rangle$ and $|i\rangle$ states are again coherently mixed by the Ramsey field of same as in the first zone Rabi frequency Ω_R. Finally, the atom falls on a state-selective field-ionization detector D. The detector can distinguish between whether the atom was in $|g\rangle$ or $|i\rangle$ state. Atom in $|g\rangle$ state points to storing one photon in the cavity and the population of $|i\rangle$ state means no photons. In order to understand why this binary outcome is uniquely related to energy of the cavity field and why the energy is not altered while probing by the atom, we shall follow [9] and present below a simple theoretical study.

3.2   theory

Three atomic states, two transitions, and three electromagnetic fields (two Ramsey fields and the cavity field) will be revelevant to our theoretical study. In the interaction representation with respect to the unperturbed atomic and field Hamiltonians, the "atom + field" system wave function can be expanded as

$$|\Psi\rangle = c_i(t)|i, n, \Omega_R\rangle + c_g(t)|g, n, \Omega_R\rangle + c_e(t)|e, n, \Omega_R\rangle, \qquad (9)$$

where $c_i(t)$, $c_g(t)$, and $c_e(t)$ are time-dependent probability amplitudes, i, g, and e are notations of atomic states made appropriate to Fig. 2, n stands for the photon number (0 or 1) in the cavity, and Ω_R is Rabi frequency of each Ramsey field. The last is assumed to be strong enough so that Ω_R can be approximated as c-number.

Initial conditions for the probability amplitudes after the stage of preparation and before the atom enters first Ramsey zone R_1, are as following:

$$c_i(0) = 0, \qquad (10)$$

$$c_g(0) = i, \qquad (11)$$

$$c_e(0) = 0, \qquad (12)$$

i.e. the atom is in state g with probability equal to unity. While the atom is transiting the first Ramsey zone only two levels are coupled by the field, and the amplitudes evolve according to the following set of Schrödinger equations:

$$\dot{c}_i(t) = -i\frac{\Omega_R}{2}\, c_g(t), \qquad (13)$$

$$\dot{c}_g(t) = -i\frac{\Omega_R}{2}\, c_i(t), \qquad (14)$$

$$\dot{c}_e(t) = 0. \qquad (15)$$

The evolution ends after transit time τ_R. Solutions to above equations at $t = \tau_R$ with initial conditions Eqs. (10)-(12) yield for the probability amplitudes,

$$c_i(\tau_R) = \sin \frac{\Omega_R \tau_R}{2}, \tag{16}$$

$$c_g(\tau_R) = i \cos \frac{\Omega_R \tau_R}{2}, \tag{17}$$

$$c_e(\tau_R) = 0. \tag{18}$$

After a short period of free evolution accompanied with no changes in the probability amplitudes, the atom enters the cavity zone and passes through the cavity transverse mode. During the passage the atom interacts with the field of Rabi frequency $\Omega_n = 2g\sqrt{n}$ where g is the atom-field coupling constant and n can be either 0 or 1. In contrast to the Ramsey fields, quantum properties of the signal field are essential for our analysis and we shall treat this field fully quantum-mechanically. The interaction Hamiltonian Hamiltonian allows only transitions with absorption of one photon accompanied with simultaneous promotion of the atom from g to e, or reverse processes - with depositing a photon into the cavity mode at the expense of getting the atom downwards from e to g. Before entering the cavity containing n photons the atom has zero probability to be in state e, as regulated by Eq. (18). So, relevant states compatible with this situation are $|g, n\rangle$ and $|e, n - 1\rangle$. In the first spot we indicate the atomic state and in second one - state of the signal field. Probability amplitudes associated with these two states evolve in the cavity zone as given by equations

$$\dot{c}_g(n, t) = -i \frac{\Omega_n}{2} c_e(n - 1, t), \tag{19}$$

$$\dot{c}_e(n - 1, t) = -i \frac{\Omega_n}{2} c_g(n, t). \tag{20}$$

The lower state $|i\rangle$ is not altered. Solutions to these equations for times $t > \tau_R$ yield

$$c_g(n, t) = c_g(n, \tau_R) \cos \frac{\Omega_n(t - \tau_R)}{2}, \tag{21}$$

$$c_e(n - 1, t) = -i c_g(n, \tau_R) \sin \frac{\Omega_n(t - \tau_R)}{2}. \tag{22}$$

Initial conditions for the evolution in the cavity are set by the outcome state emerging from the first Ramsey zone, Eqs. (16)-(18). Solutions Eqs. (21) and (22) already take account of the fact that the upper state is not populated before the atom enters the cavity zone. We now complete the substitution of the initial conditions Eqs. (16)-(18) and write

$$c_g(n,t) = i \cos \frac{\Omega_R \tau_R}{2} \cos \frac{\Omega_n(t - \tau_R)}{2}, \qquad (23)$$

$$c_e(n-1,t) = \cos \frac{\Omega_R \tau_R}{2} \sin \frac{\Omega_n(t - \tau_R)}{2}. \qquad (24)$$

At this point it is instructive to recall that the goal of experimental arrangement under discussion is to measure energy of the cavity field in a non-destructive way. Formally, this requirement implies that the probability of finding n photons in the cavity after completing interaction with the atom is equal to that before the interaction, that is unity. In general, the probability $p(n,t)$ of having n photons in the field at time t is obtained by taking the trace over the atomic states:

$$p(n,t) = |c_i(n,t)|^2 + |c_g(n,t)|^2 + |c_e(n,t)|^2. \qquad (25)$$

The last term is zero because it is incompatible to get e state excited and simultaneously still have n photons in the cavity. Explicit expressions for first two terms in Eq. (25) follow from solutions Eqs. (16) and (23). So,

$$p(n,t) = \sin^2 \frac{\Omega_R \tau_R}{2} + \cos^2 \frac{\Omega_R \tau_R}{2} \cos^2 \frac{\Omega_n(t - \tau_R)}{2}. \qquad (26)$$

In order to fulfill the QND requirement the interaction time τ_{int} between the atom and the cavity field should be tuned so that to satisfy $p(n, \tau_R + \tau_{int}) = p(n,0) = 1$. With $t = \tau_R + \tau_{int}$ in the last term of Eq. (26) the QND condition reads

$$\Omega_n \tau_{int} = 2\pi, \qquad (27)$$

Note that 0 or any multiple of 2π on the right side of Eq. (27) also satisfies the QND condition.

The arrangement when the atom moves through the motionless cavity can be alternatively imagined as a pulsed field interacting with fixed atom. Then the interaction time τ_{int} becomes the duration of the pulse. The two descriptions are formally equivalent. The second one is a little more preferable as providing a direct reference to visualization of dynamics of the two-level system on the Bloch sphere [10]. The pulse with the property expressed by Eq. (27) is called 2π-pulse as it corresponds to rotation of the Bloch vector on 2π angle. The tip of the Bloch vector undergoes full circle on the Bloch sphere thus experiencing the full excursion with coming back to the starting point. Under action of the first half of the resonant 2π-pulse the atomic population is fully transferred from the lower state g to upper state i. Second half of the pulse reverses the evolution returning all population back to the lower level.

Returning the population back assures that energy of the cavity field was not absorbed by the atom. In case if one photon was in the cavity then it remained there. In case of zero photons present initially, the cavity remained empty. So, the intra-cavity field was not *demolished* by probing by the atom. Now, the atom should be "asked" whether it had interaction with the photon or the cavity was empty. For this purpose, the atom is directed through the second Ramsey zone where it evolves according to equations (13)-(15), the same as in the first stage. After interaction time τ_R, the probability amplitudes end up with

$$c_i(2\tau_R + \tau_{int}) = \sin\frac{\Omega_R\tau_R}{2}\cos\frac{\Omega_R\tau_R}{2}\left(1 + \cos\frac{\Omega_n\tau_{int}}{2}\right), \quad (28)$$

$$c_g(2\tau_R + \tau_{int}) = i\cos^2\frac{\Omega_R\tau_R}{2}\cos\frac{\Omega_n\tau_{int}}{2} - i\sin^2\frac{\Omega_R\tau_R}{2}, \quad (29)$$

$$c_e(2\tau_R + \tau_{int}) = \cos\frac{\Omega_R\tau_R}{2}\sin\frac{\Omega_n\tau_{int}}{2}. \quad (30)$$

It is convenient to adjust the transit time through the Ramsey zones such that the fields Ω_R act as $\pi/2$ pulses, i.e. $\Omega_R\tau_R = \pi/2$. Then, solutions Eqs. (28)-(30) are simplified to

$$c_i(2\tau_R + \tau_{int}) = \frac{1}{2}\left(1 + \cos\frac{\Omega_n\tau_{int}}{2}\right), \quad (31)$$

$$c_g(2\tau_R + \tau_{int}) = \frac{1}{2}i\left(\cos\frac{\Omega_n\tau_{int}}{2} - 1\right), \quad (32)$$

$$c_e(2\tau_R + \tau_{int}) = \frac{\sqrt{2}}{2}\sin\frac{\Omega_n\tau_{int}}{2}. \quad (33)$$

Two cases are to be considered. Empty cavity implies $\Omega_n = 0$ and with this expression in Eqs. (31)-(33) we get

$$c_i(2\tau_R + \tau_{int}) = 1, \quad (34)$$

$$c_g(2\tau_R + \tau_{int}) = 0, \quad (35)$$

$$c_e(2\tau_R + \tau_{int}) = 0. \quad (36)$$

On the other hand, if the cavity contained a photon, we should substitute $\Omega_n = 2g$ and after imposing the 2π-pulse condition Eq. (27) get from Eqs. (31)-(33)

$$c_i(2\tau_R + \tau_{int}) = 0, \quad (37)$$

$$c_g(2\tau_R + \tau_{int}) = -i, \quad (38)$$

$$c_e(2\tau_R + \tau_{int}) = 0. \quad (39)$$

The two situations are distinguished rather clearly. For empty cavity the atom is with unit probability in $|i\rangle$ state: $|c_i|^2 = 1$. Otherwise, it appears in $|g\rangle$ state:

$|c_g|^2 = 1$. In order to distinguish between these two cases experimentally, the atom is now directed to a state-selective detector D. Depending on its outcome, the experimenter can infer whether the cavity contained (and still contains) one photon or none.

3.3   discussion

The QND measurement is thus completed. It is important to emphasize that as any measurement inevitably disturbs the system under observation, here as well, the cavity field is altered by the interaction with the atom. The alteration does not however penetrate the energy and is fully loaded on variables non-commutative with the energy. One can identify such a variable and makes sure that its uncertainty indeed increases, at least to the extent which is necessary to satisfy Heisenberg's uncertainty principle. This exercise is however far from trivial.

Some remarks are in order. Two Ramsey fields used in the experimental setup are auxiliary fields. They serve as a part of the detection apparatus adjusting the measurement outcome to the method of state-selective detection. In absence of the fields the outcome would read $c_i(2\tau_R + \tau_{int}) = c_e(2\tau_R + \tau_{int}) = 0$ and

$$c_g(2\tau_R + \tau_{int}) = i \quad \text{for} \quad n = 0, \tag{40}$$

$$c_g(2\tau_R + \tau_{int}) = -i \quad \text{for} \quad n = 1, \tag{41}$$

following Eqs. (28)-(30) with $\Omega_R = 0$. Though the atomic state shows distinct difference for $n = 0$ and $n = 1$ there is no experimental device capable of direct measurements of the atomic phase phase. The purpose of the Ramsey technique is in translation of the phase shift into directly detectable probabilities.

Note that the "full cycle" requirement $\Omega_n \tau_{int} = N \times 2\pi$ where N is an integer, cf. Eq. (27), is necessary condition for the nondestructive way of probing the field but becomes also sufficient for the purpose of the experiment only when N takes even vaues. Indeed, one can immediately see from Eqs. (28)-(30) that for an odd N two cases $n = 0$ and $n = 1$, both yield identical results. Among the family of useful $(N + 1) \times 2\pi$-pulses only the lowest order 2π-pulse is practical. The rest demand greater values of the coupling strength and/or longer interaction times.

The possibility for the atom to measure the intra-cavity energy nondestruc-tively is due to the unique reversibility of atom's evolution on the $e \leftrightarrow g$ transition for either 0 photons (no evolution at all) or 1 photon (complete cycle by a 2π-pulse). To achieve the complete reversibility for two (and more) photons and thereby non-destructively distinguish between three number states $n = 0$, $n = 1$, and $n = 2$, is however not possible. Simply because one cannot fit the interaction time to simultaneously satisfy two equalities $\cos(\Omega_1 \tau_{int}/2) = \cos(g\,\tau_{int}) = 1$ and $\cos(\Omega_2 \tau_{int}/2) = \cos(\sqrt{2}\,g\,\tau_{int}) = 1$.

The above problem is characteristic for the particular experimental arrangement. Another nondestructive technique, for example that which is based on the dispersive type of interaction between atom and cavity field can be used for QND measurements of, in principle, any number of photons, [11]. When the atomic transition is fairly far detuned from central frequency of the cavity mode, the atomic wavefunction experiences a phase shift while keeping the probability of absorption at a very low level. The value of the phase phase shift depends on the intra-cavity energy and can be sensitively detected by means of the Ramsey spectroscopy. Experimental implementation of this arrangement meets some technological difficulties associated with weakness of the off-resonant atom-field coupling. Yet another approach to preparing and detecting photon number states via so-called "trapping states" trapping of the micromaser is suggested by the group of H. Walther in Garching [12].

4. Formal QND requirements

Above example can be viewed as giving a preliminary intuitive understanding of the topic. A more extensive development needs greater formalities, see [5, 13] and particularly [14, 15]. In this section we summarize basic QND requirements and start with writing a generic form of the Hamiltonian of two interacting systems:

$$H = H_s + H_m + H_I, \qquad (42)$$

with H_s, H_m, and H_I as the Hamiltonian Hamiltonian of the signal system, meter system, and their interaction, correspondingly. In terms of example from the previous section, H_s is the Hamiltonian of the cavity field and H_m is that of the atom. The characteristic feature of the QND formalism is that interaction between the system and the meter is treated fully quantum-mechanically while the actual quantum-to-classical interface is shifted towards probing the meter by the classical apparatus, as depicted in Fig. 3. In this regard, the QND measurement is an *indirect* one.

Let our interest be in measuring A_s which is an observable of the signal system. Interaction between the signal and meter system is aimed to record the information about the value of A_s into a meter observable A_m. The readout of the wanted information from A_m is then performed by a classical detector. The detection stage will be omitted from our discussion as not carrying an essential QND content.

Evolution of the two observables (both are quanum-mechanical operators) obey Heisenberg's equations of motion

$$-i\hbar\frac{dA_s}{dt} = [H_s, A_s] + [H_I, A_s], \qquad (43)$$

$$-i\hbar\frac{dA_m}{dt} = [H_m, A_m] + [H_I, A_m]. \qquad (44)$$

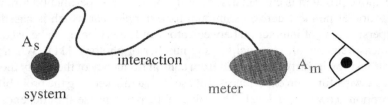

Figure 3.   A generic scheme of a QND setup. Its essential feature is that the part of the detection apparatus called meter is treated fully quantum-mechanically. The actual measurement, loosely speaking connection to the classical world, is performed on the meter system and not on the signal system as characteristic for majority of measurements in optics.

Since the goal is to record the value of A_s in A_m the interaction term in Eq. (44) first should not be zero and second, should be the function of A_s. The measurement of A_s affects the motion of its own in two ways. The first way is via the change of A_s due to $[H_I, A_s]$ term. This change should be avoided in order to ensure that free motion of A_s is not altered. Most practical is to require $[H_I, A_s] = 0$, i.e. that the interaction does not perturb the motion of A_s even deterministically. The second way of disturbance of A_s is via the signal Hamiltonian. If H_s contains an observable conjugate to A_s, then free evolution between any two subsequent measurements developes an element of uncertainty in A_s, as was demonstrated in section 2 with the example of back-action in position measurement of a free particle. In general, the QND observable requires

$$[A_s(t), A_s(t')] = 0. \tag{45}$$

More restrictive is to have A_s as constant of motion. All above requirements can be summarized as following

(a) H_I is a function of A_s;

(b) $[H_I, A_s] = 0$;

(c) $[H_I, A_m] \neq 0$;

(d) H_s is *not* a function of the conjugate observable of A_s.

5.   QND measurements via Kerr interaction

5.1   formulation of the problem

QND measurement of the e.m. energy of an optical beam can be realized via the optical Kerr effect Kerr effect, as proposed first in Ref. [15]. The signal wave is allowed to interact with another (probe) optical beam in a Kerr medium,

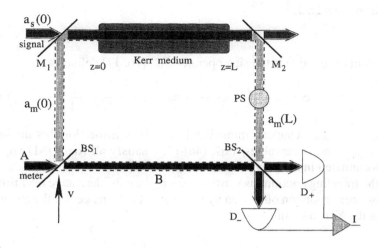

Figure 4. Schematic setup for QND measurements of the energy of a light beam using the Kerr effect. Four mirrors form the Mach-Zender interferometer for the meter beam (colored in blue): mirrors M_1 and M_2 are totally reflective while BS_1 and BS_2 are partially transparent, with η and $1/2$ as the intensity transmission coefficients for BS_1 and BS_2, correspondingly. Phase shifter PS is used for maximizing the output signal. The output is detected by photodetectors D_+ and D_- and the photocurrent difference is then taken by I. Mirrors M_1 and M_2 are fully transparent for the signal beam (colored in red).

for instance in a configuration shown in Fig. 4. Due to the effect of cross-phase modulation, the energy of the signal beam is recorded in the optical phase of the probe beam. The last serves as a meter. The photon number of the signal wave is the QND observable, A_s, and the phase of the probe beam is the readout observable, A_m.

Relevant Hamiltonians read as

$$H_m = \hbar\omega_m \left(a_m^\dagger a_m + \frac{1}{2} \right) , \qquad (46)$$

$$H_s = \hbar\omega_s \left(a_s^\dagger a_s + \frac{1}{2} \right) , \qquad (47)$$

$$H_I = \frac{\hbar^2}{2V\epsilon^2} \omega_s \omega_m \chi^{(3)} a_m^\dagger a_m a_s^\dagger a_s , \qquad (48)$$

where $a_m^\dagger$ ($a_s^\dagger$) and a_m (a_s) are the creation and annihilation operators of the probe (signal) optical wave, V is the quantization volume, $\chi^{(3)}$ is the third-order nonlinear susceptibility, and ω_m and ω_s are optical frequencies for the probe and signal beam, correspondingly. The QND observable is the photon number operator of the signal:

$$A_s = a_s^\dagger a_s \,. \tag{49}$$

The readout observable is the sine operator S_m, [16, 17] defined as

$$A_m = S_m \equiv \frac{1}{2i} \left(\frac{1}{\sqrt{n_m + 1}} \, a_m - a_m^\dagger \, \frac{1}{\sqrt{n_m + 1}} \right), \tag{50}$$

with $n_m = a_m^\dagger a_m$. One can immediately check by inspection that the set of Hamiltonians and observables, Eqs. (46)-(50), satisfy all four QND requirements formulated in the previous section.

For the traveling continuous wave we can replace the derivative over time in the Heisenberg equation of motion by space derivative and get for the evolution of a_m in the Kerr medium:

$$-i\hbar v \frac{d}{dz} a_m(z) = \frac{\hbar^2}{2V\epsilon^2} \omega_m \omega_s \chi^{(3)} \, n_s \, a_m(z) \,. \tag{51}$$

The time which is needed to traverse the Kerr medium of length L corresponds to the time during which the interaction Hamiltonian is switched on. Solution to Eq. (51) at the output of the medium reads

$$a_m(L) = e^{iX n_s} a_m(0), \tag{52}$$

with the cross-phase coefficient X defined by expression

$$X \equiv \frac{\hbar^2}{2V\epsilon^2 v} \omega_m \omega_s \chi^{(3)} L \,. \tag{53}$$

>From Eq. (52) we conclude that the goal is reached and the signal photon number is indeed recorded in the phase phase of the probe beam. Moreover, this is accomplished in the QND fashion: the signal photon number is constant of the motion and thereby is not perturbed during the interaction. In next subsection we describe how to arrange the measurement in order to extract the value of n_s from a phase-sensitive detection of the probe beam.

5.2   measurement outcome

The experiment can be arranged as suggested in [15] and depicted in Fig. 4. Three fields are essential for our consideration: the signal beam a_s, the vacuum field v whose contribution becomes relevant in the evaluation of noise, and the incoming beam A which is splitted on the beamsplitter BS_1 in two parts $a_m(0)$ and B:

$$a_m(0) = \sqrt{\eta} \, A + \sqrt{1 - \eta} \, v \,, \tag{54}$$

$$B = \sqrt{\eta} \, v - \sqrt{1 - \eta} \, A \,. \tag{55}$$

The probe beam a_m is under the measurement. We follow its evolution along the upper arm of the Mach-Zender interferometer, allow to interact with the signal beam in the Kerr medium, and finally collide on the beamsplitter BS_2 with the free propagating portion B. Finally, two fields, $a_m(L)$ and B, fall on the 50/50 beamsplitter BS_2. Two outcoming fields produce photocurrents at detectors D_+ and D_-. Subtraction one from the other yields

$$\hat{I} = i \cos{(X\hat{n}_s)} \left(\hat{v}^\dagger \hat{A} - \hat{A}^\dagger \hat{v} \right) \tag{56}$$

$$+(1 - 2\eta) \sin{(X\hat{n}_s)} \left(\hat{A}^\dagger \hat{v} + \hat{v}^\dagger \hat{A} \right)$$

$$+2\sqrt{\eta(1-\eta)} \sin{(X\hat{n}_s)} \left(\hat{A}^\dagger \hat{A} - \hat{v}^\dagger \hat{v} \right) .$$

Tracing over initial state of the system, denoted here as $\langle \ldots \rangle$, gives us the measurement outcome:

$$\langle I \rangle = 2\sqrt{\eta^{-1} - 1}\ \bar{n}_m \ \sin(X\bar{n}_s) , \tag{57}$$

where $\bar{n}_s \equiv \langle n_s \rangle$ and $\bar{n}_m \equiv \langle n_m \rangle$. Note that contribution of the vacuum field vanishes.

Without loss of generality, the nonlinear phase shift is assumed to be small: $X\bar{n}_s \ll 1$. Then, the inversion of Eq. (57) yields

$$\bar{n}_s = \frac{\langle I \rangle}{2\sqrt{\eta^{-1} - 1}\ \bar{n}_m \ X} . \tag{58}$$

As the meter beam represents a part of the detection apparatus, the knowledge of its statistics is a natural assumption. So, $\bar{n}_m$ is known, as well as the other relevant parameters, η and X. Using Eq. (58) one can now infer from the measurement the photon number of the signal beam.

5.3   discussion

Principal part of the study is completed. However, there is still a number of interesting, practical as well as fundamental, problems. One is the accuracy of the measurement. In principle, no fundamental limitation on the accuracy is imposed by the formal QND theory. In practice the situation is different. This can be conveniently illustrated on the example of measurements of a continuous variable, for instance momentum. The better one wishes to measure the momentum the longer interaction time with a meter system is required. Or alternatively, one needs to spend more energy. As both the time and the energy are limited, the precision cannot become infinitely high.

Let us support this reasoning with calculation of the error variance of the outcome of our measurement. Eq. (56) is used to evaluate $(\Delta I)^2 = \langle \hat{I}^2 \rangle - \langle \hat{I} \rangle^2$.

The result reads

$$(\Delta I)^2 = \bar{n}_m/\eta. \qquad (59)$$

An adequate measure of the accuracy is signal-to-noise ratio. It immediately follows from Eqs. (57) and (59):

$$\frac{\text{SIGNAL}}{\text{NOISE}} = 2X\sqrt{(1-\eta)\bar{n}_m}\,\bar{n}_s. \qquad (60)$$

The ratio is improved when energy (here $\bar{n}_m$ and $\bar{n}_s$) grows and/or time (here L in the expression for X) increases. Note that the photon number is a discrete and not continuous variable. Nevertheless, the energy-time limitation is manifestly present.

Another aspect of the same problem is the level of noise that separates domains with "good" and "bad" precision. The boundary between the two domains is usually called the Standard Quantum Limit (SQL). For light fields the SQL is defined as the noise level of the variable of interest in a coherent state. coherent state This is certainly no more than a convention but a very good one, as it emerges from everyday practice: an ideal laser laser generates light in a coherent state coherent state or in a state close to coherent. In this regard, the coherent state coherent state is a "standard". Achieving the accuracy below this standard is an experimental challenge.

Let us adopt the SQL as the measure of noise and estimate the accuracy of the QND measurement of the photon number in the signal beam. With the mean photon number $\bar{n}_s$ which is the outcome of our measurement we associate the noise level $\sqrt{\bar{n}_s}$ as the SQL. This suggestion does not imply that we ascribe the coherent state coherent state to the signal beam, and serves solely for ranking the precision. As pointed out in the Introduction, the quantum state of the signal in QND measurements is an irrelevant issue. So, the standard level of the signal-to-noise ratio is defned as $\bar{n}_s/\sqrt{\bar{n}_s} = \sqrt{\bar{n}_s}$. If now

$$\frac{\text{SIGNAL}}{\text{NOISE}} = 2X\sqrt{(1-\eta)\bar{n}_m}\,\bar{n}_s < \sqrt{\bar{n}_s}, \qquad (61)$$

see Eq. (60), then the precision is below the SQL and the measurement is classified as ultra-precise. For a rough numerical estimate we take $X\bar{n}_s \approx 1$ and $(1-\eta) \approx 1$, and with these values in Eq. (61) get

$$\sqrt{\frac{\bar{n}_m}{\bar{n}_s}} < 1 \qquad (62)$$

as the condition for beating the SQL. The more intense the probe beam, the better.

Typically, optical beams involved in the experiment are anyway rather energetic. They may contain millions of photons on the length of the medium.

Practically speaking, the requirement on large intensity is an inevitable consequence of rather weak nonlinearities of all known media. The Kerr nonlinearity
is particularly weak because it is originated from far-off-resonance atomic or
molecular responses.

One may draw a link between two types of QND measuremrents, of single
photons stored in the high-Q cavity and that of optical waves traveling through
the Kerr medium. Both systems are probed via interaction with atoms (can be
also molecules). In the first case the interaction is resonant, in the second -
far-off-resonant. Both, however, conserve the energy of the signal field. In the
single photon setup the atom is directly subjected to the measurement by an
atomic detector, while with the Kerr medium there is an additional mediator.
Medium's response is first mapped onto the meter beam and only then measured
by means of photodetection.

Concluding this section, we list a few interesting issues left beyond the scope
of this article and formulate them in the form of questions. The first is how
to relate the measurement outcome Eq. (57) to the measurement of the sine
operator Eq. (50). What are the approximations that should be adopted? Does
the noise introduced during the QND measurement correspond to the minimal
noise allowed by Heisenberg's uncertainty principle? If positive, what these
necessary conditions are.

6.   QND measurements of solitons in optical fibers

The study presented in this section is a natural extension of the above results.
Continuous waves are now replaced by pulses, and more specifically - solitons.
The medium, now optical fiber, is characterized not only by the Kerr nonlinearity
but also by the group velocity dispersion. It is the balance of the nonlinearity and
the dispersion that gives rise to stable formations localized in space and time,
called solitons. The nonlinear part of the Hamiltonian Hamiltonian includes
two more terms as compared to Eq. (48) and reads now as

$$H_{kerr} = \frac{\hbar^2}{2V\epsilon^2}\omega\omega\chi^{(3)}\left(a_m^\dagger a_m a_s^\dagger a_s + a_s^\dagger a_s a_s^\dagger a_s + a_m^\dagger a_m a_m^\dagger a_m\right), \quad (63)$$

where we put $\omega_s \approx \omega_m = \omega$. The first term causes the cross-phase modulation
between two interacting fields and it vanishes when one of the fields is absent.
In contrast, the second (third) term is related solely to the signal (probe) beam.
Each of the last two terms stands for the interaction of the field with itself and
causes the effect called self-phase modulation.

Let us briefly describe the experimental arrangement. When two initially
independent solitons with slightly different carrier frequencies propagate in a
fiber, they move with different velocities and can eventually collide. After
the collision occurs and the two solitons are distinctly separated, they appear
with fully restored initial shapes, velocities, and energies, [18]. The sequence of

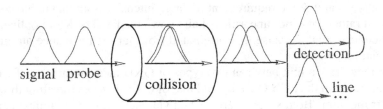

Figure 5.   An illustration for the QND measurements of optical solitons via collision in a fiber. The value of the photon number of the signal soliton (colored in red) is recorded into the phase phase of the probe soliton (colored in green)). After emerging from the fiber, the solitons are separated. The phase of the probe soliton is measured by a detector. The signal soliton is directed to a line for a further use, if necessary.

events is illustrated in Fig. 5. The only feature which records the past collision is a time and phase shift. The last depends on photon numbers and on the frequencies of the two solitons. Therefore, if the time and/or phase phase shift of one soliton (probe soliton) is measured, it becomes possible to infer the photon number and/or frequency of the other soliton (signal soliton).

As in the previous example with continuous beams, here as well, one soliton serves as the meter and the other is the signal system. The difference is that in contrast to the monochromatic wave which is described as a single mode, the soliton involves an infinite number of degrees of freedom. A nontrivial consequence of the greater complexity is the appearance of the frequency operator. The last is the constant of motion. So, one can now imagine the QND measurements of not only the photon number but also the frequency, or a linear combination of these two operators, [19, 20]. Analogously, the readout can be accomplished not only via optical phase but also via the time shift. The last is treated as a quantum-mechanical operator conjugate to the momentum.

The most adequate (from author's perspective) formalism for describing quantum solitons is based on the quantum counterpart of the classical perturbation theory for the Inverse Scattering Problem, see [21, 22, 23, 24, 6] and the latest collection of references in [20]. The quantum theory of optical solitons is by its own an elegant formalism deserving a separate lecture. The theory allows one to build a fully analytic description of the QND measurements. Quantum solitons are also promising candidates for purposes of the quantum information processing [25, 26] and quantum cryptography along the lines of the recently proposed scheme [27].

Basic principles of the QND measurements with solitons remain the same as in the previous example with continuous beams. However, the problem of accuracy becomes much more complicated. Formally, the deterioration of accuracy is caused by the effect of group velocity dispersion and self-phase modulation, for the latter see last terms in the Hamiltonian Eq. (63). The two effects contribute to the phase and time shifts. The noise accompanied this contribution washes out the useful information recorded into the phase and time shifts due to cross-phase modulation. As an illustration, for the initially coherent probe soliton, the uncertainty in phase due to self-phase modulation becomes comparable with the initial phase uncertainty already at a distance as small as $1/25$ of the soliton period. Note that at least one soliton period is needed to get well separated solitons after they collide. So, the signal-to-noise ratio is roughly 25 times above the SQL. There is a number of proposals on how to overcome the extra noise and even completely cancel the undesirable contribution. A motivated reader can find them in Refs. [23, 28, 29, 30, 20].

Acknowledgments

This lecture is a by-product of numerous discussions of the topic with J. H. Eberly while author's stay with the Rochester Theory Center, University of Rochester. The author wishes to express his deep gratitude to W. P. Schleich for his invaluable support. He is also indebted to A. S. Shumovsky for his very warm hospitality in Bilkent and Antalya.

References

[1] Quantum Theory and Measurement, Eds. J. A. Wheeler and W. H. Zurek (Princeton University, Princeton, NJ, 1983).

[2] One may also have a goal oriented more to a preparation rather than to measurement. Here the repeated measurements can be used in a feedback loop, where the variable of interest is measured and then corrected towards a desirable value by an active device. Such cases as well as any other active intervention into the quantum system are left beyond the scope of this lecture.

[3] V. B. Braginsky and Y. I. Vorontsov, "Quantum-mechanical limitations in macroscopic experiments and modern experimental technique" Usp. Fiz. Nauk **114**, 41-53 (1974) [Sov. Phys. Usp. **17**, 644-650 (1975)]; V. B. Braginsky, Y. I. Vorontsov, and F. Y. Khalili "Quantum singularities of a ponderomotive meter of electromagnetic energy" Zh. Eksp. Teor. Fiz. **73**, 1340-1343 (1977) [Sov. Phys. JETP **46**, 705-706 (1977)].

[4] P. D. Drummond, R. M. Shelby, S. R. Friberg, and Y. Yamamoto, "Quantum solitons in optical fibres", Nature (London) **365**, 307-313 (1993); P. Grangier, J. A. Levenson, and J.-P. Poizat, "Quantum non-demolition measurements in optics", Nature (London) **396**, 537-542 (1998); A. Sizmann and G. Leuchs, in "Progress in Optics" vol. XXXIX, ed. by E. Wolf, (Elsevier, NY) 1999, p.373.

[5] V. B. Braginsky and F. Ya. Khalili, Quantum measurement (Cambridge Univ. Press, 1992).

[6] H. A. Haus, Electromagnetic noise and quantum optical measurements (Springer, Berlin, 2000).

[7] G. Nogues, A. Rauschenbeutel, S. Osnaghi, M. Brune , J. M. Raimond, and S. Haroche, "Seeing a single photon without destroying it", Nature **400**, 239-242 (1999).

[8] B.-G. Englert, N. Sterpi, and H. Walther, "Parity states in the one-atom maser", Opt. Commun. **100**, 526-535 (1993).

[9] This theoretical treatment is after V.V. Kozlov and J.H. Eberly (unpublished).

[10] L. Allen and J. H. Eberly, Optical resonance and two-level atoms (Dover, New-York, 1987).

[11] M. Brune, S. Haroche, J. M. Raimond, L. Davidovich, N. Zaury, "Manipulation of photons in a cavity by dispersive atom-field coupling: quantum non-demolition measurements and generation of Schrödinger cat states", Phys. Rev. A **45**, 5193-5214 (1992).

[12] B. T. H. Varcoe, S. Brattke, M. Weidinger, and H. Walther, "Preparing pure photon number states of the radiation field", Nature (London) **403**, 743-746 (2000).

[13] K. S. Thorne, R. W. P. Drever, C. M. Caves, M. Zimmermann, and V. D. Sandberg "Quantum nondemolition measurements of harmonic oscillators" Phys. Rev. Lett. **40**, 667-671 (1978); W. G. Unruh "Quantum nondemolition and gravity-wave detection" Phys. Rev. B **19**, 2888-2896 (1979); C. M. Caves, K. S. Thorne, R. W. P. Drever, V. D. Sandberg, and M. Zimmermann, "On the measurement of a weak classical force coupled to a quantum-mechanical oscillator. I. Issues of principles" Rev. Mod. Phys. **52**, 341-392 (1980).

[14] V. B. Braginsky and F. Ya. Khalili, "Quantum nondemolition measurements: the route from toys to tools", Rev. Mod. Phys. **68**, pp. 1-11 (1996).

[15] N. Imoto, H. A. Haus, and Y. Yamamoto, "Quantum nondemolition measurement of the photon number via the optical Kerr effect", Phys. Rev. A **32**, pp. 2287-2292 (1985).

[16] P. Carruthers and M. M. Nieto, "Coherent states and the number-phase uncertainty relation", Phys. Rev. Lett. **14**, pp. 387-389 (1965).

[17] The sine operator Eq. (50) is used as a substitute for the phase operator. The definition of the last has a controversial history, see for instance a special issue on this subject in Physica Scripta, **48** (1993), "Quantum phase and phase dependent measurements", Eds. W. P. Schleich and S. M. Barnett. The sine operator is however a well defined object and fits well the context of our problem. What is equally important, the operator can be measured directly in an optical experiment, see J. W. Noh, A. Fougeres, and L. Mandel, "Measurement of the quantum phase by photon counting", Phys. Rev. Lett. **67**, 1426-1429 (1991).

[18] V. E. Zakharov and A. B. Shabat, "Exact theory of three-dimensional self-focusing and one-dimensional self-modulation of waves in nonlinear media", Soviet Physics JETP **34**, 62-69 (1972) [Zh. Eksp. Teor. Fiz. **61**, 118-134 (1971)].

[19] see H. A. Haus, K. Watanabe, and Y. Yamamoto, "Quantum-nondemolition measurement of optical solitons", J. Opt. Soc. Am. **B6**, 1138-1148 (1989), for the first proposal of the QND measurements of optical solitons. See also S. R. Friberg, S. Machida, and Y. Yamamoto, "Quantum-nondemolition measurement of the photon number of an optical soliton", Phys. Rev. Lett. **69**, 3165-3168 (1992), for the first QND experiment with optical solitons.

[20] V. V. Kozlov and D. A. Ivanov, "Accurate quantum nondemolition measurements of optical solitons", Phys. Rev. A **65**, 023812 (2002).

[21] D. J. Kaup, J. Math. Phys. **16**, 2036 (1975).

[22] H. A. Haus and Y. Lai, "Quantum theory of soliton squeezing: a linearized approach", J. Opt. Soc. Am. **B7**, 386-392 (1990).

[23] P.D. Drummond, J. Breslin, R.M. Shelby, "Quantum-nondemolition measurements with coherent soliton probes", Phys. Rev. Lett. **73**, 2837-2840 (1994).

[24] V. V. Kozlov and A. B. Matsko, "Second-quantized models for optical solitons in nonlinear fibers: Equal-time versus equal-space commutation relations", Phys. Rev. A **62**, 033811 (2000).

[25] V. V. Kozlov and A. B. Matsko, "Einstein-Podolsky-Rosen paradox with quantum solitons in optical fibers", Europhys. Lett., **54**, 592-598 (2001).

[26] V. V. Kozlov and M. Freyberger, "High-bit-rate quantum communication", Opt. Commun. **206**, 287-294 (2002).

[27] Ch. Silberhorn, T. C. Ralph, N. Lütkenhaus, and G. Leuchs, "Continuous variable quantum cryptography: beating the 3 dB loss limit", Phys. Rev. Lett. **89**, 167901 (2002).

[28] J.-M. Courty, S. Spälter, F. König, A. Sizmann, and G. Leuchs, "Noise-free quantum-nondemolition measurement using optical solitons", Phys. Rev. A **58**, 1501 (1998).

[29] V. V. Kozlov and A. B. Matsko, "Cancellation of the Gordon-Haus effect in optical transmission system with resonant medium", J. Opt. Soc. Am. B. **16**, 519-522 (1999).

[30] A. B. Matsko, V. V. Kozlov, and M. O. Scully, "Back-action cancellation in quantum nondemolition measurement of optical solitons", Phys. Rev. Lett. **82**, 3244-3247 (1999).

CAVITY QUANTUM ELECTRODYNAMICS WITH SINGLE ATOMS

Herbert Walther
Max-Planck-Institut für Quantenoptik and
Sektion Physik der Universität München,
85748 Garching, Germany

1.   Introduction

Experiments with single atoms got routine. In this lecture two groups of those experiments will be reviewed with special emphasis on applications to study quantum phenomena in the atom-radiation interaction. The first one deals with the one-atom maser and the second one with another cavity quantum electrodynamic device on the basis of trapped ions. The latter device has interesting applications in quantum computing Quantum computing and quantum information.

2.   Review of the One-Atom Maser

The most basic avenue to study the generation process of radiation in lasers laser and masers is to drive a maser or laser consisting of a single-mode cavity by single atoms. This system, at first glance, seems to be another example of a Gedanken experiment, but such a one-atom maser [1] really exists and can in addition be used to study the basic principles of radiation-atom interaction. The advantages of the system are:

(1) it is the only system which allows to study the periodic photon exchange between a single atom and a single cavity mode,

(2) it leads to the first maser which sustains oscillations with much less than one atom on the average in the cavity,

(3) this setup allows one to study in detail the conditions necessary to obtain nonclassical radiation, especially radiation with sub-Poissonian photon statistics and even Fock states in a maser system directly, even with Poissonian pumping, and

(4) it is possible to study a variety of phenomena of quantum physics including the quantum measurement process.

A.S. Shumovsky and V.I. Rupasov (eds.), Quantum Communication and Information Technologies, 125–144.
© 2003 *Kluwer Academic Publishers.*

What are the tools that make this device work? It is the enormous progress in constructing superconducting cavities together with the laser laser preparation of highly excited atoms – Rydberg atoms – that have made the realization of such a one-atom maser possible. Rydberg atoms Rydberg atom have quite remarkable properties [2] which make them ideal for such experiments: The probability of induced transitions between neighbouring states of a Rydberg atom scales as n^4, where n denotes the principle quantum number. Consequently, a single photon is enough to saturate the transition between adjacent levels. Moreover, the spontaneous lifetime of a highly excited state is very large. We obtain a maser by injecting these Rydberg atoms into a superconducting cavity with a high quality factor. The injection rate is such that on the average there is much less than one atom present inside the resonator at any time. A transition between two neighbouring Rydberg levels is resonantly coupled to a single mode of a cavity field. Due to the high quality factor of the cavity, the radiation decay time is much longer than the characteristic time of the atom-field interaction, which is given by the inverse of the single-photon Rabi-frequency. Therefore it is possible to observe the dynamics [3] of the energy exchange between atom and field mode leading to collapse and revivals in the Rabi oscillations [4, 5]. Moreover a steady state field is built up inside the cavity when the mean time between the atoms injected into the cavity is shorter than the cavity decay time.

The one-atom maser or micromaser setup used for the experiments is shown in Fig. 1; for a detailed description of the present setup see Ref. [6].

A ^{3}He-^{4}He dilution refrigerator houses the closed superconducting microwave cavity. A rubidium oven provides two collimated atomic beams: the main beam passing directly into the cryostat and the second being used to stabilize the laser frequency [6]. A frequency-doubled dye laser ($\lambda = 297$ nm) is used to excite rubidium (^{85}Rb) atoms to the Rydberg $63P_{3/2}$ state starting from the $5S_{1/2}$(F = 3) ground state ground state. In the excitation process a particular velocity subgroup of the atoms is populated. In this way the interaction time of the atoms with the cavity is controlled. The cavity is tuned to the 21.456 GHz transition from the $63P_{3/2}$ state to the $61D_{5/2}$ state, which is the lower or ground state of the maser transition. For this experiment a cavity with a Q-value of 4×10^{10} was used, this corresponding to a photon lifetime of 0.3 s. This Q-value is the largest ever achieved in this type of experiments and the photon lifetime is more than two orders of magnitude higher than that of related setups, see e.g Ref. [7]. This cavity is used to study micromaser operation in great detail. The cavity is carefully shielded against magnetic fields by several layers of cryoperm. In addition, three pairs of Helmholtz coils are used to compensate the earth's magnetic field to a value of several mG in a volume of $(10 \times 4 \times 4)$ cm^3. This is necessary in order to achieve the high quality factor and prevent the different magnetic substates of the maser levels from mixing during the

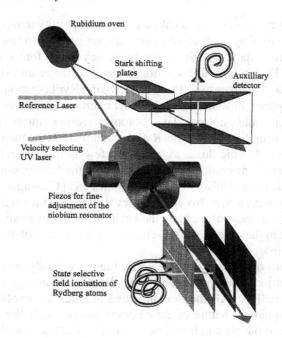

Figure 1. The atoms leaving the rubidium oven are excited into the $63P_{3/2}$ Rydberg state by means of a UV laser laser at an angle of $11°$. After the cavity the atoms are detected by state-selective field ionisation. The cavity is tuned with two piezo translators. An auxiliary atomic beam (not shown) is used to stabilize the laser frequency. The laser is locked to a Stark-shifted atomic resonance of the auxiliary beam, thus allowing the velocity subgroup selected by excitation to be continuously changed within the range of the velocity distribution of the atoms. In this way the interaction time of the atoms with the cavity is changed

atom-field interaction time. Besides the maser transitions from the $63p_{3/2}$ level to the $61d_{5/2}$ level mentioned above also the transition to the $61d_{3/2}$ level have been studied in our experiments. The two transitions require different cavities.

The Rydberg atoms Rydberg atom in the upper and lower maser levels are detected in two geometrically different regions of a field ionisation detector. The field strength of the detector has a gradient and is adjusted so as to ensure that in the first section of the detector the atoms in the upper level are ionised, but not those in the lower level. The lower-level atoms are then ionised in the second part of the detector.

To demonstrate maser operation, the cavity is e.g. tuned over the maser transition and the flux of atoms in the excited state is recorded simultaneously. Transitions from the initially prepared $63p_{3/2}$ state to the lower level are detected by a reduction of the electron count rate.

If the rate of atoms crossing a cavity exceeds the cavity damping rate ω/Q, the photon released by each atom is stored long enough to interact with the next atom. Here ω stands for the cavity frequency and Q for the quality factor of the cavity. The atom-field coupling becomes stronger and stronger as the field builds up and evolves to a steady state. Using Rydberg atoms with a large field-atom coupling constant leads to a maser operating with exceedingly small numbers of atoms and photons. The photons corresponding to transitions between neighbouring Rydberg atoms Rydberg atom are in the microwave region at about 25 GHz. Atomic fluxes as small as a few atoms per second have generated maser action, as could be demonstrated for the first time in the group of H. Walther et al. of the University of Munich in 1985 [1] using a superconducting cavity. For such a low flux there is never more than a single atom in the resonator – in fact, most of the time the cavity is empty of atoms. It should be mentioned that in the case of such a setup a single resonant photon is sufficient to saturate the maser transition.

The one-atom maser was used to verify the complex dynamics of a single atom in a quantised field predicted by the Jaynes-Cummings model Jaynes-Cummings model[3]. All the demonstrated features are explicitly a consequence of the quantum nature of the electromagnetic field: the statistical and discrete nature of the photon field leads to a new characteristic dynamics such as collapse and revivals in the exchange of a photon between the atom and the cavity mode [4, 5]. The maser field in the cavity is measured through the number of atoms in the lower state of the maser transition. Due to the strong coupling between atom and maser field both are entangled.

The steady-state field of the micromaser shows sub-Poisson statistics. This is in contrast to regular masers and lasers where coherent fields (Poisson statistics) are observed. The reason for nonclassical radiation being produced is that a fixed interaction time of the atoms is chosen, leading to careful control of the atom-field interaction dynamics.

Under steady-state conditions, the photon statistics P(n) of the field of the micromaser is essentially determined by the pump parameter $\Theta = N_{ex}^{1/2} \Omega t_{int}$, denoting the angular rotation of the pseudospin-vector of the interacting atom. Here, N_{ex} is the average number of atoms that enter the cavity during the decay time of the cavity τ_{cav}, Ω the vacuum Rabi flopping frequency, and t_{int} is the atom-cavity interaction time. The normalised photon number of the maser $\langle \nu \rangle = \langle n \rangle / N_{ex}$ shows the following generic behaviour (see Fig. 2). It suddenly increases at the maser threshold value $\Theta = 1$ and reaches a maximum for $\Theta \approx 2$ (denoted by 1 in Fig. 2). The behaviour at threshold corresponds to the characteristics of a continuous phase transition. As Θ further increases, the normalised averaged photon number $\langle \nu \rangle / N_{ex}$ decreases to a minimum slightly below $\Theta \approx 2\pi$, and then abruptly increases to a second maximum (3a in Fig. 2). This general type of behaviour recurs roughly at integer multiples of 2π, but becomes less pronounced with increasing Θ. The reason for the periodic maxima of the average photon number is that for integer multiples of $\Theta = 2\pi$ the pump atoms perform an integer number of full Rabi flopping cycles, and start to flip over at a slightly larger value of Θ, thus leading to enhanced photon emission. The periodic maxima for $\Theta \approx 2\pi, 4\pi$, and so on may be interpreted as first-order phase transitions [8].

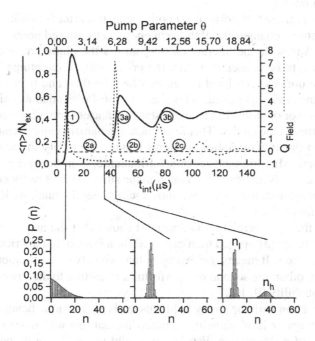

Figure 2. One-atom maser or micromaser characteristics. The upper part of the figure shows the average photon number versus interaction time (solid curve) and the photon number fluctuations represented by the Q-factor (dotted line). Both curves are determined by the photon exchange dynamics between atom and field. The pump parameter gives the angular rotation of the pseudospin vector of the interacting atom (e.g. 2π corresponds to a full rotation, i.e. the atom is again in the upper level). In the lower part of the figure the steady-state photon number distribution $P(n)$ is shown for three values of t_{int}. The distribution on the left side corresponds to the maser threshold, the one on the right gives an example of the double-peaked distribution associated with a bistable bistability behaviour leading to quantum jumps in the maser field. In the situation around n_l, the atoms are back in the excited state and can therefore emit a photon, leading to a higher steady-state photon number n_h. With increasing t_{int} this part will grow and n_l will decrease and disappear. A new jump occurs in the region close to 3b.

The photon statistics of the maser radiation is usually characterised by the Q-pamameter introduced by Mandel:

$$Q_{field} = \{(\langle n^2 \rangle - \langle n \rangle^2)/\langle n \rangle\} - 1.$$

It can be seen that $Q_{field} = 0$ corresponds to a Poissonian photon distribution. Q_{field} for the micromaser is plotted as a dotted line in Fig. 2. The value drops below zero in the region 2a, 2b, etc. This shows the highly sub-Poissonian character of the one-atom-maser field being present over a large range of parameters.

The reason for the sub-Poissonian atomic statistics is the following: A changing flux of atoms changes the Rabi frequency via the stored photon number in the cavity. Adjusting the interaction time allows the phase phase of the Rabi nutation cycle to be chosen such that the probability of the atoms leaving the cavity in the upper maser level increases when the flux, and hence the photon number in the cavity, is raised. This leads to sub-Poissonian atomic statistics since the number of atoms in the lower state decreases with increasing flux and photon number in the cavity. This feed-back mechanism can be neatly demonstrated when the anticorrelation of atoms leaving the cavity in the lower state is investigated. Measurements of this 'antibunching' antibunching phenomena could be made. It is interesting to note that the situation at $\Theta \approx 2\pi$ corresponds to a quantum-non-demolition case introducing less fluctuations leading to a sub-Poissonian field.

The fact that anticorrelation is observed shows that the atoms in the lower state are more equally spaced than expected for a Poissonian distribution of the atoms in the beam. It means, for example, that when two atoms enter the cavity close to each other, the second one performs a transition to the lower state with reduced probability [9, 10].

The interaction with the cavity field thus leads to an atomic beam with atoms in the lower maser level showing number fluctuations which are up to 40 % below those of a Poissonian distribution found in usual atomic beams. This is interesting because atoms in the lower level have emitted a photon to compensate for cavity losses inevitably present. Although this process is induced by dissipation giving rise to fluctuations, the atoms still obey sub-Poissonian statistics. sub-Poissonian statistics

The field strongly fluctuates at the positions where phase phase transitions occur, these being caused by the presence of two maxima in the photon number distribution P(n) at photon numbers n_l and $n_h (n_l < n_h)$; see Fig. 2. The phenomenon of the two coexisting maxima in P(n) was also studied in a semi-heuristic Fokker-Planck Fokker-Planck approach (FP) approach. There, the photon number distribution P(n) is replaced by a probability function $P(\nu, \tau)$ with continuous variables $\tau = t/\tau_{cav}$ and $\nu = n/N_{ex}$, the latter replacing the photon number n. The steady-state solution obtained for $P(\nu, \tau), \tau \gg 1$, can be constructed by means of an effective potential $V(\nu)$ showing minima at positions where maxima of the probability function $P(\nu, \tau)$ are found. Close to $\Theta = 2\pi$

and multiples thereof, the effective potential $V(\nu)$ exhibits two equally attractive minima located at stable gain-loss equilibrium points of maser operation (see Fig. 2). The mechanism at the phase transitions mentioned is always the same: A minimum of $V(\nu)$ loses its global character when Θ is increased, and is replaced in this role by the next one. This reasoning is a variation of the Landau theory of first-order phase transitions, with $\sqrt{\nu}$ being the order parameter. This analogy actually leads to the notion that in the limit $N_{ex} \rightarrow \infty$ the change of the micromaser field around integer multiples of $\Theta = 2\pi$ can be interpreted as first-order phase phase transitions [8].

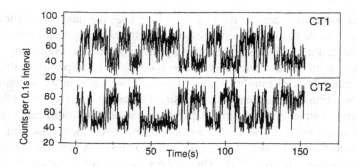

Figure 3.   Quantum jumps between two equally stable operation points of the one-atom-maser field. CT1 is the measurement of the upper state atoms and BT2 that of the lower state atoms. It can be seen that coth signals show counterphase behaviour.

Close to first-order phase transitions long time constants of the field evolution are expected. This phenomenon was experimentally demonstrated, as well as related phenomena such as spontaneous quantum jumps between equally attractive minima of $V(\nu)$, bi-stability, bistability and hysteresis. Some of these phenomena are also predicted in the two-photon micromaser, for which qualitative evidence of first-order phase transitions and hysteresis has been reported.

In the following, we would like to discuss one example of this switching behaviour. The maser was operated under steady-state conditions close to a first-order phase transition. The two maxima in P(n) are manifested in spontaneous jumps of the maser field between the two maxima with a time constant of 10 s. This fact, and the relatively barge pump rate, led to the clearly observable field jumps shown in Fig. 3. The two discrete values of the counting rates correspond to the meta-stable operating points of the maser, these being ca. 70 and ca. 140 photons. In the description, the two values correspond to two equally attractive minima in the FP potential $V(\nu)$. If one considers, for instance, the counting rate of lower-state atoms (CT2 in Fig. 3), the lower (higher) plateaus correspond to time intervals in the low (high)-field meta-stable operating point. If the actual photon number distribution is averaged over a time interval containing many spontaneous field jumps, the steady-state result P(n) of the micromaser theory is recovered [8].

The measurements performed with the one-atom maser in our group have been discussed in several review papers [9, 10]. In the following we will focus on the recent results with this device. They deal with the generation of Fock states of the radiation field.

3.   Generation of Fock States in the One-Atom Maser

The quantum treatment of the radiation field uses the number of photons in a particular mode to characterise the quantum states. the ground state ground state of the quantum field is represented by the vacuum state consisting of field fluctuations with no residual energy. The states with fixed photon number are usually called Fock or number states. They are used as a basis on which any general radiation field state can be expressed. Fock states thus represent the most basic quantum states and maximally differ from what one would call a classical field. Although Fock states of vibrational motion are routinely observed, e.g. in ion traps Ion-trap[11], Fock states of the radiation field are very fragile and very difficult to produce and maintain. They are perfectly number-squeezed, extreme sub-Poissonian states in which intensity fluctuations completely vanish.

The special case of single-photon fields are necessary for secure quantum communication [12, 13, 14] and quantum cryptography [15] and are required in some cases for quantum computing Quantum computing [16]. Photon fields with fixed photon numbers are also interesting from the point of view of fundamental physics since they represent, as pointed out above, the ultimate non-classical limit of radiation. When the photon number state is produced by strong coupling of excited-state atoms, a corresponding number of ground-state atoms is simultaneously populated. Such a system therefore produces a fixed number of atoms in the lower state as well. This type of atom source is a long sought after *gedanken* device as well [17]. Single photons have been generated by several processes such as single-atom fluorescence [18, 19, 20], single-molecule fluorescence [21], two-photon down-conversion [22], and Coulomb blockade of electrons [23], and one- and two-photon Fock states have been generated in the micromaser [24, 6]. As these sources do not produce photons on demand, they are better described as "heralded" photon sources, because they are stochastic either in the emission direction or in the time of creation. A source of single photons or, more generally, of Fock states generated on demand has just recently been demonstrated in our laboratory [32, 33]. Cavity quantum electrodynamics (QED) provides us with the possibility of generating a photon both at a particular time and with a predetermined direction. To this end there have been several proposals making use of high-Q cavities, which are basically capable of serving as sources of single photons [14, 25, 26, 27]. The current paper reviews our work based on the one-atom maser or micromaser; it allows a specified photon Fock state ($n \geq 1$) to be generated *on demand*, without need of conditional measurements, thus making it independent of detector efficiencies.

As discussed above steady-state operation of the one-atom maser has been extensively studied both theoretically [28] and experimentally. Sub-Poissonian photon statistics has been observed [29] earlier. More recently, two experiments demonstrated that Fock states (i.e. states with a fixed photon number) can readily be created in normal operation of the maser by means of either state reduction state reduction [24] or steady-state operation of the micromaser in a trapping state [6]. The trapping states trapping in the micromaser are the quantum states of the radiation field produced in the maser cavity. They are described in detail below. State reduction is possible owing to entanglement entanglement between the state of the outgoing atoms and the cavity field; detection of a lower state atom means that a field originally in an n-photon Fock state is projected onto the $n + 1$ state [30]. As a source of single photons, such a source can be compared to two-photon down-conversion, in which an idler beam is used to prove the creation of a photon in the signal beam. Both are subject to the same limitation in that the creation of the Fock state is unpredictable, and imperfect detectors further reduce the probability of the preparation of such a state. In contrast, it is shown here that the micromaser can be used to prepare Fock states *on demand* with small photon numbers in the cavity. Simultaneously, an equal number of ground state ground state atoms are produced with an efficiency of up to 98%.

Trapping states are a feature of low-temperature operation of the micromaser, for which the steady-state photon distribution closely approximates a Fock state under certain conditions. The trapping states show up when the maser is operated at temperatures at which the blackbody photons in the cavity practically disappear. They would otherwise contribute to the field fluctuations washing out the trapping state minima. They are typical of strongly coupled systems. They occur when atoms perform an integer number k of Rabi cycles under the influence of a fixed photon number n:

$$\sqrt{n + 1}gt_{\text{int}} = k\pi, \tag{1}$$

where g is the effective atom-field coupling constant and t_{int} is the interaction time. Trapping states are characterised by the number of photons n and the number of Rabi cycles k. The trapping state $(n, k)=(1, 1)$ therefore refers to the one-photon, one-Rabi-oscillation trapped field state. In other words, trapping states occur when the interaction time is chosen such that the emission probability becomes zero for certain operating parameters of the maser. When the trapping state is reached in steady-state operation, the micromaser field will become stabilised. The particular Fock state is known and is determined by the interaction time between the atom and cavity as given by the trapping state formula (Eq. 1). The Fock state, once prepared, is preserved owing to the trapping condition with a minimum probability of photon emission. Following preparation of the state, the beam of pump atoms can be turned off and the Fock state remains in the cavity for the duration of the cavity decay time. For simplicity, we concentrate in the following on preparation of a one-photon Fock state, but the method can also be generalised to generation of fields of higher photon numbers. Fig. 4 shows the trapping states for small photon numbers.

To realise the Fock state, it is necessary to switch the excitation of the Rydberg atoms Rydberg atom on and off in a predefined pulse sequence; this was achieved by means of an intensity-modulating electro-optical modulator triggered by control software. The pulse duration and pulse separation can both be tailored to the conditions required for the particular experiment.

To demonstrate the principle of this source, Fig. 5 shows a sequence of twenty successive pulses obtained by Monte Carlo simulation [31] of the micromaser operating in the (3, 1) trapping state. In each pulse there is a single emission event, producing a single lower-state atom and leaving a single photon in the cavity. In the case of loss of a photon by dissipation, one of the next incoming excited-state atoms will restore the single-photon Fock state. This condition was observed in [6] when sub-Poissonian atom statistics was measured with the maser operating in a trapping state. The influence of thermal photons and variations in interaction time or cavity tuning further complicates this picture, resulting in reduced visibility of steady-state Fock states [6]. Pulsed excitation as discussed here, however, reduces these perturbations present in the case of continuous operation of the atomic beam and the Fojk state is maintained with a high probability [33].

Figure 6 shows three curves, again obtained from a computer simulation, that illustrate the dynamical behaviour of the maser under pulsed excitation as a function of the interaction time for more ideal (but achievable) experimental parameters. The simulations show the probability of finding no ground-state atom per pulse ($P^{(0)}$) and exactly one ground-state atom per pulse ($P^{(1)}$); and the conditional probability of finding a second ground-state atom in a pulse, if one has already been detected ($P^{(>1;1)}$). The latter plot of the conditional probability, $P^{(>1;1)}$, has the advantage of being especially suitable for comparing theory and experiment since it is relatively insensitive to the detection probability for atoms in the upper and lower maser levels [33].

From the simulations it follows, with an interaction time corresponding to the (1, 1) trapping state, that both one-photon in the cavity and a single atom in the lower state are produced with a 98% probability. In order to maintain an experimentally verifiable quantity, the simulations presented relate to the production of lower-state atoms rather than to the Fock state left in the cavity.

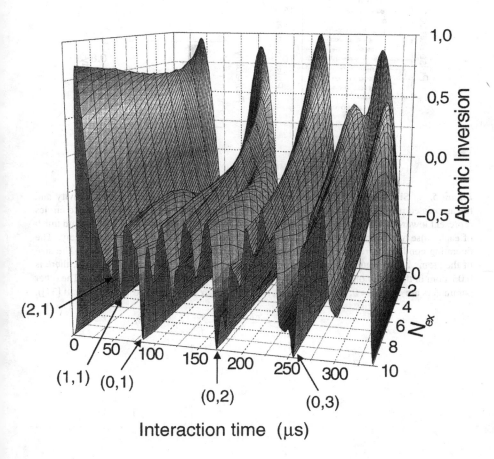

Figure 4. Dynamic behaviour of the micromaser at low temperature showing the trapping states, manifested as valleys along the N_{ex} axis. For the designation of the trapping states see text.

Since the duration of the pumping pulses can be kept rather short ($0.01\tau_{\text{cav}} \leq \tau_{\text{pulse}} \leq 0.1\tau_{\text{cav}}$), the dissipation in the cavity can be neglected.

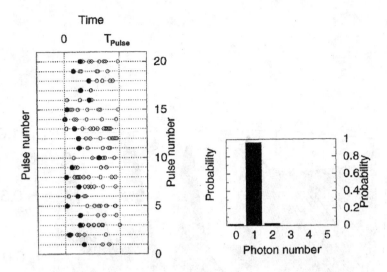

Figure 5.   Simulation of a subset of twenty successive atom bunches after the cavity and the associated probability distribution for photons or lower-state atom production (solid circles represent lower-state atoms and blank circles represent excited-state atoms). The start and finish of each pulse are indicated by the vertical dotted lines marked 0 and τ_{pulse}, respectively. The operating conditions are the (1,1) trapping state trapping ($gt_{int} = 2.2$) conditions. The size of the atoms in this figure is exaggerated for clarity. With the real atomic separation, there is 0.06 atom in the cavity on average (i.e. the system operates in the one-atom regime). The other parameters are $\tau_{pulse} = 9,92 \times \tau_{cav}$, $n_{th} = 10^{-4}$ and $N_a = 7$ (see also Refs. [32] and [37]).

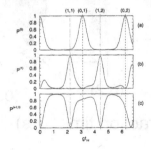

Figure 6.   The probability of finding (a) no lower-state atoms per pulse, $P^{(0)}$, (b) exactly one lower-state atom per pulse, $P^{(1)}$, and (c) a second lower-state atom, if one has already been detected, $P^{(>1;1)}$. The parameters are $\tau_{pulse} = 0.02_{cav}$, $N_a = 7$ atoms, and $n_{th} = 10^{-4}$. The maximum value of P^1 is 98% for the (1,1) trapping state.

The variation of the time when an emission event occurs during an atom pulse in Fig. 5 is due to the variable time spacing between the atoms as a consequence of Poissonian statistics of the atomic beam and the stochasticity of the quantum process. The atomic rate therefore has to be high enough to ensure a sufficient number of excited atoms per laser laser pulse, so as to maintain the 90% probability of an atom emitting. Bo guarantee single-atom single-photon operation, the duration of the preparation pulses must be short in relation to the cavity decay time. For practical purposes, the pulse duration should be smaller than $0.1\tau_{cav}$ for dissipative losses to be less than 10%.

For s large range of operating conditions, the production of Fock states of the field and single lower-state atoms is remarkably robust against the influence of thermal photons, variations of the velocity of atoms, and other influences such as mechanical vibrations of the cavity; much more so than the steady-state trapping states, for which highly stable conditions with low thermal photon numbers pre required [6, 32].

An obvious side-effect of production of a single photon in the mode is, as already mentioned, that a single atom in the lower state is produced. This atom is in a different state when it leaves the cavity, and is therefore distinguishable from the pump atoms. Under these operating conditions, the micromaser thus also serves as a source of single atoms in a particular state, a requirement for many experiments proposed [17, 34].

Our present micromaser setup was specifically designed for steady-state operation and is therefore not ideal for the parameter vange presented here. However, the setup does permit comparison between theory and experiment in a relatively small parameter range. With the setup it could be demonstrated that the cavity field is correctly prepared in 83.2% of the pulses [33]. By improving the experimental parameters, we can achieve conditions leading to the expected theoretical result.

One interesting application of the single photon states produced in the micromaser is the determination of the Wigner function Wigner function of a single photon state giving the phase space distribution of that state. Recently, a new method using the Rabi nutation of atoms in the micromaser was proposed [35]. The method allowed us to use earlier measurements of a single-photon state for evaluation [24]. The value at the origin $W_{|1\rangle}(0) = -1.9$ we got is very close to the theoretically expected value -2.

4.   Cavity Quantum Electrodynamics with Trapped Ions

The second part of this paper reports the progress of our work on ions in optical cavities. The interaction of a single atom with a single field mode of a high-finesse cavity has been the subject of a number of experiments in the field of cavity QED (see, for example, Ref. [27]). However, most of these investigations suffer from a lack of control over the position of the atom, which results in non-deterministic fluctuations of the coupling between atoms and the field. In this context, the strong localisation and position control available when an ion trap Ion-trap is combined with an optical cavity would be a big step forward and would become a key technology for future progress in cavity QED in the optical range. We are now implementing two experiments exploiting localisation of an ion in a cavity. Pulsed excitation of a maximally coupled ion allows single-photon wave packets to be emitted from the cavity on demand [26, 27, 36] (single-photon gun). Under conditions of strong coupling, a single calcium ion in the cavity provides sufficient gain to build up a laser field [25]. Like a single ion in free space, which was previously shown to be an excellent source of antibunched light [19, 20], radiation from a single-ion laser laser has nonclassical photon statistics and correlations.

An equally attractive goal in the area of cavity QED is the simultaneous interaction of two or more ions with a single-cavity mode. Due to the linear geometry of our trap several ions can be stored within the mode volume. As a first test, we placed an array of two ions in the cavity field and observed the total fluorescence. We succeeded in matching the ion crystal to the two maxima of the TEM_{01} mode of the cavity. In such a configuration the cavity field may be used to entangle the two ions [37, 38]. This is a promising alternative to schemes involving the ions' motional degrees of freedom, since there is no need for cooling the vibrational modes of the string below the Doppler temperature. Using a cavity to perform quantum operations on adjacent pairs of ions in a long string is a viable route to a scalable quantum computer.

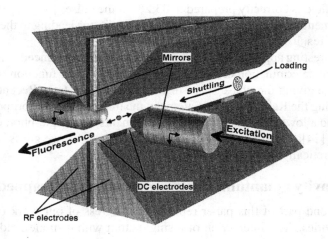

Figure 7.   Experimental arrangement of trap electrodes and cavity mirrors. The ion is loaded at the rear end of the trap and shuttled to the mirror region. Fluorescence is observed from the side of the cavity. For scans in the direction of the trap axis, the ion is moved with DC electrodes. In all other directions, the cavity is translated relatively to the ion's position, as indicated by arrows.

In the following, we now give a progress report of our experiment. (For details see also Ref. [39].) We are using a linear trap with Ca ions (see Fig. 7). As an initial test of the setup for the above-mentioned cavity QED experiments, we used the trapped Ca ions to probe the optical field in the cavity. The Ca ion is sensitive to radiation close to the resonance line $4^2S_{1/2} - 4^2P_{1/2}$ at a wavelength of $\lambda = 397$ nm. The fluorescent light emitted by the ion is collected and detected with a photomultiplier tube. The observed fluorescence rate R is proportional to the local intensity of the optical field at the position $\vec{r}$ of the ion, i.e. $R \propto I(\vec{r})$, provided there is no saturation of the atomic transition. By scanning the position of the ion in the field and detecting the fluorescence rate at each point, a high-resolution map of the optical intensity distribution is obtained. It should be noted that a single ion can also probe the amplitude distribution $\vec{E}(\vec{r})$ of the light field and hence measure its phase. To this end, heterodyne detection of the fluorescent light must be used, with the exciting laser as a local oscillator [18, 20].

With the single ion as a probe, we investigated the eigenmodes of a Fabry-Perot resonator formed by two mirrors (radius of curvature = 10 mm) at a distance of L = 6 mm (Fig.7). The transverse mode pattern is described by Hermite-Gauss functions with a beam waist $w_0 \approx 24 \mu$m, while in the direction of the cavity axis a standing wave builds up. In the experiment, a particular cavity mode is excited by a laser beam with a power of a few hundred nanowatts at 397 nm. The length of the cavity is actively stabilised to this mode.

An ion is loaded in the trap after electron-impact ionisation of calcium atoms. Since the electron beam and the calcium beam would degrade the optical mirrors and make stable trapping difficult, we use a linear trap and load it in a region spatially separated from the observation zone (Fig. 7). Subsequently, DC electrodes along the axis are employed to shuttle the ion over a distance of 25 mm to the uncontaminated end of the trap, where the cavity is located, oriented at right angles to the trap axis. Residual DC fields in the radial direction must be carefully compensated with correctional DC voltages to place the ion precisely on the nodal line of the RF field (coinciding with the trap axis), so as to prevent the trapping field from exciting the micromotion of the ion.

In the direction of the trap axis, the ion is confined in a DC potential well which is approximately harmonic with an oscillation frequency of $\omega_z \approx 300$ kHz. By applying asymmetric voltages, the minimum of the potential well, and thus the equilibrium position of the ion, is moved along the trap axis. By simultaneously monitoring the fluorescence, we sampled one-dimensional cross-sections of the cavity mode. The width of the ion's wave function in the axial potential well is a few hundred nanometres, which provides sufficient resolution to map the transverse mode pattern with an intensity distribution varying on a scale given by the cavity waist w_0.

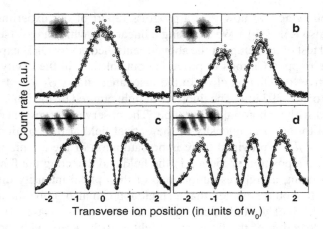

Figure 8.   Transverse profiles of the Hermite-Gauss modes of the cavity, obtained by monitoring the ion's fluorescence while scanning over a range of 120 μm. The solid line is a fit including saturation of the transition. The inset shows the calculated intensity distribution of the mode and indicates the scan path. The modes are a) TEM_{00}, b) TEM_{01}, c)TEM_{02}, d) TEM_{03}.

Figure 8 shows scans of the first four TEM_{0n} modes of the cavity obtained in this way. The fluorescence data are not entirely symmetric because of a small displacement and rotation of the cavity eigenmodes with respect to the trap axis. In each plot, an inset indicates the path along which the ion is scanned. The solid curves in Fig. 8 are obtained from a fit using Hermite-Gauss functions and take into account saturation of the ion's transition. The influence of saturation is apparent in Fig. 8c, where a slightly higher intensity was injected into the cavity. In all cases, the correspondence with the measured fluorescence is excellent.

The ion's motion must be restrained to the trap axis, since off the axis the radio–frequency field of the trap would lead to micromotion. To scan other dimensions of the field, the sample must be moved. In our experiment this is done by piezoelectrically translating the entire cavity assembly perpendicularly to the trap axis. In this way complete three-dimensional mapping of the mode field can be obtained [39].

RF confinement of the ion perpendicular to the trap axis is also harmonic, but the corresponding oscillation frequency $\omega_r \approx 1.1$ MHz is larger than the axial frequency so that field structures in the radial direction of the trap are better resolved. The resolution achieved by our method may be most accurately determined by probing the standing-wave field created between the cavity mirrors, which varies on a scale of $\lambda/2$. To this end, the cavity was moved parallel to its axis while keeping the ion stationary and monitoring its fluorescence. Figure 9 shows the mapping of the cavity field obtained in this way. A pronounced

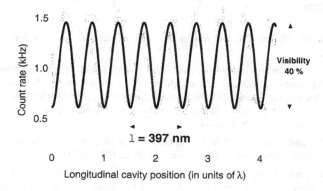

Figure 9. Single-ion mapping of the longitudinal structure of the cavity field. The visibility is determined by the residual thermal motion of the Doppler-cooled ion. It corresponds to a resolution of 60 nm. The localisation of the ion's wave packet in this measurement is 16 nm.

standing-wave pattern is observed with a visibility of 40 %. For details see also Ref. [39].

Taking advantage of the excellent localisation in ion traps Ion-trap, we performed the hitherto most precise measurement of a three-dimensional spatial structure of an optical field over a range of up to 100 μm. As a demonstration, we scanned modes of a low-loss optical cavity. The precise positioning we achieve implies deterministic control of the coupling between the ion and field. At the same time, the field and the internal states of the ion are not affected by the trapping potential. What we have realised, therefore, is an ideal system for cavity QED with a single particle.

5. Conclusion

This paper reviews our work on the generation of photon number states on demand by means of the micromaser. In addition, it describes first experiments using a single trapped ion in conjunction with an optical cavity. The two systems described show great promise and will afford a series of interesting new applications [33, 39].

Acknowledgments

This work was supported by the European IST/FET programme QUBITS.

References

[1] Meschede, D., Walther, H., and Müller, G. (1985) One-atom maser, *Phys. Rev.Lett.* **54**, 551–554.

[2] For a review, see the articles by Haroche, S. and Raimond, J.M. (1985) Radiative properties of Rydberg states in resonant cavities, *Advances in Atomic and Molecular Physics* **20**, 347–411; Gallas, J.A., Leuchs, G., Walther, H., and Figger, H. (1985) Rydberg atoms: high resolution spectroscopy and radiation interaction - Rydberg molecules, *Advances in Atomic and Molecular Physics* **20**, 413–466.

[3] Jaynes, E.T. and Cummings, F.W. (1963) Comparison of quantum and semiclassical radiation theory with application to the beam maser, Proc. IEEE **51**, 89–.

[4] See, for example: Eberly, J.H., Narozhny, N.B., and Sanchez-Mondragon, J.J. (1980) Periodic spontaneous collapse and revival in a simple quantum model, *Phys. Rev. Lett.* **44**, 1323–1326, and references therein.

[5] Rempe, G., Walther, H., and Klein, N. (1987) Observation of quantum collapse and revival in a one-atom maser, *Phys. Rev. Lett.* **58**, 353–356.

[6] Weidinger, M., Varcoe, B. T. H., Heerlein, P., and Walther, H. (1999) Trapping states in the micromaser, *Phys. Rev. Lett.* **82**, 3795–3798.

[7] Nogues, G., Rauschenbeutel, A., Osnaghi, S., Brune, M., Raimond, J. M., and Haroche, S. (1999) Seeing a single photon without destroying it, *Nature* **400**, 239–242.

[8] Meystre, P. (1992) Cavity quantum optics and the quantum measuement process, *Progress in Optics* **30**, 261–355.

[9] Raithel, G., Wagner, C., Walther, H., Narducci, L.M., and Scully, M.O. (1994) The micromaser: a proving ground for quantum physics, in P. Berman (ed.), *Cavity Quantum Electrodynamics*, Academic Press, New York, pp. 57-121

[10] Englert, B., Löffler, M., Benson, O., Weidinger, M., Varcoe, B., and Walther, H. (1998) Entangled atoms in micromaser physics, *Fortschr. Phys.* **46**, 897–926.

[11] Leibfried, D., Meekhof, D.M., King, B.E., Monroe, C., Itano, W. M., and Wineland, D. J. (1996) Experimental determination of the motional quantum state of a trapped atom, *Phys. Rev. Lett* **77**, 4281–4285.

[12] Zbinden, H., Gisin, N., Huttner, B., and Tittel, W. (2000) Practical aspects of quantum cryptographic key distribution, *J. Cryptol.* **13**, 207–220; Lo, H.-K. and Chau, H. F. (1999) Unconditional security of quantum key distribution over arbitrarily long distances, *Science* **283**, 2050–2056.

[13] Gheri, K. M., Saavedra, C., Törmä, P., Cirac, J. I., and Zoller, P. (1998) Entanglement engineering of one-photon wave packets using a single-atom source, *Phys. Rev. A* **58**, R2627–R2630; van Enk, S. J., Cirac, J. I., and Zoller, P. (1997) Ideal quantum communication over noisy channels: a quantum optical implementation, *Phys. Rev. Lett.* **78**, 4293–4296; van Enk, S. J., Cirac, J. I., and Zoller, P. (1998) Photonic channels for quantum communication, *Science* **279**, 205–208.

[14] Cirac, J. I., Zoller, P., Kimble, H. J., and Mabuchi, H. (1997) Quantum state transfer and entanglement distribution among distant nodes in a quantum network, *Phys. Rev. Lett.* **78**, 3221–3224; Parkins, A. S., Marte, P., Zoller, P., and Kimble, H. J. (1993) Synthesis of arbitrary quantum states via adiabatic transfer of Zeeman coherence, *Phys. Rev. Lett.* **71**, 3095–3098; Parkins, S. and Kimble, H. J. (1999) Quantum state transfer between motion and light, *J. Opt. B.* **1**, 496–504.

[15] Jennewein, T., Simon, C., Weihs, G., Weinfurter, H., and Zeilinger, A. (2000) Quantum cryptography with entangled photons, *Phys. Rev. Lett.* **84**, 4729–4732; Naik, D. S., Peterson, C. G., White, A. G., Berglund, A. J., and Kwiat, P. G. (2000) Entangled state quantum cryptography: eavesdropping on the Ekert protocol, *Phys. Rev. Lett.* **84**, 4733–4736; Tittel, W., Brendel, J., Zbinden, H., and Gisin, N. (2000) Quantum cryptography using entangled photons in energy-time Bell states, *Phys. Rev. Lett.* **84**, 4737–4740.

[16] Gottesman, D. and Chuang, I. L. (1999) Demonstrating the viability of universal quantum computation using teleportation and single-qubit operations, *Nature* **402**, 390–393. See also Preskill, J. (1999) Plug-in quantum software, *Nature* **402**, 357–358; Johnathon, D. and Plenio, M. B. (1999) Entanglement-assisted local manipulation of pure quantum states, *Phys. Rev. Lett.*, **83**, 3566–3569.

[17] Sources of single atoms, such as that described in this paper, are routinely employed for hypothetical tasks such as creation of an atomic beam with arbitrary timing sequence or for stabilisation of cavity states. See, for example, Vitali, D., Tombesi, P., and Milburn, G. (1998) Quantum-state protection in cavities, *Phys. Rev. A* **57**, 4930–4944.

[18] Höffges, J. T., Baldauf, H. W., Lange, W., and Walther, H. (1997) Heterodyne measurement of the resonance fluorescence of a single ion, *J. Mod. Opt.* **44**, 1999–2010.

[19] Diedrich, F. and Walther, H. (1987) Nonclassical radiation of a single stored ion, *Phys. Rev. Lett.* **58**, 203–206.

[20] Höffges, J. T., Baldauf, H. W., Eichler, T., Helmfrid, S.R., and Walther,H. (1997) Heterodyne measurement of the fluorescent radiation of a single trapped ion, *Opt. Comm.* **133** 170–174.

[21] Brunel, C., Lounis, B., Tamarat, P., and Orrit, M. (1999) Triggered source of single photons based on controlled single molecule fluorescence, *Phys. Rev. Lett.* **83**, 2722–2725.

[22] Hong, C. K. and Mandel, L. (1986) Experimental realisation of a localized one-photon state, *Phys. Rev. Lett.* **56**, 58–60.

[23] Kim, J., Benson, O., Kan, H., and Yamamoto, Y. (1999) Single-photon turnstile device, *Nature* **397**, 500–503.

[24] Varcoe, B. T. H., Brattke, S., Weidinger, M., and Walther, H. (2000) Preparing pure photon number states of the radiation field, *Nature* **403**, 743–746.

[25] Meyer, G. M., Briegel, H.-J., and Walther, H. (1997) Ion-trap laser, *Europhys. Lett.* **37**, 317–322.

[26] Law, C. K. and Eberly, J. H. (1996) Arbitrary control of a quantum electromagnetic field, *Phys. Rev. Lett.* **76**, 1055–1058; Law, C. K. and Kimble, H. J. (1997) Deterministic generation of a bit-stream of single-photon pulses, *J. Mod. Opt.* **44**, 2067–2074; Domokos, P., Brune, M., Raimond, J. M., and Haroche, S. (1998) Photon-number-state generation with a single two-level atom in a cavity: a proposal, *Eur. Phys. J. D* **1**, 1–4.

[27] See Kuhn, A., Hennrich, M., Rempe, G. (2002) Deterministic single-photon source for distributed quantum networking *Phys. Rev. Lett.* **89**, 067901 and references therein, e.g. Kuhn, A., Hennrich, M., Bondo, T., and Rempe, G. (1999) Controlled generation of single photons from a strongly coupled atom-cavity system, *Appl. Phys B* **69**, 373–377; Pinkse, P. W. H., Fischer, T., Maunz, P., and Rempe, G. (2000) Trapping an atom with single photons, *Nature* **404**, 365–368; Ye, J., Vernooy, D. W., and Kimble, H. J. (2000) Trapping of single atoms in cavity QED, *Phys. Rev. Lett.* **83**, 4987–4990; Hood, C. J., Lynn, T. W., Doherty, A. C., Parkins, A. S., and Kimble, H. J. (2000) The atom-cavity microscope: single atoms bound in orbit by single photons, *Science* **287**, 1447–1453.

[28] See, for example, Scully, M.O. and Zubairy, M. S. (1997) *Quantum Optics* (Cambridge University Press.

[29] Rempe, G., Schmidt-Kaler, F., and Walther, H. (1990) Observation of sub-Poissonian photon statistics in a micromaser, *Phys. Rev. Lett.* **64**, 2783–2786.

[30] Krause, J., Scully, M.O., and Walther, H. (1987) State reduction and $|!!n\rangle$-state preparation in a high-Q micromaser, *Phys. Rev. A* **36**, 4547–4550.

[31] A detailed account of the simulations used in this paper and a comparison with ideal micromaser theory are given in Brattke, S., Englert, B.-G., Varcoe, B. T. H., and Walther, H. (2000) Fock states in a cyclically pumped one-atom maser, *J. Mod. Opt.* **47**, 2857–2867.

[32] Brattke S., et al. (2001) Preparing Fock states in the micromaser, *Optics Express* **8**, 131-144.

[33] Brattke, S., Varcoe, B. T. H., and Walther, H. (2001) Generation of photon number states on demand via cavity quantum electrodynamics, *Phys. Rev. Lett.* **86**, 3534–3537.

[34] Proposals such as teleportation of an atomic state using multiple atomic beams would be substantially enhanced if atoms arrived on demand rather than by chance. See, for example, Davidovich, L., Zagury, M., Brune, M., Raimond, J. M., and Haroche, S. (1994) Teleportation of an atomic state between two cavities using nonlocal microwave fields, *Phys. Rev. A* **50**, R895–R898; Cirac, J. I. and Parkins, A S. (1994) Schemes for atomic-state teleportation, *Phys. Rev. A* **50**, R4441–R4444; Moussa, M. H. Y. (1997) Teleportation with identity interchange of quantum states, *Phys. Rev. A* **55**, R3287–R3290.

[35] Lougowski, P., Solano, E., Zhang, Z.M., Walther, H., Mack, H., and Schleich, W.P. (2002) Fresnel transform: an operational definition of the Wigner function, quant-ph/0206083v1.

[36] Hennrich, M., Legero, T., Kuhn, A., and Rempe, G. (2000) Vacuum-stimulated Raman scattering based on adiabatic passage in a high-finesse optical cavity, *Phys. Rev. Lett.* **85**, 4872–4875.

[37] Pellizzari, T., Gardiner, S. A., Cirac, J. I., and Zoller, P. (1995) Decoherence, continuous observation, and quantum computing: a cavity QED model, *Phys. Rev. Lett.* **75**, 3788–3791.

[38] Zheng, S. B. and Guo, G. C. (2000) Efficient scheme for two-atom entanglement and quantum information processing in cavity QED, *Phys. Rev. Lett.* **85**, 2392–2395.

[39] Guthöhrlein, G. R., Keller, M., Hayasaka, K., Lange, W., and Walther, H. (2001) A single ion as a nanoscopic probe of an optical field, *Nature* **414**, 49–52.

MULTIPOLE RADIATION IN QUANTUM DOMAIN

Alexander S. Shumovsky
Physics Department, Bilkent University
Bilkent, 06800 Ankara, Turkey

M. Ali Can
Physics Department, Bilkent University
Bilkent, 06800 Ankara, Turkey

Öney Soykal
Physics Department, Bilkent University
Bilkent, 06800 Ankara, Turkey
shumo@fen.bilkent.edu.tr

Abstract   We review recent results based on the application of spherical wave representation to description of quantum properties of multipole radiation generated by atomic transitions. In particular, the angular momentum Angular momentum of photons including the angular momentum entanglement, the quantum phase of photons, and the spatial properties of polarization are discussed.

Keywords: Angular momentum, atoms, entanglement, polarization, quantum phase, qutrit, spherical waves of photons.

Introduction

It is well known since the end of XIX century that the time-varying classical electromagnetic (EM) field can be expanded in vector spherical waves and that this representation is convenient for electromagnetic boundary-value problems possessing spherical symmetry and for the discussion of multipole radiation from a local sources (e.g., see [1, 2]). Since both plane and spherical waves form complete sets of orthonormal functions, they are equivalent, so that the use of either representation of classical electromagnetic radiation is caused by the usability reasons.

A.S. Shumovsky and V.I. Rupasov (eds.), Quantum Communication and Information Technologies, 145–169.
© 2003 *Kluwer Academic Publishers.*

The underlying motive for consideration of quantum EM radiation in terms of spherical waves of photons is the fact that the atomic and molecular transitions create the multipole photons, in other words, the photons with given angular momentum and parity rather than plane photons specified by the linear momentum and polarization polarization [3, 4].

It should be stressed that, unlike the classical domain, the plane and spherical wave representations of photons are not completely equivalent. The point is that the vector potential of classical EM field is defined in the three-dimensional Euclidian space $\mathcal{R}^3$, while the operator vector potential of quantum EM radiation is defined in the space

$$\mathcal{H} = \mathcal{R}^3 \otimes \mathcal{H}_{ph}, \tag{1}$$

where $\mathcal{H}_{ph}$ denotes the Hilbert space of photons. According to Wigner's approach [5], *the general properties of a quantum mechanical system are specified by the dynamic symmetry of the corresponding Hilbert space.* The Hilbert spaces of plane and spherical photons have different symmetry properties. Viz, the former manifests the $SO(2)$ symmetry caused by invariance with respect to rotations in xy-plane whose positive normal coincides with direction of propagation $\vec{k}/k$. In turn, the latter has the $SU(2)$ symmetry agreed upon the invariance with respect to rotations in three dimensions about a local source (say, atom or molecule).

In particular, the symmetry reasons imply the different sets of quantum numbers, specifying the photons in the two representations [3, 4, 6]. Any photon in the plane wave representation (PWR) is specified by given energy, linear momentum, and polarization. In turn, a photon in the spherical wave representation (SWR) has given energy, angular momentum Angular momentum, and parity that corresponds to the type of radiation, either electric or magnetic.

SWR was introduced by Heitler [7] and then discussed in his famous monograph [6]. Unfortunately, the most of books on quantum optics leave this representation aside, so that it is discussed in very few monographs (e.g., see [3, 4]). Meanwhile, a number of modern experiments with trapped atoms interacting with photons corresponds to the interatomic distances that are much shorter than the wave length [8, 9]. It should be stressed that the difference between the properties of photons in PWR and SWR is particularly strong just in the near and intermediate zones.

Another reason to use SWR is connected with the problem of use of the angular momentum (AM) of photons in quantum information processing that has attracted recently a great deal of interest [10, 11, 12, 13, 14]. In the usual treatment, AM of photons is discussed in terms of PWR [10, 15]. Unfortunately, this representation leads to the wrong commutation relations for spin spin components as well as for orbital angular momentum (OAM) components [10, 16].

The great importance of true commutation relations in quantum mechanics is obvious to everyone. In fact, just existence of nontrivial commutation relations for observables is the distinguishing feature of quantum mechanics in comparison with classical physics. In particular, the commutation relations are responsible for the quantum noise that influences the precision of measurements in quantum domain. Unlike PWR, the use of SWR leads to the true commutation relations for spin and OAM operators [17].

Besides that, SWR can be used to define the inherent quantum phase phase of photons [18, 19, 20, 21], to describe the quantum properties of polarization [22, 23], and to discuss emission of photons with high AM [24].

The main aim of present course is to give a brief review of SWR, its application, and recent results related to the problem of quantum communications and information processing.

The course is arranged as follows. In Sec. I we introduce SWR and discuss its general properties. In Sec. II we consider representation of AM in SWR. In particular, we examine AM entanglement entanglement of two photons emitted by a cascade decay of an electric quadrupole transition. In Sec. III inherent quantum phase of multipole photons is discussed. In Sec. IV we consider the polarization of quantum multipole radiation. Finally in Sec. V we briefly summarize the results and discuss their applications.

1. Quantization of Multipole Radiation

Following [6], let us construct a representation of photons with given AM and parity. This means that we have to represent the vector potential in terms of a superposition of states with given AM and parity.

As for any other particle, AM of a photon consists of the spin and OAM contributions

$$\vec{J} = \vec{S} + \vec{L}, \tag{2}$$

where $\vec{S}$ and $\vec{L}$ denote the spin spin and OAM, respectively. Since rest mass of photons is equal to zero, the spin is defined to be the minimum possible value of AM. From the atomic spectroscopy we know that minimum $j = 1$ (in the units of $\hbar$). Hence, spin of a photon is $s = 1$. Nevertheless, the requirement of Poincaré invariance on the light cone allows only two spin degrees of freedom.

The eigenfunctions of the operators J_z and $\vec{J}^2$ are the *vector spherical harmonics* [2, 3, 25]

$$\vec{J}^2 \vec{Y}_{j\ell m} = j(j+1)\vec{Y}_{j\ell m}, \qquad J_z \vec{Y}_{j\ell m} = m\vec{Y}_{j\ell m}. \tag{3}$$

The eigenstates of spin 1 are the columns

$$\vec{\epsilon}_+ = \begin{pmatrix} 1 \\ 0 \\ 0 \end{pmatrix}, \quad \vec{\epsilon}_0 = \begin{pmatrix} 0 \\ 1 \\ 0 \end{pmatrix}, \quad \vec{\epsilon}_- = \begin{pmatrix} 0 \\ 0 \\ 1 \end{pmatrix},$$

that can be associated with the base vectors in $\mathcal{R}^3$ as follows

$$\vec{\epsilon}_\pm = \mp\frac{\vec{e}_x \pm i\vec{e}_y}{\sqrt{2}}, \qquad \vec{\epsilon}_0 = \vec{e}_z. \tag{4}$$

In fact, the vectors (4) form the so-called helicity basis [2, 26]. In particular, the vectors $\vec{\epsilon}_\pm$ can be associated with unit vectors of polarization with either positive or negative helicity. Let us stress that quantum electrodynamics interprets the polarization as a given spin state of photons [3].

In turn, the eigenstates of quantum mechanical OAM operator $-i(\vec{r} \times \vec{\nabla})$ are the spherical harmonics $Y_{\ell m}(\vec{k}/k)$. Thus, the vector spherical harmonics (3) can be constructed as the linear combinations of spin states and spherical harmonics

$$\vec{Y}_{j\ell m} = \sum_\mu \langle 1\ell\mu, m-\mu|jm\rangle\vec{\epsilon}_\mu Y_{\ell,m-\mu}, \tag{5}$$

where $\langle \cdots | \cdots \rangle$ denotes the Clebsch-Gordon coefficients of vector addition of spin spin and OAM. Taking into account the properties of Clebsch-Gordon coefficients [27], it is easy to conclude that the quantum numbers j and ℓ connected in the following way

$$j = \ell + 1, \ell, |\ell - 1|. \tag{6}$$

Thus, for each value of AM j, there are three different states specified by the vector spherical harmonics (5) under the condition (6).

Since under inversion vector $\vec{\epsilon}_\mu$ changes sign and function $Y_{\ell,m-\mu}$ is multiplied by $(-1)^\ell$, the vector spherical harmonics have given parity $(-1)^{\ell+1}$. Thus, the functions $\vec{Y}_{jjm}$ have the parity $(-1)^{j+1}$, while the parity of functions $\vec{Y}_{j,j\pm1,m}$ is $(-1)^j$.

The vector spherical harmonics (5) form a complete orthonormal set of functions:

$$\int_0^{2\pi} d\phi \int_0^\pi \vec{Y}_{j\ell m}^+ \cdot \vec{Y}_{j'\ell'm'} \sin\theta d\theta = \delta_{jj'}\delta_{\ell\ell'}\delta_{mm'}. \tag{7}$$

It is seen that $(\vec{k}/k) \cdot \vec{Y}_{jjm}(\vec{k}/k) = 0$. This function $\vec{Y}_{jjm}$ is usually cold the *transversal vector spherical harmonics of magnetic type*. Another transversal function can be constructed as a combination of the functions with $\ell = j \pm 1$

$$\vec{Y}_{jm}^E \equiv \frac{1}{\sqrt{2j+1}}(\sqrt{j}\vec{Y}_{j,j+1,m} + \sqrt{j+1}\vec{Y}_{j,j-1,m}), \tag{8}$$

which is called the *transversal spherical harmonics of electric type*. It is seen that $\vec{Y}_{jjm}$ and $\vec{Y}_{jm}^E$ are mutually orthogonal for the same $\vec{k}/k$.

The functions with $\ell = j \pm 1$ can also be used to construct the *longitudinal vector spherical function*

$$\vec{Y}_{jm}^{L} = -\sqrt{\frac{j+1}{2j+1}}\vec{Y}_{j,j+1,m} + \sqrt{\frac{j}{2j+1}}\vec{Y}_{j,j-1,m},$$

which is orthogonal to both $\vec{Y}_{jjm}$ and $\vec{Y}_{jm}^{E}$. Nevertheless, this longitudinal function should be discarded because the Poincaré invariance on the light cone.

Thus, the states of the field with given AM and parity can be obtained by expansion of vector potential over the transversal vector spherical harmonics. Taking into account the expansion [2]

$$e^{i(\vec{k}\cdot\vec{r} - \omega_k t)} = 4\pi \sum_{\ell,m} (i)^{\ell} j_{\ell}(kr) Y_{\ell m}^{*}(\vec{k}/k) Y_{\ell m}(\vec{k}/k) e^{-i\omega_k c},$$

where $j_{\ell}(kr)$ denotes the spherical Bessel functions, we can conclude that the positive-frequency part of the vector potential has the form

$$\vec{A}_{Mkjm} = N_M j_{\ell} \vec{Y}_{jjm} a_{Mkjm}, \tag{9}$$

$$\vec{A}_{Ekjm} = N_E[\sqrt{j} j_{j+1}(kr)\vec{Y}_{j,j+1,m} - \sqrt{j+1} j_{j-1}(kr)\vec{Y}_{j,j-1,m}] a_{Ekjm} \tag{10}$$

in the case of parity $(-1)^{j+1}$ and $(-1)^j$, respectively. Here N_λ denotes the normalization factor. In order to have vector potential with discrete values of k, the right-hand sizes in (9) and (10) should be defined inside an ideal spherical cavity of big radius R. Then, the spectrum is defined by the roots of equation

$$j_{\ell}(kR) = 0.$$

In this case, it is convenient to renormalize the spherical Bessel functions by the condition

$$\forall \ell \qquad \int_0^R j_{\ell}(kr) j_{\ell}(k'r) r^2 dr = \frac{4\pi R^3}{3}\delta_{kk'}.$$

In Eqs. (9) and (10), the complex amplitudes $a_{\lambda kjm}$ specify the amount of the corresponding multipole field. The harmonic time dependence is usually included into these amplitudes. In classical electrodynamics, the amplitudes $a_{\lambda kjm}$ are determined by the properties of the source of radiation (harmonically varying current or intrinsic magnetization) [2]. Within the quantum picture, the amplitudes $a_{\lambda kjm}$ are supposed to be the annihilation operators of multipole photons [6, 25] that obey the commutation relations

$$[a_{\lambda kjm}, a_{\lambda'k'j'm'}^{+}] = \delta_{\lambda\lambda'}\delta_{kk'}\delta_{jj'}\delta_{mm'}. \tag{11}$$

Hence, they form a representation of the Weyl-Heisenberg algebra of multipole photons. In this case, (9) and (2) should be considered as the positive frequency parts of the *operator vector potential* of the magnetic-type radiation (with $\lambda = M$ and parity $(-1)^{j+1}$) and of the electric-type radiation (with $\lambda = E$ and parity $(-1)^j$), respectively.

Hereafter, we consider expressions (9) and (10) as the quantum operators.

Let us now note that the operators (9) and (10) can be represented in $\mathcal{R}^3$ as follows

$$\vec{A}_{\lambda kjm} = \sum_{\mu}(-1)^{\mu}\epsilon_{-\mu}\mathcal{A}_{\lambda kjm\mu}a_{\lambda kjm},$$

where $\mathcal{A}_{\lambda kjm\mu}$ denotes the *mode function* of the multipole field. By construction, this function obey the homogeneous Helmholtz wave equation

$$\nabla^2 \mathcal{A}_{\lambda kjm\mu} + \omega_k^2 \mathcal{A}_{\lambda kjm\mu} = 0.$$

In fact, the vector $\vec{A}_{\lambda kjm}$ can be considered as a function from $\mathcal{R}^3$ to the Hilbert space $\mathcal{H}$ of complex linear functions on $\mathcal{R}^3$ in (1). The operators (9) and (10) obey the same wave equations but assumes values in the Hilbert space $\mathcal{H} \times \mathcal{H}$, where the second factor $\mathcal{H}$ comes from the spin states.

In view of the wave equation, the mode functions $\vec{A}_{\lambda kjm}$ can be interpreted as the *wave functions of multipole photons* [3].

It should be emphasized that under rotations the vector spherical functions are transformed along an irreducible representations of the $O^+(3)$ group. Thus, they are the irreducible tensors of rank j rather than vectors.

It is useful to show that the operators (9) and (10) are invariant with respect to the $SU(2)$ group. Consider first the electric-type multipole radiation and introduce an auxiliary operator

$$\vec{\mathbf{A}}_\ell(\vec{r}/r) = \sum_{\mu,m} Y_{1\mu}Y_{\ell m}\vec{\epsilon}_\mu \otimes a_{\ell m} \qquad (12)$$

For simplicity, we drop here all other indexes. Because rotations do not influence the radial dependence in (10) provided by the spherical Bessel functions, the auxiliary function (12) depends only on the direction $\vec{r}/r$ in $\mathcal{R}^3$.

Let φ be an arbitrary transformation belonging to the $SU(2)$ group. Then [28]

$$
\begin{aligned}
\vec{\mathbf{A}}_\ell(\varphi\vec{r}/r) &= \sum_{\mu,m} Y_{1\mu}(\varphi\vec{r}/r)Y_{\ell m}(\varphi\vec{r}/r) \\
&= \sum_{\mu,\mu'}\sum_{m,m'} Y_{1\mu'}(\vec{r}/r)\varphi_{\mu\mu'}Y_{\ell m'}(\vec{r}/r)\vec{\epsilon}_\mu \otimes \varphi_{mm'}a_{\ell m} \\
&= \sum_{\mu,\mu'}\sum_{m,m'} Y_{1\mu'}(\vec{r}/r)Y_{\ell m'}(\vec{r}/r)[\varphi_{\mu\mu'}\vec{\epsilon}_\mu] \otimes [\varphi a_{\ell m}] \\
&= \sum_{mu',m'} Y_{1\mu'}(\vec{r}/r)Y_{\ell m'}(\vec{r}/r)[\varphi\vec{\epsilon}_{\mu'}] \otimes [\varphi a_{\ell m}] = \varphi\vec{\mathbf{A}}_\ell(\vec{r}/r).
\end{aligned}
$$

Thus, the auxiliary operator (12) is invariant with respect to the $SU(2)$ group.

Since the spherical harmonics form a basis of an irreducible representation $\mathbf{M}_\ell$ of the $SU(2)$ group, the product $Y_{1\mu}Y_{\ell m}$ in (12) form a basis of

$$\mathbf{M}_1 \otimes \mathbf{M}_\ell = \mathbf{M}_{\ell-1} \oplus \mathbf{M}_\ell \oplus \mathbf{M}_{\ell+1}. \tag{13}$$

The operator (10) is defined just in (13).

Let $(Y_{1\mu}Y_{\ell m})_s$ $(s = \ell, \ell \pm 1)$ be the component (projection) of $Y_{1\mu}Y_{\ell m}$ in $\mathbf{M}_s$. Then the vector operator

$$\vec{\mathbf{A}}_{\ell s} = \sum_{\mu,m}(Y_{1\mu}Y_{\ell m})_s \vec{\epsilon}_\mu \otimes a_{\ell m}$$

is also invariant with respect to the $SU(2)$ group. This implies the invariance of (10), because rotations do not influence the radial dependence. The $SU(2)$ invariance of (9) can be proven in the same way.

2. Angular Momentum of Multipole Photons

A classical distribution of electromagnetic field in vacuum carries AM of the form

$$\vec{J} = \frac{1}{4\pi c} \int \vec{r} \times (\vec{E} \times \vec{B}) d^3 r, \tag{14}$$

where

$$\vec{E} = \frac{1}{c}\frac{\partial \vec{A}}{\partial t}, \qquad \vec{B} = \vec{\nabla} \times \vec{A}$$

are the electric and magnetic fields. For the fields produced a finite time in the past and so localized to a finite region, this expression can be rewritten in the form

$$\vec{J} = \frac{1}{4\pi c} \int \left[\vec{E} \times \vec{A} + \sum_\mu E_\mu (\vec{r} \times \vec{\nabla}) B_\mu \right] d^3 r. \tag{15}$$

The first term is usually identified with the spin contribution, while the second term represents OAM because of the presence of the quantum mechanical angular momentum Angular momentum operator $-i(\vec{r} \times \vec{\nabla})$ [15]. Let us stress that Eq. (15) is obtained within the classical picture, so that the use of the notions spin and OAM has here a conditional meaning.

Within the quantum domain, both terms in the right-hand side of (15) are represented by the bilinear forms in the photon operators (11).

Consider first PWR, when the operator vector potential has the form

$$\vec{A}(\vec{r}) = N \sum_{\mu=\pm} \sum_k \vec{\epsilon}_{k\mu}(e^{i\vec{k}\cdot\vec{r}} a_{k\mu} + H.c.)$$

because the third direction ϵ_{k0} is forbidden in this case [15]. Averaging the first term in the right-hand side of (15) over time to eliminate the rapidly oscillation terms with a^2 and $(a^+)^2$ and changing summation over k by integration, we get for the spin spin operator

$$\vec{S} = \frac{N^2}{2\pi c} \int \frac{d^3k}{2\pi}^3 \vec{k}(a^+_{k+}a_{k+} - a^+_{k-}a_{k-}). \tag{16}$$

Thus, the components of "spin" (16) obey the commutation relations [10, 16]:

$$[S_\alpha, S_\beta] = 0, \qquad \alpha, \beta = x, y, z. \tag{17}$$

Employing the same procedure to the second (OAM) term in (15) then gives the following commutation relations

$$[L_\alpha, L_\beta] = i\epsilon_{\alpha\beta\gamma}(L_\gamma + S_\gamma), \qquad [L_\alpha, S_\beta] = i\epsilon_{\alpha\beta\gamma}S_\gamma,$$

so that neither spin nor OAM obey the true commutation relations. In fact, the definition (2) assumes the structure $SU(2) \times SU(2)$ of the total AM. At the same time, the above commutation relations correspond to the structure $\mathbf{A} \times SU(2)$, where $\mathbf{A}$ denotes the Abelian group of translations caused by the $SO(2)$ symmetry of the Hilbert space.

The use of SWA leads to a different result. According to the results of Sec. II, the total AM (2) has the structure $SU(2) \times SU(2)$ by construction.

Consider first a single-mode photon emitted by an electric dipole (E1) transition in a two-level atom two-level atom located at the center of an ideal spherical cavity. Let us stress that E1 photons represent the most frequently encountered type of EM radiation in visible and IR regions. If AM of the excited atomic state is $j = 1$, then this state is triple degenerated with respect to the quantum number $m = 0, \pm 1$. The Jaynes-Cummings Hamiltonian Hamiltonian of such a system has the form [18]

$$H = H_0 + H_{int}, \tag{18}$$
$$H_0 = \sum_m (\omega a^+_m a_m + \omega_0 R_{mm}),$$
$$H_{int} = \gamma \sum_m (R_{mg}a_m + a^+_m R_{gm}).$$

Here ω is the cavity mode frequency, ω_0 is the atomic transition frequency, γ is the coupling constant, a_m, a^+_m are the E1 photon operators (11), and R denotes the atomic operators:

$$R_{mm'} = |j = 1, m\rangle\langle j = 1, m|, \qquad R_{mg} = |j = 1, m\rangle\langle j' = 0, 0|.$$

Since the angular momentum is conserved in the atom-photon interaction [4], the total angular momentum

$$\vec{J} = \vec{J}^{(a)} + \vec{J}^{(ph)} \tag{19}$$

should be an integral of motion with the Hamiltonian (18). It is clear that the photon, created by the atom, takes away AM of the excited atomic state. The latter is specified by the operators

$$J_x^{(a)} = \frac{1}{\sqrt{2}}(R_{0+} + R_{0-} + H.c.), \quad J_y^{(a)} = \frac{1}{\sqrt{2}}(R_{0+} - R_{-0} - H.c.), \quad (20)$$

$$J_z^{(a)} = R_{++} - R_{--}, (21)$$

that obey the commutation relations

$$[J_\alpha^{(a)}, J_\beta^{(a)}] = i\epsilon_{\alpha\beta\kappa} J_\kappa^{(a)}, \quad \alpha, \beta, \kappa = x, y, z. \quad (22)$$

It is now a straightforward matter to arrive at conclusion that the photon operator, complementing (20) with respect to the integral of motion with the Hamiltonian (18), has the components

$$J_x^{(ph)} = \frac{1}{\sqrt{2}}\{a_0^+(a_+ + a_-) + H.c.\},$$

$$J_y^{(ph)} = \frac{i}{\sqrt{2}}\{a_0^+(a_+ - a_-) - H.c.\},$$

$$J_z^{(ph)} = a_+^+ a_+ - a_-^+ a_-. \quad (23)$$

It follows from (11) that the operators (22) obey the same commutation relations as (21), that are the true commutation relations for the components of AM operator. By construction, the operators (22) define AM carried away by the photon from the atom. Thus, the use of SWR leads to the true commutation relations for AM of photons.

Let us stress the principle difference in the operator structure of Eqs. (15) and (22). In the former, the symbols $\pm$ denote the circular polarization polarization, while in the latter, the subscripts $m = 0, \pm 1$ correspond to the projection of angular momentum Angular momentum $j = 1$ on the quantization axis.

Assume that the atom emits E1 photon with given m. Then, for the mean values of AM operators (21) we get

$$\forall m \qquad \langle 1_m | J_{x,y}^{(ph)} | 1_m \rangle = 0, \quad \langle 1_m | J_z^{(ph)} | 1_m \rangle = m.$$

In turn, the variances are

$$\langle 1_m | (\Delta J_{x,y}^{(ph)})^2 | 1_m \rangle = \begin{cases} \frac{1}{2}, & \text{at } m = \pm 1 \\ 1, & \text{at } m = 0 \end{cases}$$

and

$$\langle 1_m | (\Delta J_z^{(ph)})^2 | 1_m \rangle = |m|.$$

Thus, the Fock number state of E1 photon manifests strong quantum fluctuations of AM, and the quantum fluctuations of AM in the state with $m = 0$ are stronger than those in the states with $m = \pm 1$.

Eqs. (21) can be used to specify mean values and variances of AM of many E1 photons as well. Assume for example that the local source emits E1 photons in coherent state coherent state $|\alpha_m\rangle$ with given m. Then

$$\langle \alpha_m | J_{x,y}^{(ph)} | \alpha_m \rangle = 0, \quad \langle \alpha_m | J_z^{(ph)} | \alpha_m \rangle = m |\alpha_m|^2$$

and

$$\langle \alpha_m | (\Delta J_{x,y}^{(ph)})^2 | \alpha_m \rangle = \begin{cases} \frac{1}{2} |\alpha_\pm|^2, & m = \pm 1 \\ |\alpha_0|^2, & m = 0 \end{cases}, \tag{24}$$

$$\langle \alpha_m | (\Delta J_z^{(ph)})^2 | \alpha_m \rangle = |m| |\alpha_m|^2.$$

In PWR, the physical quantities related to the operators $J_{x,y}^{(ph)}$ do not manifest quantum fluctuations at all.

Let us now establish a contact with the definitions of AM given by Eqs. (13) and (14). Consider first the spin density operator

$$\vec{S}(\vec{r}) = \frac{1}{4\pi c} \vec{E}(\vec{r}) \times \vec{A}(\vec{r}) \tag{25}$$

in the case of E1 monochromatic radiation. Using (10), one can see that the components of (23) contain all possible bilinear combinations of photon operators (11). Taking into account the property of spherical Bessel functions that $j_0(kr) \to 1$ and $j_2(kr) \to 0$ at $r \to 0$, we can conclude that the components of the spin operator (23) have the same structure in the photon operators as (21). Moreover, it is seen that the integrand of the second term in (14) vanish in the same limit. Thus, in a certain vicinity of the origin (atom), AM of photons consists of spin, while OAM contribution arises with distance from the source.

Taking into account that the photon localization appears in the form of a wavefront [29], we should integrate (23) over a spherical shall of radius r together with averaging over time to calculate the amount of spin carried by E1 photon at any distance r from the source. Performing straightforward but tedious calculations, we can conclude that

$$\vec{\mathcal{S}}(r) \equiv \int_0^{4\pi} d\phi \int_0^\pi \vec{S}(\vec{r}) \sin\theta d\theta = f(kr) \vec{J}^{(ph)}, \tag{26}$$

$$f(kr) \sim \frac{1}{3} [2j_0^2(kr) - \frac{1}{2} j_2^2(kr)].$$

In turn, OAM of E1 photons at distance r from the source can be calculated in the same fashion as $\vec{\mathcal{S}}(r)$:

$$\vec{\mathcal{L}}(r) \equiv \int_0^{4\pi} d\phi \int_0^\pi \vec{L}(\vec{r}) \sin\theta d\theta \sim \frac{1}{2} j_2^2(kr) \vec{j},$$

so that

$$\int_0^R [f_S(kr) + f_L(kr)]r^2 dr = 1$$

as all one can expect.

The fact that OAM has the same operator structure as the spin and total AM reflects the known property of electric-type photons [3, 25]. Viz, in the states described by the vector spherical harmonics of electric type (8), OAM does not have a given value but is a superposition of states with $\ell = j \pm 1$. Thus, in these states, the total AM *cannot be divided into spin and OAM contributions.*

A more detailed examination shows that, unlike the energy of electromagnetic field, AM is not contained in the pure wave zone, and the main contribution to AM comes from the near and intermediate zones. At far distances where

$$j_\ell(x) \sim [\sin(x - \ell\pi/2)]/x, \qquad x \gg \ell,$$

we get $\vec{S}(kr) = \vec{L}(kr)$, so that the spin and OAM contribute equally into the total AM at far distances.

Consider now the radiation by atom in free space, when the continuum mode distribution, corresponding to the natural line breadth, should be taken into account [17]. In this case, we should extend the model Hamiltonian Hamiltonian (17) on the multi-mode case by adding integration over k and apply the Markov approximation Markov approximation, which is similar to the Wigner-Weisskopf approach [30, 31]. Then, the time-dependent wave function of the atom-photon system can be written in the form

$$|\psi(t)\rangle = C(t)|\psi^{(a)}\rangle + \int B(k,t)|\psi(k)\rangle dk \qquad (27)$$

with the initial conditions

$$C(0) = 1, \qquad \forall k \quad B(k,0) = 0.$$

Here $|\psi^{(a)}\rangle$ corresponds to the excited atomic state and vacuum for photons, while $|\psi(k)\rangle$ gives the ground atomic state and single E1 photon with given k and m. Employing the standard analysis [30, 31] than gives

$$C(t) = e^{-i\omega_0 t - \Gamma t}, \quad B(k,t) = \frac{-k^{3/2}}{\omega_k - \omega_0 + i\Gamma}\left(1 - e^{i(\omega_k - \omega_0) - \Gamma t}\right),$$

where Γ is the radiative decay width.

Carrying out the averaging of z components of spin and OAM contributions in (14) over the state (24), we get [17]

$$\langle S_z(t)\rangle = \langle L_z(t)\rangle = \frac{1}{2}(1 - e^{-2\Gamma t}). \qquad (28)$$

Since the Markov approximation Markov approximation corresponds to the "rough" scale $t \gg \Gamma^{-1}$ [31], Eq. (25) shows that spin and OAM contribute equally into the total AM of E1 photons at the distances $r \geq c/\Gamma \gg c/\omega_0$, again corresponding to the wave zone.

The obtained results can also be applied to the problem of entanglement of photon twins created by an electric quadrupole (E2) transition between the states $|j = 2, m = 0\rangle$ and $|j' = 0, m' = 0\rangle$ (Fig. 2) [17]. The cascade decay of this state gives rise to the two E1 photons propagating in the opposite directions. Because of the AM conservation, the state of the radiation field has the form

$$|\psi\rangle = \frac{1}{\sqrt{3}}(|1_+, 1_-\rangle + |1_-, 1_+\rangle + |1_0, 1_0\rangle), \qquad (29)$$

where the subscripts correspond to the quantum numbers m and $|1_m, 1_{m'}\rangle$ is the product of number states of "left" and "right" photons. Let us stress that photons with $m = 0$ may have the most probable direction of propagation different from that for the photons with $m = \pm 1$ because of the structure of the radiation pattern.

We now prove that (26) represents the maximum entangled qutrit state. It was shown in Refs. [32, 33] that the maximum entangled states entangled states of a composite system obey the following criterion. Local measurement *The local measurements at all subsystems have maximum uncertainty in comparison with the other states allowed for a system under consideration.* The complete set of local measurements is defined by the dynamic symmetry group of the Hilbert state of the composite system [33]. In the case of qutrit system, this is the $SU(3)$ group. Then, the local ("left" and "right") measurements in the system under consideration are described by the nine Hermitian generators of the $SU(3)$ subalgebra in the Weil-Heisenberg algebra of E1 photons (11):

$$M = \begin{cases} a_+^+ a_+ - a_0^+ a_0, & a_0^+ a_0 - a_-^+ a_-, & a_-^+ a_- - a_+^+ a_+, \\ \frac{1}{2}(a_+^+ a_0 + a_0^+ a_+), & \frac{1}{2}(a_0^+ a_- + a_-^+ a_0), & \frac{1}{2}(a_-^+ a_+ + a_+^+ a_-), \\ \frac{1}{2i}(a_+^+ a_0 - a_0^+ a_+), & \frac{1}{2i}(a_0^+ a_- - a_-^+ a_0), & \frac{1}{2i}(a_-^+ a_+ - a_+^+ a_-). \end{cases} \qquad (30)$$

It is easily seen that

$$\langle \psi | M_n | \psi \rangle = 0$$

for all $n = 1, \cdots, 9$ in (27). Thus, the uncertainties of the measurements (27)

$$\langle (\Delta M_n)^2 \rangle \equiv \langle (M_n)^2 \rangle - \langle M_n \rangle^2$$

achieve the maximum value $\langle (\Delta M_n)^2 \rangle = \langle (M_n)^2 \rangle$ in the case of averaging over the state (26).

Let us stress that similar qutrit states have been considered in the context of quantum information processing and quantum cryptography [34, 35].

It is easily seen that AM operators (22) can be constructed as the linear combinations of the generators (27). Thus, $\langle J_\alpha^{(ph)} \rangle = 0$ in the state (26). At the same time, this state provide the maximum quantum fluctuations of the components of AM

$$\langle (\Delta J_\alpha^{(ph)})^2 \rangle = \frac{2}{3}, \quad \alpha = x, y, z,$$

as well as the maximum correlation of measurements at the opposite sides of the "quantum information channel" provided by the state (26):

$$\langle [J_\alpha^{(ph)}]_{left}, [J_\alpha^{(ph)}]_{right} \rangle = \frac{2}{3}.$$

Here $\langle A, B \rangle \equiv \langle AB \rangle - \langle A \rangle \langle B \rangle$. This results illustrates the idea that the entangled states entangled states carry information in the form of correlations between the local measurements Local measurement [36].

3.   Quantum Phase of E1 Photons

The problem of quantum phase was discussed in quantum optics for a long time (for review, see Refs. [37, 38, 39]). Among the results in the field, the two should be mentioned, first of all. One is the so-called Pegg-Barnett approach [40, 41] (for further references, see [38]). Their method is based on a contraction of the infinite-dimensional Hilbert space of photons $\mathcal{H}$. Viz, the quantum phase phase is first defined in an arbitrary s-dimensional subspace in $\mathcal{H}$. The formal limit $s \rightarrow \infty$ is taken only after the averaging of the operators, describing the physical quantities. A weak spot of the approach is that any restriction of dimension of the Hilbert-Fock space Fock space of photons leads to an effective violation of the algebraic properties of the photon operators. This, in turn, can lead to an inadequate picture of quantum fluctuations.

Another approach has been proposed by Noh, Fougères, and Mandel [42, 43]. It is based on the *operational* definition of the quantum phase (in terms of what can be measured). The main result of the approach is that *there is no unique quantum phase variable, describing universally the measured phase properties of the light*. This very strong statement has obtained a totally convincing confirmation in a number of experiments.

The use of SWR permits us to define the quantum phase phase of photons, corresponding to the azimuthal phase of AM [18, 21], in the whole Hilbert space without any contraction. The approach proposed in Ref. [18] complements, in a sense, the Noh-Fougères-Mandel approach. In fact, it defines the quantum phase in terms of what can be emitted by a source.

Let us use again the Jaynes-Cummings Hamiltonian (18) and the atomic AM (20). The latter can be specified by the operators

$$J_+^{(a)} = \sqrt{2}(R_{+0} + R_{0-}), \quad J_-^{(a)} = (J_+^{(a)})^+, \quad J_z^{(a)} = R_{++} - R_{--}, \quad (31)$$

forming a representation of the $SU(2)$ algebra in the three-dimensional Hilbert space of excited atomic states:

$$[J_+^{(a)}, J_-^{(a)}] = 2J_z^{(a)}, \qquad [J_z^{(a)}, J_\pm^{(a)}] = \pm J_\pm^{(a)}. \tag{32}$$

Since the enveloping algebra of (28)-(29) contains the uniquely defined Casimir operator

$$(\vec{J}^{(a)})^2 = 2 \sum_{m=-1}^{1} R_{mm} = 2 \times \mathbf{1},$$

where $\mathbf{1}$ denotes the unit operator in the three-dimensional Hilbert space, a dual phase-dependent representation of (28) can be constructed through the use of method proposed by Vourdas [44]. Viz, the rising and lowering operators in (28) can be represented in the "polar" form

$$J_+^{(a)} = J_r^{(a)} E, \qquad J_-^{(a)} = E^+ J_r^{(a)},$$

where $J_r^{(a)}$ is the Hermitian "radial" operator and E is the unitary ($EE^+ = 1$) "exponential of the phase" operator. It is easily seen that

$$E = R_{+0} + R_{0-} + e^{i\psi} R_{-+}, \tag{33}$$

where ψ denotes an arbitrary real reference phase. Using (30), one can define the cosine and sine of the atomic AM azimuthal phase operators

$$C^{(a)} = \frac{1}{2}(E + E^+), \qquad S^{(a)} = \frac{1}{2i}(E - E^+), \tag{34}$$

such that

$$[C^{(a)}, S^{(a)}] = 0 \quad \text{and} \quad (C^{(a)})^2 + (S^{(a)})^2 = 1.$$

The phase states of the atomic AM are then defined to be the eigenstates of the operator (30)

$$E|\phi_m\rangle = e^{i\phi_m}|\phi_m\rangle,$$

which leads

$$|\phi_m\rangle = \frac{1}{\sqrt{3}} \sum_{m'=-1}^{1} e^{-im'\phi_m}|j = 1, m\rangle, \qquad \phi_m = \frac{\psi + 2m\pi}{3}, \tag{35}$$

where m acquires the values 0 and ± 1 as above. Through the use of the phase states (32), it is easy to define the following dual representation of the $SU(2)$ algebra (28)-(29):

$$\mathcal{J}_\pm^{(a)} = \sum_m \sqrt{2 - m(m \pm 1)}|\phi_{m\pm1}\rangle\langle\phi_m|, \qquad \mathcal{J}_z^{(a)} = \sum_m m|\phi_m\rangle\langle\phi_m|. \tag{36}$$

It should be stressed that the $SU(2)$ phase states can be constructed for an arbitrary number of two-level atoms. In particular, it can be shown that the $SU(2)$ phase states form the set of maximum entangled $2N$-qubit states [32].

The representation of the $SU(2)$ subalgebra in the Weyl-Heisenberg algebra of E1 photons (11) has the form

$$J_+^{(ph)} = \sqrt{2}(a_+^+ a_0 + a_0^+ a_-), \quad J_-^{(ph)} = (J_+^{(ph)})^+, \quad J_z^{(ph)} = \sum_m m a_m^+ a_m. \ (37)$$

This expressions can be obtained directly from (22). The operators (34) complement the atomic operators (28) with respect to an integral of motion with the Hamiltonian (18). Unfortunately, there is no *isotype* representation of the $SU(2)$ subalgebra in the Weyl-Heisenberg algebra [45]. In other words, there is no uniquely defined Casimir operator in the enveloping algebra of (34). Therefore, Vourdas' [44] approach cannot be directly used here to describe the phase properties of AM of E1 photons.

At the same time, we again can use the conservation of AM in the process of radiation, which is independent of whether we use the standard form of AM operators or their dual representation. In particular, it is seen that [18]

$$\cdot[(E + \varepsilon), H] = 0,$$

where

$$\varepsilon = a_+^+ a_0 + a_0^+ a_- + e^{i\psi} a_-^+ a_+ \tag{38}$$

is the photon counterpart of the exponential of the phase operator (30). In contrast to (30), Eq. (35) does not determine a unitary operator. At the same time, (35) represents the *normal* operator

$$[\varepsilon, \varepsilon^+] = 0,$$

commuting with the total number of photons

$$[\varepsilon, \sum_m a_m^+ a_m] = 0.$$

The quantum phase phase properties of E1 photons can now be described in terms of the dual representation of photon operators that has been introduced in Ref. [21]. Let us use the following Bogolubov-type [46] canonical transformation

$$a_m = \frac{1}{\sqrt{3}} \sum_{m'=-1}^{1} e^{-im'\phi_m} \mathbf{a}_{m'}, \quad \mathbf{a}_m = \frac{1}{\sqrt{3}} \sum_{m'=-1}^{1} e^{im'\phi_m} a_{m'}, \quad [\mathbf{a}_m, \mathbf{a}_{m'}^+] = \delta_{mm'}. \ (39)$$

Here ϕ_m represents the same phase angle as above. It is seen that the operator (35) takes the diagonal form in the representation (36):

$$\varepsilon_\phi = \sum_{m=-1}^{1} e^{i\phi_m} \mathbf{a}_m^+ \mathbf{a}_m. \qquad (40)$$

Let us note that the atomic operator (30) is also diagonal in the representation of phase states (32)

$$E_\phi = \sum_m e^{i\phi_m} |\phi_m\rangle\langle\phi_m|$$

and that

$$[(E_\phi + \varepsilon_\phi), H] = 0.$$

Thus, all one can conclude is that the operators $\mathbf{a}_m$ and $\mathbf{a}_m^+$ (36) provide a representation of E1 photon operators of annihilation and creation with *given quantum phase* that, by construction, is the azimuthal phase of AM of photons.

In particular, the annihilation operators in the phase representation (36) obey the stability condition

$$\forall m \qquad \mathbf{a}_m|0\rangle = 0,$$

where $|0\rangle$ is the vacuum state. Thus, the conjugated creation operator can be used to construct the Fock number states in the phase representation in usual way:

$$|\nu_m\rangle = \frac{1}{\sqrt{\nu_m}}(\mathbf{a}_m^+)^{\nu_m}|0\rangle, \qquad \mathbf{a}_m^+\mathbf{a}_m|\nu_m\rangle = \nu_m|\nu_m\rangle, \qquad \nu_m = 0,1,\cdots, \quad (41)$$

such that

$$\langle\nu_m|\nu_{m'}\rangle = \delta_{mm'}, \qquad \bigotimes_{m=-1}^{1} \sum_{\nu_m} |\nu_m\rangle\langle\nu_m| = \mathbf{1}. \qquad (42)$$

Thus, the photon phase states $|\nu_m\rangle$ form a complete orthonormal denumerable set of states of E1 photons, spanning the "phase" Hilbert-Fock space Fock space. This space is dual to the conventional space of states of E1 photons. It is seen that

$$\varepsilon_\phi|\nu_m\rangle = \nu_m e^{i\phi_m}|\nu_m\rangle.$$

Therefore, (37) can be interpreted as the non-normalized exponential of the phase operator. In turn, the cosine and sine of the photon phase phase operators can be defined as follows

$$C_\phi^{(ph)} = K \sum_m \mathbf{a}_m^+\mathbf{a}_m \cos\phi_m, \qquad S_\phi^{(ph)} = K \sum_m \mathbf{a}_m^+\mathbf{a}_m \sin\phi_m, \qquad (43)$$

where the normalization coefficient K is determined by the condition

$$\langle (C_\phi^{(ph)})^2 + (S_\phi^{(ph)})^2 \rangle = 1 \tag{44}$$

for the averaging over an arbitrary state of the radiation field. Similar coefficient was used in the Noh-Fougères-Mandel operational approach as well [43].

The phase representation (36) can be used to describe the azimuthal phase of AM of E1 photons [18, 19, 20, 21, 39]. In particular, it is possible to show that the eigenvalues of the azimuthal quantum phase of AM of photons have a discrete spectrum, depending on the number of photons. All eigenvalues lie in the interval $[0, 2\pi]$. In the classical limit, provided by the high-intensity coherent state coherent state of radiation, the phase eigenvalues are distributed uniformly over the interval $[0, 2\pi]$ as all one can expect in classical domain [21, 39].

The comparison with the Pegg-Barnett approach shows the qualitative coincidence of results for mean value of the cosine and sine operators. At the same time, there is a striking difference in the behavior of variance of quantum phase phase in the case of very few photons, corresponding to the quantum domain [19, 39].

4. Polarization of Multipole Photons

The polarization is usually defined to be the measure of transversal anisotropy of electromagnetic field with respect to the direction of propagation provided by the Poynting vector $\vec{P}$ [26]. Quantum electrodynamics interprets the polarization as given spin state of photons [3]. In spite of the fact that spin spin is equal to 1 and hence has three states, the photons have only two polarizations because of the absence of the rest mass.

In PWR, direction of $\vec{P}$ always coincides with $\vec{k}/k$, and the polarization is uniquely defined everywhere. However, this is no longer a case for SWR, where $(\vec{r} \times \vec{P})$ is not necessary equal to zero, at least in a certain vicinity of the source.

As a matter of fact, E1 radiation obey the condition $(\vec{r} \cdot \vec{B}) = 0$, while the electric field $\vec{E}$ is not orthogonal to the radial direction [2]. Therefore, if we consider the radiation in the "laboratory frame" spanned by the basis (4) with the origin at the atom location, the three polarizations should be taken into account [22, 23].

This fact can be illustrated in the following way. Within the relativistic picture, the field is described by the field-strength tensor

$$F_{\alpha\beta} = \partial_\alpha A_\beta - \partial_\beta A_\alpha = \begin{pmatrix} 0 & E_x & E_y & E_z \\ -E_x & 0 & -B_z & B_y \\ -E_y & B_z & 0 - B_x \\ -E_z & -B_y & B_x & 0 \end{pmatrix}. \tag{45}$$

Since the anisotropy of the field is specified by the magnitudes of the components and by the phase angles between the components, consider the following (4×4) matrix

$$
\mathcal{R} = F^+ F = \begin{pmatrix} (\vec{E}^+ \cdot \vec{E}) & \vec{P} \\ \vec{P}^+ & \mathcal{P} \end{pmatrix}.
$$

Here F (F^+) denotes the positive- (negative-)frequency part of (42), $\vec{P}$, apart from an unimportant factor, coincides with the positive-frequency part of the Poynting vector, and $\mathcal{P}$ is the Hermitian (3×3) matrix additive with respect to contributions coming from electric and magnetic fields

$$
\mathcal{P} = \mathcal{P}_E + \mathcal{P}_B, \tag{46}
$$

where

$$
\mathcal{P}_{E\alpha\beta} = E_\alpha^+ E_\beta, \quad \alpha, \beta = x, y, z \tag{47}
$$

and

$$
\mathcal{P}_{B\alpha\beta} = \begin{cases} \vec{B}^+ \cdot \vec{B} - B_\alpha^+ B_\alpha & \text{at } \alpha = \beta \\ -B_\alpha^+ B_\beta & \text{otherwise} \end{cases}. \tag{48}
$$

Thus, the matrix (43)-(45) specifies the magnitudes of the components and the phase angles between the components of the complex field strengths in the "laboratory frame" with the origin at the source location. We chose to interpret (43) as the general polarization polarization matrix.

To justify this choice, consider first the case of plane waves propagating in the z-direction. Then, because of the relations $B_x = -E_y$ and $B_y = E_x$, both terms in (43) are reduced to the same (2×2) matrix of the form

$$
\begin{pmatrix} E_x^+ E_x & E_x^+ E_y \\ E_y^+ E_x & E_y^+ E_y \end{pmatrix},
$$

that is, to the conventional polarization matrix [26]. In the case of multipole radiation, the matrices (44) and (45) can also be reduced to the (2×2) conventional polarization matrices by a local unitary transformation, rotating the z-axis in the direction of Poynting vector $\vec{P}(\vec{r})$ at any point $\vec{r}$.

The diagonal terms in (44) give the radiation intensities of the components. The off-diagonal terms give the "phase information" described by the phase differences

$$
\Delta_{\alpha\beta} = \arg E_\alpha - \arg E_\beta, \quad \Delta_{xy} + \Delta_{yz} + \Delta_{zx} = 0.
$$

The polarization matrices (44) and (45) can also be expressed in the helicity basis (4) through the use of the unitary transformation

$$
U \begin{pmatrix} E_x \\ E_y \\ E_z \end{pmatrix} = \begin{pmatrix} E_+ \\ E_0 \\ E_- \end{pmatrix}, \quad U = \frac{1}{\sqrt{2}} \begin{pmatrix} 1 & i & 0 \\ 0 & 0 & \sqrt{2} \\ -1 & i & 0 \end{pmatrix}
$$

and similar transformation for $\vec{B}$. Then, $U\mathcal{P}_E U^+$ coincides, to within the transposition of columns, with the polarization matrix with elements

$$\tilde{\mathcal{P}}_{\mu\mu'} = E_\mu^+ E_{\mu'}, \qquad \mu, \mu' = 0, \pm 1, \tag{49}$$

have been introduces in [22] for E1 radiation.

Since $\vec{E}(\vec{r}) \cdot \vec{B}(\vec{r})$ at any point $\vec{r}$ and $\vec{r} \cdot \vec{B} = 0$ for the electric-type radiation, the complete information about the phase differences is provided by the matrix (44) or by the equivalent matrix (46) in this case. In the case of magnetic-type radiation with $\vec{r} \cdot \vec{E} = 0$, the matrix (45) should be used instead of (44) [47].

Consider now the quantum E1 radiation. In this case, the field amplitudes should be changed by corresponding operators. Let us note that, in addition to (46), defined in terms of *normal product* of photon operators, we can also construct the anti-normal ordered polarization matrix

$$\tilde{\mathcal{P}}_{\mu\mu'}^{(an)} = E_{\mu'} E_\mu^+.$$

Then, the difference

$$\tilde{\mathcal{P}}_{\mu\mu'}^{(0)} \equiv \tilde{\mathcal{P}}_{\mu\mu'}^{(an)} - \tilde{\mathcal{P}}_{\mu\mu'} = [E_\mu, E_{\mu'}^+] \tag{50}$$

defines the elements of the *vacuum polarization matrix*, in other words, the zero-point oscillations (ZPO) of polarization [48]. It is easily seen that ZPO of polarization depend on the distance from the source r and have a uniform angular distribution. Consider first the z-direction when $\theta = 0$ and

$$Y_{j\pm1,m-\mu}(0, \phi) = \sqrt{\frac{2(j \pm 1) + 1}{4\pi}} \delta_{m\mu}$$

for all ϕ. Then, taking into account the definition of vector spherical harmonics (5), operator vector potential (10), and commutation relations (11), for the right-hand side of (47) we get

$$\tilde{\mathcal{P}}_{\mu\mu'}^{(0)}(r, 0, \phi) = \begin{cases} N_E^2 \left[j_2(kr)\langle 12\mu 0|1\mu\rangle\sqrt{\frac{5}{4\pi}} - j_0(kr)\sqrt{\frac{1}{2\pi}} \right]^2, & \text{at } \mu = \mu' \\ 0, & \text{otherwise} \end{cases} \tag{51}$$

Because of the $SU(2)$ invariance of the operator vector potential, proven in Sec. II, there is a local unitary transformation $V(\vec{r})$, transforming (47) into (48) at any point. For explicit form of V see Ref. [39]. Since $\langle 1210|11\rangle = \langle 12, -1, 0|1, -1\rangle = 1/\sqrt{10}$ and $\langle 1200|10\rangle = -\sqrt{(2/5)}$, the transversal (with respect to $\vec{r}$) elements $\tilde{\mathcal{P}}_{\pm\pm}^{(0)}$ in (48) have equal magnitude. In view of definition of spherical Bessel functions, it is seen that ZPO of polarization are strong enough in the near and intermediate zones, while vanish at $r \to \infty$.

Let us now apply the above unitary transformation to the components of the operator vector potential of E1 field $V(\vec{r})\vec{A}_{E1k\mu}(\vec{r})$ and calculate the commutator

$$[V(\vec{r})\vec{A}_{E1k\mu}(\vec{r}), \vec{A}^{+}_{E1k\mu'}(\vec{r})V^{+}(\vec{r})] = \delta_{\mu\mu'} \times \begin{cases} \tilde{\mathcal{P}}^{(0)}_{++}(r) & \text{at } \mu = \pm 1 \\ \tilde{\mathcal{P}}^{(0)}_{00}(r) & \text{at } \mu = 0 \end{cases}$$

It is seen that these relations coincide with (11) to within the distance-dependent factors, describing ZPO of polarization. In turn, the normalized operators

$$b_{k\mu}(\vec{r}) = \frac{V(\vec{r})\vec{A}_{E1k\mu}(\vec{r})}{\sqrt{\tilde{\mathcal{P}}^{(0)}_{\mu\mu}(r)}} \tag{52}$$

obey the commutation relations (11) at any point $\vec{r}$ and hence form a local representation of the Weyl-Heisenberg algebra of E1 photons. Instead of the global index m, specifying AM, they depend on the coordinates μ. In other words, they specify the field oscillations in the "laboratory frame" spanned by the helicity basis (4) and hence can be interpreted as the *local operators of E1 photons with given polarization* [48]. This means that the eigenstates of the number operators $b^{+}_{\mu}(\vec{r})b_{\mu}(\vec{r})$ give the number states of E1 photons with given polarization μ at any point $\vec{r}$.

In fact, Eq. (49) represents a local Bogolubov-type canonical transformation from the photon operators with given m to the photon operators with given μ

$$b_{k\mu}(\vec{r}) = \frac{1}{\sqrt{\tilde{\mathcal{P}}^{(0)}_{\mu\mu}(r)}} \sum_{m}\sum_{\mu'}(-1)^{\mu'}V_{\mu\mu'}(\vec{r})\mathcal{A}_{km\mu'}(\vec{r})a_m \equiv \sum_{m}\mathcal{B}_{m\mu}(\vec{r})a_m, \tag{53}$$

where $\mathcal{A}$ denotes the mode function (see Sec. II). Since $\det[\mathcal{B}] \neq 0$, there is an inverse transformation, representing operators a_m in terms of b_{μ}.

It is a straightforward matter to show that, in the representation of local operators (49)-(50), the polarization matrix (46), apart from an unimportant constant factor, takes the form

$$\tilde{\mathcal{P}}_{\mu\mu'}(\tilde{r}) = b^{+}_{\mu}(\vec{r})b_{\mu'}(\vec{r}). \tag{54}$$

Assume now that the two-level E1 transition have been discussed in Sec. III emits a single photon in the state $|1_m\rangle$. Then, the averaging of (51) over this state gives the polarization at any point $\vec{r}$ described by the matrix with elements

$$\langle\tilde{\mathcal{P}}_{\mu\mu'}(\vec{r})\rangle = \frac{\mathcal{B}^{*}_{m\mu}(\vec{r})\mathcal{B}_{m\mu'}(\vec{r})}{\tilde{\mathcal{P}}^{(0)}_{\mu\mu}(r)}. \tag{55}$$

Taking again into account that the properties of a multipole photon correspond to a spherical shall of radius r, we should integrate (52) over $\sin\theta d\theta d\phi$ to get the polarization matrix of the photon at any distance r from the atom like we did in Sec. III for the spin carried by photon. In particular, it can be seen that the photons with any m have only two circular polarizations $\mu = \pm 1$ in the wave zone [23] even in the "laboratory frame".

It should be emphasized that the polarization can also be described in terms of Stokes parameters stokes parameters that become the Stokes operators in quantum domain [49]. Since in the "laboratory frame" we have three degrees of freedom, the set of Hermitian Stokes operators is provided by the generators of the $SU(3)$ subalgebra in the local Weyl-Heisenberg algebra of operators (49). In other words, the Stokes operators in "laboratory frame" coincide with (27) with the substitution of $b_\mu(\vec{r})$ instead of a_m [39].

In this way, the quantum properties of polarization of E1 photons can be described, including the quantum fluctuations of polarization [23, 39].

5. Summary

We have reviewed some recent results concerning the quantum radiation by multipole transitions in atoms and molecules. It is shown that because of the difference in the dynamic symmetry group of the Hilbert space, the use of SWR leads to an adequate picture of AM in quantum domain, based on the structure of the type $SU(2) \times SU(2)$. In particular, SWR permits us to evaluate the quantum fluctuations of AM of photons. It is also shown that spin and OAM contribute equally into the total AM of photons in the wave zone, while spin prevail over OAM in the near and intermediate zones.

It is also shown that the cascade decay of E2 transition can lead to creation of E1 photon twins in the qutrit state. This state obey the criterion of maximum entanglement of Refs. [32, 33] and manifests maximum correlation of local measurements of AM.

The use of SWR permits us to define the inherent quantum phase of multipole photons that is the azimuthal phase of AM. This definition develops the operational approach by Noh, Fougères and Mandel [42, 43]. The approach based on the polar decomposition of AM does not violate the algebraic properties of photon operators and leads to a qualitatively different picture of quantum fluctuations of phase from that obtained within the Pegg-Barnett approach [38, 40, 41].

It should be stressed that the approach based on the consideration of the $SU(2)$ phase states has shown its efficiency in the problem of definition of maximum entangled N-qubit states in atomic systems [32] of the type proposed in Ref. [50, 51], and in quantum cryptography [52]. In particular, it can be used to specify the qubit multipartite states in three-level atoms [53]. It can also be applied to classification of maximum entangled states entangled states that can be obtained through the use of strong-driving-assisted processes in cavity QED [54].

Finally, the use of SWR permits us to describe the quantum properties of polarization at any distance from the source. Since the polarization is a local property of multipole radiation, the representation of the Weyl-Heisenberg algebra of photons with given polarization at any distance from the source can be constructed through the use of SWR. It should be stressed that usually the problem of photon localization is discussed in terms of wave functions (see [55] and references therein). At first sight, there is no principle contradiction between the approaches based on the use of operators and wave functions. A connection between the approaches deserves a more detailed investigation.

In fact, the most of results reviewed in this paper were obtained by mapping of the operators from the finite-dimensional atomic Hilbert space to the infinite-dimensional Hilbert space of photons through the use of AM conservation. This procedure reflects an old idea that the properties of quantum radiation are specified by the source [56].

The SWR can be successfully used to describe the quantum properties of multipole radiation, generated by real atomic transitions. In many cases, the use of SWR permits us to avoid the difficulties peculiar to PWR. The approach based on the SWR has also manifested its efficiency in the number of problems connected with investigation of photon band-gap materials [57, 58].

Although the results discussed in this paper are mostly connected with E1 radiation, they can also be generalized on the case of an arbitrary quantum multipole radiation.

References

[1] Bateman, H. (1955). *The Mathematical Analysis of Electric and Optical Wave-Motion.* New York: Dover

[2] Jackson, J.D. (1975). *Classical Electrodynamics.* New York: Wiley.

[3] Berestetskii, V.B., E.M. Lifshitz, and L.P. Pitaevskii. (1982). *Quantum Electrodynamics.* Oxford: Pergamon Press.

[4] Cohen-Tannouji, C., J. Dupont-Roc, and G. Grinberg. (1992). *Atom-Photon Interaction.* New York: Wiley.

[5] Wigner, E.P. (1959). *Group Theory and Its Application to the Quantum Mechanics of Atomic Spectra.* New York: Academic Press.

[6] Heitler, W. (1954). *The Quantum Theory of Radiation,* 3rd edition. London: Clarendon Press.

[7] Heitler, W. (1936). Proc. Camb. Phil. Soc. **37**, 112.

[8] Bederson, B. and H. Walther. (2000). *Advances in Atomic, Molecular, and Optical Physics,* Vol. 42. New York: Academic Press.

[9] Raymond, J.M., M. Brune, and S. Haroche. (2001). Rev. Mod. Phys. **73**, 565.

[10] Arnavut, H.H. and G.A. Barbosa. (2000). Phys. Rev. Lett. **85**, 286.

[11] Mair, A., A. Vazari, G. Weihs, and A. Zeilinger. (2001). Nature **412**, 313.

[12] Padgett, M.J., S.M. Barnett, S. Franke-Arnold, and J. Courtial. (2002). Phys. Rev. Lett. **88**, 257901.

[13] Vazari, A., G. Weihs, and A. Zeilinger. (2002). Phys. Rev. Lett. **89**, 240401.

[14] Muthukrishnan, A. and C.R. Stroud. (2002). J. Opt. B **4**, S73.

[15] Mandel, L. and E. Wolf. (1995). *Optical Coherence and Quantum Optics*, New York: Cambridge University Press.

[16] Van Enk, S.J. and G. Nienhuis. (1994). J. Mod. Opt. **41**, 963.

[17] Cakır, Ö. and A.S. Shumovsky. (2003). E-print quant-ph

[18] Shumovsky, A.S. (1997). Opt. Commun. **136**, 219.

[19] Shumovsky, A.S. and Ö. Müstecaplıoğlu.(1998).*Opt.Commun.***146**, 124

[20] Shumovsky, A.S. and *Ö.* Müstecaplıoğlu. (1998). In *Causality and Locality in Modern Physics*, edited by G. Hunter, S. Jeffers, and J.-P. Vigier, Dordreht: Kluwer.

[21] Shumovsky A.S. (1999). J. Phys. A **32**, 6589.

[22] Shumovsky, A.S. and *Ö.* Müstecaplıoğlu. (1998). Phys. Rev. Lett. **80**, 1202.

[23] Can, M.A. and A.S. Shumovsky. (2002). J. Mod. Opt. **49**, 1423.

[24] Allen, L., M.J. Padgett, and M. Babiker. (1999). In *Progress in Optics XXXIX*, edited by E. Wolf. New York: Elsevier Science.

[25] Davydov, A.S. (1976). *Quantum Mechanics* Oxford: Pergamon Press.

[26] Born, M. and E. Wolf. (1970). *Principles of Optics* Oxford: Pergamon Press.

[27] Thompson, W.J. (1994). *Angular Momentum* New York: Wiley.

[28] Klyachko, A.A. (2000). Private communication.

[29] Acharya, R. and E.C.G. Sudarshan. (1960). J. Math. Phys. **1**, 532.

[30] Davies, B. (1972). *Quantum Theory of Open Systems* New York: Academic Press.

[31] Gardiner, C.W. and P. Zoller. (2000). *Quantum Noise* Berlin: Springer.

[32] Can, M.A., A.A. Klyachko, and A.S. Shumovsky. (2002). Phys. Rev. A **66**, 022111.

[33] Klyachko, A.A. and A.S. Shumovsky. (2003), J. Opt. B **5** (in print); preprint quant-ph/0302008.

[34] Burlakov, A.V., M.V. Chekhova, O.A. Karabutova, D.N. Klyshko, and S.P. Kulik. (1999). Phys. Rev. A **60**, R4209.

[35] Bechmann-Pasquinucci, H. and A. Peres (2000). Phys. Rev. Lett. **85**, 3313.

[36] Brukner, Č., M. Žukowski, and A. Zeilinger. (2001). E-print quant-ph/0106119.

[37] Lynch, R. (1995). Phys. Rep. **256**, 367.

[38] Tanaś, R., A. Miranowitz, and T. Gantsog. (1996). Prog. Opt. **35**, 1.

[39] Shumovsky, A.S. (2001). In *Modern Nonlinear Optics*, Pat 1, *Advances in Chemical Physics, Vol.* **119**, edited by M.W. Evans, Series editors I. Prigogine and S.A. Rice, New York: Wiley.

[40] Pegg, D.T. and S.M. Barnett. (1988). Europhys. Lett. **6**, 483.

[41] Pegg, D.T. and S.M. Barnett. (1997). J. Mod. Opt. **44**, 225.

[42] Noh, J.W., A. Fougères, and L. Mandel. (1991). Phys. Rev. Lett. **67**, 1426.

[43] Noh, J.W., A. Fougères, and L. Mandel. (1992). Phys. Rev. A **45**, 242.

[44] Vourdas, A. (1990). Phys. Rev. A, **41**, 1653.

[45] Serre, J.-P. (1977). *Linear Representations of Finite Groups* New York: Springer.

[46] Bogolubov, N.N., A.M. Kurbatov, and A.S. Shumovsky (Eds.). (1991). *N.N. Bogolubov Selected Works, Part II* New York: Gordon and Breach.

[47] Wolf, E. (1954). Nuovo Cimento **12**, 884.

[48] Klyachko, A.A. and A.S. Shumovsky. (2001). Laser Phys. **11**, 57.

[49] Jauch, J.M. and F. Rohrlich. (1959). *The Theory of Photons and Electrons* Reading MA: Addison-Wesley.

[50] Plenio, M.B., S.F. Huelga, A. Beige, and P.L. Knight. (1999). Phys Rev. A **59**, 2468.

[51] Beige, A., W.J. Munro, and P.L. Knight. (2000). Phys. Rev. A **62**, 052102.

[52] Vourdas, A. (2002). Phys. Rev. A **64**, 04321.

[53] Can, M.A., A.A. Klyachko, and A.S. Shumovsky. (2002). Appl. Phys. Lett. **81**, 5072.

[54] Solano, E., G.S. Agarwal, and H. Walther. (2002). Preprint quant-ph/0202071.

[55] Can, K.W., C.K. Law, and J.H. Eberly. (2002). Phys. Rev. Lett. **88**, 100402.

[56] Koo, I.C. and J.H. Eberly. (1976). Phys. Rev. A **14**, 2174.

[57] Rupasov V.I. and M. Singh. (1996). Phys. Rev. Lett. **77**, 338.

[58] John. S. and V.I. Rupasov. (1997). Phys. Rev. Lett. **79**, 821.

THEORY OF ATOMIC HYPERFINE STRUCTURE

Anatoly V. Andreev

Physics Department, M.V.Lomonosov Moscow State University, Moscow 119992 Russia E-mail
andreev@sr1.phys.msu.su

Abstract   The new equation for ½ spin relativistic particle is proposed. In comparison with Dirac equation the proposed equation (i) is completely symmetric with respect to particle and its antiparticle, (ii) predicts the hyperfine structure of electron energy levels in Coulomb field of nucleus with zero magnetic moment, (iii) explains the Lamb shift and electron anomalous magnetic moment without secondary quantization either the matter or the radiation field. The analytically tractable solutions of the proposed equation are found for the problems of electron motion in the Coulomb field, in uniform magnetic field and in superposition of Coulomb and uniform magnetic fields.

1.   Introduction

The hydrogen atom spectrum calculated with the help of Dirac equation is in good agreement with results of its experimental measurements. This is one the main evidence of the fundamental nature of the Dirac theory. Nevertheless the real energy-level diagram of hydrogen atom is still wealthy than this predicted by Dirac equation. In addition to fine structure the hydrogen spectrum includes the shift in fine structure (Lamb shift [1]), hyperfine structure of levels, anomalous value of electron magnetic moment [2], etc. At the present time all of these specific features of the hydrogen atom spectrum have the reasonable explanations. The hyperfine structure of levels is explained by interaction of electron spin with nucleus spin spin [2]. The Lamb shift [3] and anomalous magnetic moment [4] are explained by interaction of electron with vacuum fluctuations of electromagnetic field (radiation corrections) and their calculated values are in good agreement with the experimental data. It should be noted that a number of attempts were made to explain the Lamb shift and anomalous magnetic moment avoiding quantization of the field. Jaynes and co-workers have constructed nonquantized- field "neoclassical" theory [5] based on the old Schrodinger interpretation of quantum mechanics. Eberly and co-wokers [6] have developed the radiation reaction or source-field approach in quantum electrodynamics. Barut and co-workers [7] have developed the quantum electrodynamics based

A.S. Shumovsky and V.I. Rupasov (eds.), Quantum Communication and Information Technologies, 171–193.

on self-fields. The complex structure of spectrum of hydrogen atom, which is one of the simplest physical objects, makes evident the necessity for search of ways to revise the Dirac theory. The form of Hamiltonian of Dirac equation, including the first degree of electron momentum, gives additional arguments for revision of this equation. Indeed, the energy of free particle should not depend on the direction in which the particle moves.

It is well known that the Dirac equation can be produced in result of splitting of the second order differential equation into the two first order differential equations

$$\frac{1}{c^2}\frac{\partial^2}{\partial t^2} - \Delta + \frac{m_0^2 c^2}{\hbar^2} = \left[\frac{1}{c}\frac{\partial}{\partial t} + \vec{\alpha} \cdot \vec{\nabla} + \frac{im_0 c}{\hbar}\beta\right]\left(\frac{1}{c}\frac{\partial}{\partial t} - \vec{\alpha} \cdot \vec{\nabla} - \frac{im_0 c}{\hbar}\beta\right),$$

where α and β are Dirac matrices. One of these first order operators is then associated with the negatively charged particle and another with the positively charged. On the other hand it seems that this splitting could be associated with the splitting of two counter propagating waves that appear when we solve the second order differential equation. Indeed the Hamiltonian Hamiltonian of monodirectional propagating component should depend on the momentum but not on its magnitude. In such a case we shall not meet any problems in analysis of the free motion of particle, but the serious problems may arise when we apply this equation for the study of bound states of the particle in the external field.

Recently [8] we proposed the equation for action for spin $1/2$ relativistic particle interacting with the electromagnetic field. The proposed action leads to the second order differential equation for the bispinor electron wave function. The obtained equation is symmetric with respect to the negatively and positively charged particles. This symmetry reflects the equivalence of physical properties of particle and its antiparticle. The present paper consists of two main parts. In the first part we study the general properties of the proposed equation and then we discuss a number solutions of this equation that were obtained in analytically tractable form. The main distinctive features of the proposed equation can be easily understood from the comparison with the Dirac equation. One of the most important features that follows from this comparison is in the fact that the all solutions of the first order differential Dirac equation are at the same time the solutions of the proposed second order equation. But the second order equation have the additional solutions that are not the solutions of Dirac equation. There is a technique in quantum electrodynamics (see for example [9]) when we solve initially the second order differential equation and then we retain only those solutions that satisfy the first order Dirac equation. However there is a principle difference between these two equations when we study the interaction of particle with electromagnetic field. The proposed equation for action result not only in the second order differential equation for bispinor wave function of particle, but it generates also the new equations for charge and current density.

The transformation properties of charge and current density with respect to CPT transformations differ from that in Dirac theory. As a result the atomic response on the external electromagnetic field is significantly different from that calculated on the basis of Dirac theory. We show that the equations for the charge and current density of Dirac theory follow from the proposed theory under some additional suppositions. The analytically tractable solutions for the problem on the electron motion in the Coulomb field are found. The calculated energy level spectrum of hydrogen atom is extremely close to that calculated from Dirac equation ($\Delta E/E = 10^{-4} - 10^{-5}$) but differs from it by the presence of the hyperfine structure. The levels with the zero orbital moment split into the two sublevels and the levels with the non-zero orbital moment split into the three sublevels. The hyperfine splitting has its origin in the interaction of electron spin with electrostatic field of nucleus.

The energy-level diagram for electron in uniform magnetic field is calculated. The calculated spectrum is different from the previous calculations because it depends on the sum $(n + M)$, where n is the radial quantum number and $\hbar M$ is the projection of the total angular momentum Angular momentum $\vec{j} = \vec{\ell} + \vec{s} = -i\hbar[\vec{r} \times \vec{\nabla}] + \hbar\vec{\sigma}/2$. Therefore the levels with $n + M = const$ are degenerated. It is extremely important result because it means that electron with arbitrary value of angular moment could be in the ground state ground state, i.e. $n + M = 0$. It opens up the new possibilities for interpretation of a number of different phenomena that is due to the interaction of electron with magnetic field.

The analytically tractable solutions for the problem on the electron motion in superposition of the Coulomb and uniform magnetic fields are found. It enables us to determine the structure of energy levels for hydrogen atom in uniform magnetic field and calculate the electron magnetic moment at different states of obtained spectrum. It is shown that the value of magnetic moment is different at different states of hyperfine structure.

2. Equation for Spin $1/2$ Particle Interacting with Electromagnetic Field

2.1 The principle of action

Let the action for the spin $1/2$ particle be

$$S = \frac{1}{16\pi} \int F_{\mu\nu} F_{\nu\mu} dV dt$$

$$-\frac{1}{2m_0} \int \left[\left(i\hbar\nabla_\mu \bar{\Psi} - \frac{q}{c} A_\mu \bar{\Psi} \right) \left(-i\hbar\nabla_\mu \Psi - \frac{q}{c} A_\mu \psi \right) + m_0^2 c^2 \bar{\Psi}\Psi - \frac{iq\hbar}{2c} \bar{\Psi}\gamma_\mu \gamma_\nu F_{\nu\mu}\Psi \right] dV dt, \quad (1)$$

where $F_{\mu\nu} = \Delta_\mu A_\nu - \Delta_{nu} A_\mu$ is the field tensor, $\nabla_\mu = \partial/\partial x_\mu$, $x_\mu = (\vec{r}, ict)$, $A_\mu = (\vec{A}, i\varphi)$ is the 4-vector of field, $\bar{\Psi} = \Psi^+ \gamma_4$ is the Dirac conjugate wave function, q and m_0 are the charge and mass of particle, respectively. The matrices $\vec{\gamma}$ and γ_4 are related to Dirac matrices $\vec{\alpha}$ and β by the well known equations: $\gamma_4 = \beta$, $\vec{\gamma} = -i\beta\vec{\alpha}$. The field part of action (1) has standard form. The first two terms in the second item are similar to action for spinless particle and the last term describes the interaction of particle spin with electromagnetic field. In contrast to the case of spinless particle the action (1) includes not the Hermitian conjugated but the Dirac conjugated wave function. It is necessary because only in this case equation (1) became the self-conjugated equation.

The Euler- Lagrange equations, when S is varied with respect to $\bar{\Psi}$, is

$$\left[(h\partial/\partial t + iq\varphi)^2 + (-ihc\vec{\nabla} - q\vec{A})^2 + m_0^2 c^4\right]\bar{\Psi} = -iqhc\vec{\alpha}\vec{E}\Psi + qhc\vec{\Sigma}\vec{H}\Psi. \quad (2)$$

Variation of S with respect to A_μ results in

$$\nabla_\nu^2 A_\mu - \nabla_\mu \nabla_\nu A_\nu = -(4\pi/c)j_\mu, \quad (3)$$

where

$$j_\mu = \frac{q}{m_0}\left[\frac{ih}{2}\left(\frac{\partial\bar{\Psi}}{\partial x_\mu}\Psi - \bar{\Psi}\frac{\partial\Psi}{\partial x_\mu}\right) - \frac{q}{c}\bar{\Psi}A_\mu\Psi - \frac{ih}{4}\frac{\partial}{\partial_\nu}(\bar{\Psi}(\gamma_\mu\gamma_\nu - \gamma_\nu\gamma_\mu)\Psi)\right]. \quad (4)$$

The current density 4-vector j_μ satisfies the continuity equation

$$\nabla_\mu j_\mu = 0.$$

The spatial and temporal components of the 4-vector j_μ are the current density

$$\vec{j}(\vec{r}, t) = \frac{q}{m_0}\left[\frac{ih}{2}(\vec{\nabla}\bar{\Psi} \cdot \Psi - \bar{\Psi} \cdot \vec{\nabla}\Psi) - \frac{q}{c}\bar{\Psi}\vec{A}\Psi + \frac{h}{2}\vec{\nabla} \times (\bar{\Psi}\vec{\Sigma}\Psi) - \frac{ih}{2c}\frac{\partial}{\partial t}(\bar{\Psi}\vec{\alpha}\Psi)\right] \quad (5)$$

and charge density

$$\rho(\vec{r}, t) = \frac{q}{m_0 c}\left[-\frac{ih}{2c}\left(\frac{\partial\bar{\Psi}}{\partial t}\Psi - \bar{\Psi}\frac{\partial\Psi}{\partial t}\right) - \frac{q}{c}\bar{\Psi}\varphi\Psi + \frac{ih}{2}\vec{\nabla}(\bar{\Psi}\vec{\alpha}\Psi)\right], \quad (6)$$

respectively. It is seen that the current density four-vector is completely different from that in Dirac theory. But there is some similarity of the equation (4b) for charge density with that for spinless particle. Indeed, if we substitute the Dirac conjugated wave function by Hermitian conjugated wave function then the first two terms in (4b) will coincide with the charge density of spinless particle. The last terms in (4a) and (4b) have no analogy in Klein- Gordon or Dirac theory and they depend on the spin variables. We shall use the standard form of bispinor wave function [9]

$$\Psi = \begin{pmatrix} \varphi \\ \chi \end{pmatrix}. \quad (7)$$

It is seen from (4a) that if we substitute the bispinor wave function (5) by the spinor wave function $\bar{\Psi} \to \varphi^+$ and $\Psi \to \varphi$ then the first three terms in (4a) will coincide with the equation for current density that could be obtained in the first approximation under expansion of Dirac equation into series over $1/c$. The interpretation of equation (4) will be given in Sec.IV.

2.2 Connections with Dirac equation

By adding the equation (1) the term

$$\frac{i}{2}\left[h\frac{\partial}{\partial t}(\bar{\Psi}\gamma_4\Psi) - ihc\vec{\nabla}(\bar{\Psi}\gamma\Psi)\right],$$

we can transform the action to the following form

$$S = \frac{1}{16\pi}\int F_{\mu\nu}F_{\nu\mu}dV\,dt$$

$$-\frac{1}{2m_0}\int\left\{\bar{\Psi}\left[\gamma_4\left(\frac{h}{c}\frac{\partial}{\partial t} - iq\varphi\right) + \vec{\gamma}\left(ih\vec{\nabla} - \frac{q}{c}\vec{A}\right) + im_0c\right]\right\}$$

$$\times\left\{\left[\gamma_4\left(-\frac{h}{c}\frac{\partial}{\partial t} - iq\varphi\right) + \vec{\gamma}\left(-ih\vec{\nabla} - \frac{q}{c}\vec{A}\right) - im_0c\right]\Psi\right\}dV\,dt \tag{8}$$

The operators in the left and right braces act on $\bar{\Psi}$ and Ψ respectively. The variation of action (6) with respect to $\bar{\Psi}$ results in the following equation

$$(\gamma_\mu p_\mu + im_0c)(\gamma_\mu p_\mu - im_0c)\Psi = 0, \tag{9}$$

where $p_\mu = -ih\partial/\partial x_\mu - qA_\mu/c$. It is well known that the addition to action of the total derivative does not change the resulting equations of motion. Indeed it is seen that the equation (7) coincides identically with the equation (2). At the same time we can see now in an explicit form that any solution of the Dirac equation

$$(\gamma_\mu p_\mu - im_0c)\Psi = 0 \tag{10}$$

is simultaneously a solution of equation (2). But among the solutions of the equation (2) there are solutions not satisfying the Dirac equation (8).

The variation of action (6) with respect to results in the following equation for field

$$\frac{\partial^2 A_\mu}{\partial x_\nu^2} - \frac{\partial}{\partial x_\mu}\left(\frac{\partial A_\nu}{\partial x_\nu}\right) = -\frac{2\pi q}{m_0c}$$

$$\times\left\{\bar{\Psi}\gamma_\mu\left[\left(\gamma_\nu\left(-ih\frac{\partial}{\partial x_\nu} - \frac{q}{c}A_\nu\right) - im_0c\right)\Psi\right] - \left[\bar{\Psi}\left(\gamma_\nu\left(-ih\frac{\partial}{\partial x_\nu} + \frac{q}{c}A_\nu\right) - im_0c\right)\right]\gamma_\mu\Psi\right\} \tag{11}$$

Again the equation (9) coincides identically with the equation (3) because the terms proportional to m0 are mutually cancelled. But the right-hand-side of the equation (9) has the form, which is convenient to compare the current density four-vector (4) with that of Dirac theory.

It should be noted that, in Dirac theory, the equation conjugated to the equation (8) has the form

$$\bar{\Psi}\left[\gamma_\mu\left(-ih\frac{\partial}{\partial x_\mu} + \frac{q}{c}A_\mu\right) + im_0c\right] = 0. \tag{12}$$

Applying the equations (8) and (10) to the right-hand-side of equation (9), we get the following equation for the 4-vector of current density

$$j_\mu = icq\bar{\Psi}\gamma_\mu\Psi. \tag{13}$$

Thus if the wave function satisfies the Dirac equation (8) the current density four-vector (4) takes the form of that in Dirac theory

$$j_\mu = (\vec{j}, ic\rho) = (cq\Psi^+\vec{\alpha}\Psi, icq\Psi^+\Psi).$$

If the function Ψ obeys the equation

$$(\gamma_\mu p_\mu + im_0c)\Psi = 0, \tag{14}$$

then the current density four-vector takes the form

$$j_\mu = -icq\bar{\Psi}\gamma_\mu\Psi.$$

Taking into account that is a positively defined value we can see that the equations (8) and (12) correspond to particles with the opposite signs of charge. It should be noted that the sign of charge depends on choice of sign that we put before the second integral in the equation (1) for action. Thus, it is seen that the particle describing by equation (2) interacts with electromagnetic field in different way than the particle describing by Dirac equation (8) or (12). In spite of the fact that in some specific cases the current density four-vector (4) coincides with that for Dirac particle in general case the current density four-vectors (4) and (11) are significantly different in their CPT transformation properties. Indeed the first two terms in (4) are transformed as a product spinor of the second rank. The last term in (4) can be transformed to the following form

$$j_\mu^{(T)} = -\frac{gh}{2m_0}\frac{\partial T_{\mu\nu}}{\partial x_\nu},$$

where second rank bispinor $T_{\mu\nu}$ is a bilinear form of wave function components transformed as antisymmetric 4-tensor

$$T_{\mu\nu} = \frac{i}{2}\bar{\Psi}(\gamma_\mu\gamma_\nu - \gamma_\nu\gamma_\mu)\Psi.$$

The symmetry of equations with respect to CPT transformations plays an extremely important role in quantum electrodynamics [9].

2.3   Energy-momentum tensor

Following [9], we define the energy-momentum tensor as follows

$$T_{\mu\nu} = \frac{\partial A_\lambda}{\partial x_\mu}\frac{\partial L}{\partial(\partial A_\lambda/\partial x_\nu)} + \frac{\partial \bar\Psi}{\partial x_\mu}\frac{\partial L}{\partial(\partial \bar\Psi/\partial x_\nu)} + \frac{\partial L}{\partial(\partial \Psi/\partial x_\nu)}\frac{\partial \Psi}{\partial x_\mu} - L\delta_{\mu\nu}.$$

The Lagrange function L of equation (1) yields the following expression for the energy-momentum tensor

$$T_{\mu\nu} = \frac{1}{4\pi}\frac{\partial A_\lambda}{\partial x_\mu}F_{\lambda\nu} - \frac{1}{16\pi}F_{\alpha\beta}F_{\beta\alpha}\delta_{\mu\nu} - \frac{h^2}{2m_0}\left(\frac{\partial \bar\Psi}{\partial x_\mu}\frac{\partial \Psi}{\partial x_\nu} + \frac{\partial \bar\Psi}{\partial x_\nu}\frac{\partial \Psi}{\partial x_\mu}\right)$$

$$+\frac{igh}{2m_0c}\left(\frac{\partial \bar\Psi}{\partial x_\mu}A_\nu\Psi - \bar\Psi A)\nu\frac{\partial \Psi}{\partial x_\mu}\right) + \frac{1}{2m_0}\left[\left(ih\frac{\partial \bar\Psi}{\partial x_\lambda} - \frac{q}{c}A_\lambda\bar\Psi\right)\left(-ih\frac{\partial \Psi}{\partial x_\lambda} - \frac{q}{c}A_\lambda\Psi\right) + m_0^2c^2\bar\Psi\Psi\right]$$

$$+\frac{igh}{4m_0c}\left[\bar\Psi\frac{\partial A_\lambda}{\partial x_\mu}(\gamma_\lambda\gamma_\nu - \gamma_\nu\gamma_\lambda)\Psi - \bar\Psi\gamma_\alpha\gamma_\beta F_{\alpha\beta}\Psi\delta_{\mu\nu}\right] \quad (15)$$

Hence, for the energy density we get

$$T_{00} = \frac{1}{8\pi}\left[\frac{1}{c^2}\left(\frac{\partial \vec A}{\partial t}\right)^2 - (\vec\nabla\varphi)^2 + (\vec\nabla\times\vec A)^2\right] + \frac{1}{2m_0}\left[m_0^2c^2\bar\Psi\Psi + \frac{h^2}{c^2}\frac{\partial \bar\Psi}{\partial t}\frac{\partial \Psi}{\partial t}\right.$$

$$+h^2\vec\nabla\bar\Psi\vec\nabla\Psi - \frac{iqh}{c}(\vec\nabla\bar\Psi\cdot\Psi - \bar\Psi\cdot\vec\nabla\Psi)\vec A + \frac{q^2}{c^2}\bar\Psi(\vec A^2 - \varphi^2) - \frac{iqh}{c}\bar\Psi\vec\alpha\vec\nabla\varphi\Psi - \frac{qh}{c}\bar\psi\vec\Sigma\vec H\psi\right] \quad (16)$$

The integration of the second term in (14) over the whole space gives us the equation for electron energy in the external fields. If we apply the equality

$$\int \vec\nabla\bar\Psi\vec\nabla\Psi dV = -\frac{1}{2}\int(\Delta\bar\Psi\cdot\Psi + \bar\Psi\cdot\Delta\Psi)dV$$

together with the equation (2), we get

$$E = \int T_{00}dV = \frac{1}{m_0}\int\left[\frac{h^2}{c^2}\left(\frac{\partial \bar\Psi}{\partial t}\frac{\partial \Psi}{\partial t} - \frac{1}{4}\frac{\partial^2(\bar\Psi\Psi)}{\partial t^2}\right) + \frac{ihq\varphi}{2c^2}\left(\frac{\partial \bar\Psi}{\partial t}\Psi - \bar\Psi\frac{\partial \Psi}{\partial t}\right)\right.$$

$$\left.+\frac{ihq}{2c^2}\bar\Psi\vec\alpha\frac{\partial \vec A}{\partial t}\Psi\right]dV.$$

Let electron be in the steady-state external field $\varphi(\vec r)$. Then its energy of the state with the wave function $\Psi_n(\vec r, t) = \exp(-iE_nt/h)\Psi_n(\vec r)$ is

$$E = (E_n/m_0c^2)\int \bar\Psi_n(\vec r)(E_n - q\varphi(\vec r))\Psi_n(\vec r)dV.$$

By taking into account the equation (4b) for the charge density $\rho(\vec r, t)$, we can rewrite the last equation in the following form

$$E = E_x\int\left[\frac{1}{q}\rho_n(\vec r) - \frac{ih}{2m_0c}\vec\nabla(\bar\Psi_n(\vec r)\vec\alpha\Psi_n(\vec r))\right]dV. \quad (17)$$

After the integration over the whole space, the last term in (15) is cancelled and we get the following normalization condition for the wave function $\Psi_n(\vec{r})$

$$\frac{1}{q} \int \rho(\vec{r})dV = \frac{1}{m_0 c^2} \int \bar{\Psi}(\vec{r})[E_n - q\varphi(\vec{r})]\Psi_n(\vec{r})dV = 1. \qquad (18)$$

We have already mentioned that there is some arbitrariness in the choice of sign in the equation for action. Let us determine now the sign of charge as $q = -|e|$. Hence according to (16) the solutions of the equation (2) with positive energy $E_n > 0$ correspond to the negatively charged particle when $\int |\varphi_n(\vec{r})|^2 dV > \int |\chi_n(\vec{r})|^2 dV$ and positively charged particle when $\int |\varphi_n(\vec{r})|^2 dV < \int |\chi_n(\vec{r})|^2 dV$. However we can see that the equation (2) is symmetric with the respect to the spinors φ and χ (see (5)). Thus the transformations $\varphi \to \chi$ and $\chi \to \varphi$ enables us to get the solution for negatively charged particle from the solutions for positively charged one and vice versa. It should be noted finally that in the case when $E_n = m_0 c^2 + \Delta E_n$ *approxm*$_0 c^2$, the normalization condition (16) coincides with that for the wave function governed by Dirac equation.

In the case of negative energy $E < 0$ the solutions corresponding to the positively and negatively charged particles trade the places. Thus in the case of free particle the solutions with positive and negative energy are really the two degenerate solutions. This degeneracy is completely removed in the more realistic case of particle in the external field. The bound states for the particle in the external field could exist only in the case when potential energy $U(\vec{r}) = q\varphi(\vec{r})$ is negative. On the other hand, the wave functions, corresponding to the bound states of the particle in an external field, should obey the boundary condition $\Psi(\vec{r})|_{r\to\infty} = 0$. Taking into account, that in the steady state case with $A = 0$ the operator in the left-hand-side of the equation (2) became $[-(E_U)^2 - h^2 c^2 \Delta + m_0^2 c^4]$, we can see that the non-trivial solutions compatible with the above mentioned boundary conditions could exist only in the case when $E > 0$.

3.   Interaction of Electron with Electric and Magnetic Fields

3.1   Electron in Coulomb field

Let an electron be in Coulomb field

$$U(\vec{r}) = e\varphi(\vec{r}) = -\frac{Ze^2}{r}.$$

In this case the equation (2) became

$$\left[\Delta - \frac{m_0^2 c^4 - E^2}{h^2 c^2} + \frac{2EZ\alpha}{hcr} + \frac{Z^2\alpha^2}{r^2}\right]\Psi = -\frac{i}{hc}\frac{\partial U}{\partial r}\alpha_\gamma\Psi, \qquad (19)$$

where $\alpha = e^2/hc$ is the fine structure constant, $\alpha_\gamma = \vec{\alpha}\vec{e}_\gamma = \begin{pmatrix} 0 & \sigma_\gamma \\ \sigma_\gamma & 0 \end{pmatrix}$, and $\vec{e}_\gamma$ is the radial unit vector of spherical frame of reference. It is convenient to use the following equality

$$\sigma_\gamma = \sigma_+ \sin\theta \exp(-i\varphi) + \sigma_- \sin\theta \exp(i\varphi) + \sigma_z \cos\theta, \qquad (20)$$

where $\sigma_\pm$ and σ_z are the two-dimensional Pauli matrices. Due to the spherical symmetry of the problem it is natural to seek the solution as expansion into series over the spherical harmonics $Y_{\ell m}(\theta, \phi)$:

$$\Psi(\vec{r}) = \begin{pmatrix} f_{\ell_1 m_1}(r) Y_{\ell_1 m_1} \\ f_{\ell_2 m_2}(r) Y_{\ell_2 m_2} \\ g_{\ell_3 m_3}(r) Y_{\ell_3 m_3} \\ g_{\ell_4 m_4}(r) Y_{\ell_4 m_4} \end{pmatrix} \qquad (21)$$

The only spin operator of the equation (17) is the operator α_γ. Taking into account the equality (18) it is easy to show that the coupled components of the wave function (19) are

$$(\ell_1, m_1) = (\ell, m), \ (\ell_2, m_2) = (\ell, m+1), \ (\ell_3, m_3) = (\ell+1, m), \ (\ell_4, m_4) = (\ell+1, m+1) \ (22)$$

or

$$(\ell_1, m_1) = (\ell+1, m), \ (\ell_2 m_2) = (\ell+1, m+1), \ (\ell_3, m_3) = (\ell, m) \ \ell_4, m_4) = (\ell, m+1). \ (23)$$

Therefore the equation (17) becomes

$$\left[\frac{d^2}{dr^2} + \frac{2}{r}\frac{d}{dr} + \frac{Z^2\alpha^2 - \ell(\ell+1)}{r^2} + \frac{2EZ\alpha}{hc}\frac{1}{r} - \kappa^2 \right] f_{\ell, M+\nu/2} = -\frac{iZ\alpha}{r^2} \sum_{q=\pm 1} A_{pq} g_{\ell+1, M+q/2}, \ (24)$$

$$\left[\frac{d^2}{dr^2} + \frac{2}{r}\frac{d}{dr} + \frac{Z^2\alpha^2 - (\ell+1)(\ell+2)}{r^2} + \frac{2EZ\alpha}{hc}\frac{1}{r} - \kappa^2 \right] g_{\ell, M+\nu/2} = -\frac{iZ\alpha}{r^2} \sum_{q=\pm 1} B_{pq} f_{\ell+1, M+q/2}, \ (25)$$

where $M = m + 1/2$, $p, q = \pm 1$,

$$\kappa^2 = \frac{m_0^2 c^4 - E^2}{h^2 c^2},$$

and the coefficients A_{pq} have the form

$$A_1 = A(\ell, m \to \ell+1, m) = i\sqrt{\frac{(\ell+m+1)(\ell-m+1)}{(2\ell+1)(2\ell+3)}},$$

$$A_2 = A(\ell, m \to \ell+1, m+1) = -i\sqrt{\frac{(\ell+m+1)(\ell+m+2)}{(2\ell+1)(2\ell+3)}},$$

$$A_3 = A(\ell, m+1 \to \ell+1, m) = i\sqrt{\frac{(\ell-m)(\ell-m+1)}{(2\ell+1)(2\ell+3)}},$$

$$A_4 = A(\ell, m+1 \to \ell+1, m+1) = -i\sqrt{\frac{(\ell-m)(\ell+m+2)}{\ell+m+2}(2\ell+1)(2\ell+3)}.$$

There are also the following relations between the coefficients and B_{pq} and A_{pq}:

$$B(\ell+1, m \to \ell, m) = A_1, \quad B(\ell+1, m \to \ell, m+1) = A_3,$$
$$B(\ell+1, m+1 \to \ell, m) = A_2, \quad B(\ell+1, m+1 \to \ell, m+1) = A_4.$$

To solve the equations (21a-b) it is helpful to use the solution of the following equation

$$\frac{d^2 f}{dx^2} + \frac{2}{x}\frac{df}{dx} - \left(a - \frac{b}{x} - \frac{c}{x^2}\right) f(x) = 0$$

of the form

$$f(x) = \exp\left(-\sqrt{a}x + \frac{\sqrt{1-4c}-1}{2}\ln x\right) U\left(\frac{1+\sqrt{1-4c}}{2} - \frac{b}{2\sqrt{a}}, 1 + \sqrt{1-4c}, 2\sqrt{a}x\right), \quad (26)$$

where $U(p, q, z)$ is the confluent hypergeometric function. We shall omit the second linear independent solution described by confluent hypergeometric function $M(p, q, z)$ (see [10]), because this solution could not satisfy the above mentioned boundary conditions for wave function. Thus, to satisfy the equations (21a-b) we can try the following substitution

$$f_{\ell, M+n/2} = f_{0n} F(r), \quad g_{\ell, M+n/2} = g_{0n} F(r), \quad (27)$$

where

$$F(r) = \exp(\kappa r) r^{\mu-1} U\left(\mu - \frac{EZ\alpha}{hc\kappa}, 2\mu, 2\kappa r\right)$$

and

$$\mu = (1 + \sqrt{1-4\gamma})/2.$$

The parameter γ should be chosen in such a form that we could satisfy the condition of existing of nontrivial solutions for matrix of coefficients f_{0n} and g_{0n} at the terms proportional to r^{-2} in the equations (21a-b). The resulting equation is

$$(Z^2\alpha^2 - \ell(\ell+1) - \gamma)(Z^2\alpha^2 - (\ell+1)(\ell+2) - \gamma)$$
$$\times [(Z^2\alpha^2 - \ell(\ell+1) - \gamma)(Z^2\alpha^2 - (\ell+1)(\ell+2) - \gamma) - Z^2\alpha^2] = 0. \quad (28)$$

This equation has the following four roots

$$\gamma_{1,2} = Z^2\alpha^2 - (\ell+1)^2 \pm \sqrt{(\ell+1)^2 + Z^2\alpha^2}, \quad (29)$$
$$\gamma_3 + Z^2\alpha^2 - \ell(\ell+1), \quad \gamma_4 = Z^2\alpha^2 - (\ell+1)(\ell+2). \quad (30)$$

The solutions (23) are the non-divergent functions at $r \to \infty$, when the confluent hypergeometric function is a polynomial function. This condition means

$$\mu - \frac{EZ\alpha}{hc\kappa} = -n,$$

where n is a non-negative integer. As a result we get the following equation for electron energy spectrum

$$E_{n\ell}^i = \frac{m_0 c^2 (\mu_i + n)}{\sqrt{(\mu_i + n)^2 + Z^2 \alpha^2}}, \tag{31}$$

where the subscript i enumerates the roots (25a-b) of the equation (24). The matrix of coefficients f_{0n} and g_{0n} for different roots γ_i is shown in Table 1.

Table 1. *Matrix of coefficients f_{0n} and g_{0n}.*

	γ_1	γ_2		γ_3	γ_4
$f_{0,-1}$	$\sqrt{\frac{\ell+m+1}{2\ell+1}}$	$\frac{Z\alpha}{\ell+1+\Gamma}$	$\sqrt{\frac{\ell+m+1}{2\ell+1}}$	$-\sqrt{\frac{\ell-m}{(2\ell+1)}}$	0
$f_{0,+1}$	$\sqrt{\frac{\ell-m}{2\ell+1}}$	$\frac{Z\alpha}{\ell+1\Gamma}$	$\sqrt{\frac{\ell-m+1}{2\ell+3}}$	$\sqrt{\frac{\ell+m+1}{2\ell+1}}$	0
$g_{0,-1}$	$\frac{Z\alpha}{\ell+1+\Gamma}$	$\sqrt{\frac{\ell+m+2}{2\ell+3}}$	$-\sqrt{\frac{\ell+m+2}{2\ell+3}}$	0	$\sqrt{\frac{\ell+m+2}{2\ell+3}}$
$g_{0,+1}$	$-\frac{Z\alpha}{\ell+1+\Gamma}$	$\sqrt{\frac{\ell+m+2}{2\ell+3}}$	$\sqrt{\frac{\ell+m+2}{2\ell+3}}$	0	$\sqrt{\frac{\ell-m+1}{2\ell+3}}$

where $\Gamma = \sqrt{(\ell+1)^2 + Z^2\alpha^2}$. It is seen from the Table 1 that the roots $\gamma_{1,3}$ correspond to the negatively charged particle. Indeed for both roots $\int |\varphi_i(\vec{r})|^2 dV > \int |\chi_i(\vec{r})|^2 dV$, because χ_i is proportional to the fine structure constant α and $\chi_3 = 0$. On the contrary, the roots $\gamma_{2,4}$ correspond to positively charged particle. However by using the substitution and we get from these solutions the solutions for negatively charged particle, for which $\ell_1 = \ell_2 = \ell+1$ and $\ell_3 = \ell_4 = \ell$ (cf. (20a) and (20b)). As it follows from the equation (17) for free electron solution we have always $\varphi \neq 0$ and $\chi = 0$. Therefore for free particles the solutions with $\ell_1 = \ell_2 = \ell$ or $\ell_3 = \ell_4 = \ell \pm 1$ are degenerated.

Table 1 shows that there is a principle difference between the solutions associated with the roots $\gamma_{1,2}$ and $\gamma_{3,4}$. In the first case the wave function includes both spherical harmonics of the order of ℓ and $\ell' = \ell \pm 1$. It means that in this case the orbital momentum is not an integral of motion. In this case the only integral of motion is the total angular momentum $\vec{j} = \vec{\ell} + \vec{s}$, where $\vec{\ell} = -ih\vec{r} \times \vec{\nabla}$ and $\vec{s} = h\vec{\sigma}/2$. In the second case the wave function includes only spherical harmonics of the order 1. It means that in this case the orbital momentum conserves and the wave functions are the eigenfunctions of the Hamiltonian and orbital momentum operator at the same time.

It is seen that the roots $\gamma_{3,4}$ correspond to the degenerated case, when $\varphi \neq 0$ and $chi = 0$ or $\varphi = 0$ and $\chi \neq 0$, respectively. It will be recalled that the spinors φ and χ are determined by (5). Indeed it can be easily shown that the coefficients A_i satisfy the following relation

$$A_1 A_4 = A_2 A_3. \tag{32}$$

This is the condition of existence of nontrivial solutions for f_{0n} and g_{0n} separately.

It is seen from the equation (17) that the terms in the right-hand-sides of the equations (21a-b) describe the interaction of the electron spin with the gradient of the intra-atomic potential $\partial U / \partial r$. This interaction disappears when the right-hand-side of the equations became zero, i.e. $A_{pq} g_{0q} = 0$ or $B_{pq} f_{0q} = 0$. We can see that the equation (27) is exactly the condition of the existence of the non-trivial solutions of these equations. Thus the roots $\gamma_{3,4}$ correspond to the case when the electron spin does not interact with the gradient of the intra-atomic potential. In this case the equations (21a-b) take the form of the Klein-Gordon equation, which describes the spinless particle. It is this feature that explains the conservation of the orbital momentum of electron in these states.

As we have mentioned above, in order to get from γ_4 the solution, corresponding to the negatively charged particle we should use the substitution $\varphi \to \chi$ and $\chi \to \varphi$. It means that we should assume $\ell_1 = \ell_2 = \ell + 1$ and $\ell_3 = \ell_4 = \ell$. Therefore we can see that the solutions corresponding γ_3 at $\ell_1 = \ell_2 = \ell$ and electron solution, that can be obtained from solution corresponding to the root γ_4 at $\ell_3 = \ell_4 = \ell \pm 1$ with the help of the above mentioned transformation, have the same energy. It means that these two electron solutions are degenerated.

3.2   Comparison with the Dirac theory

Let us compare the obtained hydrogen atom spectrum with that described by the well known Sommerfeld formula that follows from the Dirac theory (see for example [9])

$$E_{n\ell} = \frac{m_0 c^2 (\sqrt{k^2 - Z^2 \alpha^2} + n_r)}{\sqrt{(\sqrt{k^2 - Z^2 \alpha^2} + n_r)^2 + Z^2 \alpha^2}}, \tag{33}$$

where n_r is the radial quantum number and $k = \ell, -(\ell + 1)$. The asymptotic of equation (28) at $Za \ll 1$ has the form

$$\frac{E_D}{mc^2} = 1 - \frac{Z^2 \alpha^2}{2(|k| + n_r)^2} - \frac{|k| + 4n_r}{8|k|(|k| + n_r)^4}(Z\alpha)^4, \tag{34}$$

where $|k| + n_r$ should be a positive integer [9].

On the other hand, from the equation (26) at $Z\alpha \ll 1$, we get

$$\frac{E_{n\ell}^{(1)}}{mc^2} = 1 - \frac{1}{2(k+n)^2}(Z\alpha)^2 - \frac{2k^2 + 4n + k(7+8n)}{8k(2k-1)(k+n)^4}(Z\alpha)^4, \quad (35)$$

$$\frac{E_{n\ell}^{(2)}}{mc^2} = 1 - \frac{1}{2(k+n+1)}(Z\alpha)^2 - \frac{k + 2k^2 + 8kn - 4(1+n)}{8k(2k+1)(k+n+1)^4}(Z\alpha)^4, \quad (36)$$

$$\frac{E_{n\ell}^{(3)}}{mc^2} = 1 - \frac{1}{2(\ell+n+1)^2}(Z\alpha)^2 - \frac{5 + 2\ell = 8n}{8(2\ell+1)(\ell+n+1)^4}(Z\alpha)^4. \quad (37)$$

In (30a-b), parameter k is equal to $k = \ell + 1$. Notice that the electron solutions "3" and "4" are degenerated, hence the asymptotic of solution "4" follows from (30c) after substitution $\ell \to \ell + 1$.

It should be noted here that neither Sommerfeld equation (28) or equation (26) could really describe the energy level spectra of hydrogen atom because the spin of hydrogen atom nucleus is not equal to zero. But comparing (29) and (30a-c) we can see that the Sommerfeld equation (28) and equation (26) generate the same equation for the energy level spectra of hydrogen-like ions to within $(Z\alpha)^2$. The difference between these two spectra is of the order $(Z\alpha)^2$. It means the difference could be observed only in the hyperfine structure of hydrogen-like ions.

3.3 Electron in uniform magnetic field

Let electron be in the uniform magnetic field of the strength $\vec{H} = \vec{r}_r H_0$. In this case the vector potential of the field is $\vec{A} = \vec{e}_r H_0 \rho / 2$ and in steady-state case the equation (2) became

$$\left[\Delta + \frac{E^2 - m_0^2 c^4}{h^2 c^2} - \frac{iqH_0}{hc}\frac{\partial}{\partial\varphi} - \left(\frac{qH_0}{2hc}\right)^2 \rho^2 \right] \Psi(\vec{r}) = -\frac{qH_0}{hc}\sigma_s \Psi(\vec{r}). \quad (38)$$

We can see that the Hamiltonian of the equation (31) commutes with the orbital momentum projection operator

$$L_z = -ih[\vec{r} \times \vec{\nabla}]_z = -ih\frac{\partial}{\partial\varphi}$$

and with spin projection operator

$$S_z = \frac{h}{2}\Sigma_z.$$

Hence, the Hamiltonian commutes with the projection operator J_z of the total angular momentum

$$\vec{J} = -ih\vec{r} \times \vec{\nabla} + \frac{h}{2}\vec{\Sigma}.$$

We can also see that the only spin operator of the equation (31) is the diagonal spin operator $\hat{\Sigma}$, therefore the equation (31) is degenerated with respect to the spinors φ and χ, constituting the bispinor wave function (5). Therefore we can assume without loss of generality, that $\chi = 0$, and seek the solution in the form

$$\varphi(\vec{r}) = Cf(\rho)\exp(im\varphi + ik_z z)u_\sigma,$$

where spinor u_σ is an eigenfunction of the operator σ_z:

$$\sigma_z u_\sigma = \sigma u_\sigma, \tag{39}$$

and

$$u_{+1} = \begin{pmatrix} 1 \\ 0 \end{pmatrix}, \qquad u_{-1} = \begin{pmatrix} 0 \\ 1 \end{pmatrix}.$$

For the radial wave function $f(\rho)$ we get

$$\frac{d^2 f}{d\rho^2} + \frac{1}{\rho}\frac{df}{d\rho} - \frac{m^2}{\rho^2}f + \beta_{m\sigma}f - \left(\frac{qH_0}{2hc}\right)^2 \rho^2 f = 0, \tag{40}$$

where

$$\beta_{m\sigma} = \frac{1}{4\kappa}\left(\frac{E^2 - m_0^2 c^4}{h^2 c^2} - k_z^2 - \frac{|q|H_0}{hc}(m + \sigma)\right) = \frac{h}{2m_0\omega_H}\left(\frac{E^2 - m_0^2 c^4}{h^2 c^2} - k_z^2\right) - \frac{m + \sigma}{2},$$

and

$$\omega_H = \frac{|q|H_0}{m_0 c}.$$

The solution of the equation (33) is

$$f_{m\sigma} = C(\kappa\rho^2)^{m/2}\exp(-\kappa\rho^2/2)U\left(\frac{1 + m}{2} - \beta_{m\sigma}, 1 + m, \kappa\rho^2\right), \tag{41}$$

where

$$\kappa = \frac{|q|H_0}{2hc} = \frac{m_0\omega_H}{2h}.$$

As in the case of Coulomb field, to get a solution non-divergent at $r \to \infty$, we should assume that the confluent hypergeometric function $U(p, q, z)$ is a polynomial function. It means that

$$\frac{1 + m}{2} - \beta_{m\sigma} = -n, \tag{42}$$

where n is a non-negative integer. The condition (35) determines the energy-level diagram

$$E_{nm\sigma}^2 = m_0^2 c^4 + h^2 c^2 k_z^2 + 2m_0 c^2 h\omega_H \left(n + m + \frac{1+\sigma}{2}\right). \qquad (43)$$

The normalized wave functions take the form

$$\varphi_{nm\sigma}(\vec{r}) = \sqrt{\frac{\kappa}{\pi L n!(n+m)!}} (\kappa\rho^2)^{m/2} \exp(-\kappa\rho^2/2) U(-n, 1+m, \kappa\rho^2) \exp(im\varphi + ik_z z) u_\sigma, \qquad (44)$$

where L is the length of the region available for electron motion along the direction of applied magnetic field. The confluent hypergeometric function $U(-n, 1+m, z)$ is equal to zero if $m < -n$. Therefore at a given value of quantum number n the quantum number m lays in the interval $-n \leq m \leq n$. If we introduce the projection of the total angular momentum $\vec{J}$

$$J_z \Psi_M = hM\Psi_M,$$

the equation (36) becomes

$$E_{nM} = \sqrt{m_j^2 c^4 + h^2 c^2 k_z^2 + 2m_0 c^2 \omega_H (n + M + 1/2)}. \qquad (45)$$

It is seen that the energy-level diagram of electron in magnetic field is degenerated with respect of the quantum numbers n and M. The levels with the same value $n + M$ have the same energy. If the strength of the magnetic field obeys the condition $h\omega_H \ll m_0 c^2$, we get

$$\Delta E_{nM} = E_{nM} - m_0 c^2 \approx h\omega_H \left(n + M + \frac{1}{2}\right) + \frac{h^2 k_z^2}{2m_0}.$$

We can see that the levels with the highest negative projection of the total angular momentum $M = -n - 1/2$ have the same energy $\Delta E_{n,-n-1/2} = 0$ at $k_z = 0$.

The current density that is due to the motion of electron in the external magnetic field is determined by equation (4a). By substituting the wave function (37) into the equation (4a) we get

$$\vec{j}_{nm\sigma}(\vec{r}) = \frac{qh}{m_0} \left[\left(\frac{m}{\rho} + \kappa\rho\right) f_{nm\sigma}^2(\rho) - \frac{\sigma}{2} \frac{df_{nm\sigma}^2(\rho)}{d\rho} \right] \vec{e}_\rho + \frac{qhk_z}{m_0} f_{nm\sigma}^2(\rho) \vec{e}_z.$$

If the length of the region for electron motion in the direction of the applied magnetic field is confined then the last term in this equation disappears. In this case the current density is

$$\vec{j}_{nM} = \vec{e}_\rho 4(\kappa\rho^2)^M \exp(-\kappa\rho^2) L_n^{(M+1/2)}(\kappa\rho^2) L_n^{(M-1/2)}(\kappa\rho^2) \frac{n!}{(n+M-1/2)!}. \qquad (46)$$

In the last equation we have substituted the confluent hypergeometric function $U(-n, 1+m, z)$ by the generalized Laguerre polynomial $L_n^{(m)}(z)$.

Thus we can see that the current density depends on the quantum numbers n and M only. We can also see that the induced current is equal to zero for electrons with the highest negative projection of the total angular momentum $M = -n - 1/2$. This is the most important feature of the equation (38), because it means that the motion of these electrons do not result in appearance of any magnetic field of response. This explains why these electrons have the zero energy $\Delta E_{n,-n-1/2} = 0$ in the external magnetic field.

We can easily calculate the strength of the magnetic field of response $\vec{H}$. In accordance with the Maxwell equation

$$\vec{\nabla} \times \vec{H} = \frac{4\pi}{c}\vec{j}.$$

By substituting the equation (38), we get

$$\vec{H} = \vec{e}_z H_0 \frac{2q^2 I_{nM}/L}{m_0 c^2}, \tag{47}$$

where

$$I_{nM} = \frac{n!}{(n + M - 1/2)!} \int_{\infty}^{\kappa \rho^2} L_n^{(M+1/2)}(x) L_n^{(M-1/2)}(x) \exp(-x) x^{M-1/2} dx. \tag{48}$$

The equation (39) has a very clear meaning. It is well known that the capacity of a cylindrical capacitor is proportional to its length. By taking this into account it is easily to understand that the equation

$$W_E = q^2/C_{nM} = q^2 2 I_{nM}/L \tag{49}$$

describes the energy of the electrostatic field produced by electron which is at the level with the quantum numbers n and M. Thus we can see that the strength of the magnetic field induced by the motion of electron in the external magnetic field H_0 is equal to the strength of this field multiplied by the ratio of energy of electrostatic field (41) to the electron rest energy.

3.4   Electron in Coulomb and uniform magnetic fields

Let electron be in a superposition of Coulomb and uniform magnetic fields. We assume that the strength of the magnetic field $\vec{H}_0$ is weak enough to omit the terms proportional to the square of the magnetic field strength in the Hamiltonian. In this case the equation (2) becomes

$$\left[\frac{d^2}{dr^2} + \frac{2}{r}\frac{d}{dr} + \frac{Z^2\alpha^2 - \ell(\ell+1)}{r^2} + \frac{2EZ\alpha}{hcr} - \kappa_M^2 + \frac{qH_0}{2hc}p \right] f_{\ell,M+p/2}$$

$$= -\frac{iZ\alpha}{r^2} \sum_{q=\pm 1} A_{pq} g_{\ell+1,M+q/2}, \qquad (50)$$

$$\left[\frac{d^2}{dr^2} + \frac{2}{r}\frac{d}{dr} + \frac{Z^2\alpha^2 - (\ell+1)(\ell+2)}{r^2} + \frac{2EZ\alpha}{hcr} - \kappa_M^2 + \frac{qH_0}{2hc}p \right] g_{\ell,M+p/2}$$

$$= -\frac{iZ\alpha}{r^2} \sum_{q=\pm 1} B_{pq} f_{\ell+1,M+q/2}, \qquad (51)$$

where

$$\kappa_M^2 = \frac{m^2c^4 - E^2}{h^2c^2} - \frac{qH_0}{hc}\left(m + \frac{1}{2}\right).$$

To solve the equations (42a-b) we can again use the function (22). The solution of the equations (21a-b) we have found by varying the coefficient c of the function (22). To satisfy the equations (42a-b) we should vary both coefficients c and a of this function. In result the coefficients γ_i are again determined by the equations (25a-b) and the coefficients a_i are

$$a_i = \kappa_M^2 - \frac{qH_0}{2hc}\nu_i = \kappa_M^2 - \frac{qH_0}{2hc} \frac{f_{0,-1}^{(i)} - f_{0,+1}^{(i)} + g_{0,-1}^{(i)} - g_{0,+1}^{(i)}}{f_{0,-1}^{(i)} + f_{0,+1}^{(i)} + g_{0,-1}^{(i)} + g_{0,+1}^{(i)}}.$$

The energy-level diagram then is

$$E_{n\ell m}^{(i)} = \frac{m_0 c^2 (\mu_i + n)}{\sqrt{(\mu_i + n)^2 + Z^2\alpha^2}} \sqrt{1 + \frac{h\omega_H}{m_0 c^2} m_i}, \qquad (52)$$

where

$$\omega_H = \frac{|e|H_0}{m_0 c}, \qquad m_i = m + \frac{1 + \nu_i}{2}.$$

In accordance with the equation (14) the magnetic moment of electron is determined by

$$\vec{m} = -\frac{qh}{2m_0 c} \int \bar{\Psi}(\vec{r}) \vec{\Sigma} \Psi(\vec{r}) dV.$$

The magnetic moment is different for different sublevels of hyperfine structure. From the experimental verification point of view the most interesting are the equations for magnetic moment in s-states, because for these states the energy structure is not overcomplicated by the dependency on the angular moment projection. For the states describing by the roots $\gamma_{1,3}$ we have

$$\vec{m}_1 = \vec{e}_z \frac{|e|\hbar}{2m_0 c} \frac{2m+1}{2\ell+1} \frac{(n+\mu_1)^2[(\ell+1+\Gamma)^2 + Z^2\alpha^2(2\ell+1)/(2\ell+3)]}{[(n+\mu_1)^2 + Z^2\alpha^2][(\ell+1+\Gamma)^2 - Z^2\alpha^2]}, \tag{53}$$

$$\vec{m}_3 = -\vec{e}_z \frac{|e|\hbar}{2m_0 c} \frac{2m+1}{2\ell+1} \frac{m_0 c^2}{E_{n\ell m}^{(3)}} \frac{(n+\mu_3)^2}{(n+\mu_3)^2 + Z^2\alpha^2}. \tag{54}$$

As it could be anticipated, only z projections of magnetic moment are not equal to zero.

The obtained equations enable us to understand the origin of the hyperfine splitting. Indeed for s-states the equations (44a-b) become

$$\vec{m}_1 \approx \vec{e}_z \mu_B, \qquad \vec{m}_3 \approx -\vec{e}_z \mu_B,$$

where μ_B is Bohr magneton. It is seen that the root γ_1 corresponds to the positive projection of electron spin and γ_3 to the negative one. Thus we can see that the interaction of electron spin with the electrostatic field of nucleus results in the splitting of states with the different projection of spin.

3.5   Comparison with experimental data

As we have mentioned above, the Sommerfeld equation (28) and equation (26) produce the different values for energy of hydrogen atom levels. The maximum difference is $7.310^{-4}eV$ for $1s$ state, $9.0610^{-5}eV$ for $2s$ state, and $1.510^{-5}eV$ for $2p$ state. We can see that the difference is small, $\Delta E/E = 10^{-4} - 10^{-5}$. The experimental data for energy of 1s and 2s states of hydrogen atom (see for example [11]) are $E(1s) = 13.5984eV$, $E(2s) = 3.39959eV$, and $E(2p3/2) = 3.39955eV$. The Sommerfeld equation gives for these levels data that are closer to the above mentioned than the equation (26). However the spin of nucleus in hydrogen atom is not equal to zero therefore the electron in hydrogen atom moves in non-zero magnetic field. The strength of the magnetic field produced by the hydrogen nucleus can be estimated as $H_0 \approx 2g|e|\hbar/M_n cr^3$, where M_n and g are the proton mass and g-factor, respectively. Thus, at a distance $r = a_B$, where a_B is the Bohr radius, the strength of the magnetic field is $0.5kOe$. Let us take into account the magnetic field of nucleus and will make the calculations with the help of the equation (43). It is seen that to compensate the above mentioned difference we need in the magnetic field with the strength of a few hundreds of Oe for $2p$ state, a few tens of kOe for $2s$ state, and a few hundreds of kOe for $1s$ state. It seems that the last figure exceeds significantly the estimation for strength of magnetic field in hydrogen atom. However the wave function of 1s state has maximum at $r = 0$, while the strength of the magnetic field decreases with distance as $1/r^3$. If we take into account that the equation (43) was obtained for uniform magnetic field we can see that the coincidence is quit good. The probability of $2s$ electron to be at the point $r = 0$ is twice smaller than that for $1s$ electron, and the wave function of 2p electron is equal to zero at $r = 0$. Thus we can see that the required magnitudes of the magnetic field strength are in good agreement with experimental data for all three levels.

4.   Discussion

Let us turn now to the set of coupled equations (2)-(4) describing the interaction of electron with the electromagnetic field. It is natural to suppose that the term of current density four-vector (4) including the spin operators is responsible for the magnetic properties of a particle. Indeed this term depends on the spatial or temporal derivatives of bilinear combinations of wave functions. Thus in steady-state case this term will not contribute into the electric charge and current, because this term integration over the whole space results in zero. Therefore the rest part of current density four-vector

$$\rho_e(\vec{r}, t) = \frac{q}{m_0 c} \left[-\frac{ih}{2c} \left(\frac{\partial \bar{\Psi}}{\partial t} \Psi - \bar{\psi} \frac{\partial \Psi}{\partial t} \right) - \frac{q}{c} \bar{\Psi} \varphi \Psi \right],$$

$$\vec{j}_e(\vec{r}, t) = \frac{q}{m_0} \left[\frac{ih}{2} (\vec{\nabla} \bar{\Psi} \cdot \Psi - \bar{\Psi} \cdot \vec{\nabla} \Psi) - \frac{q}{c} \bar{\Psi} \vec{A} \Psi \right]. \tag{55}$$

is responsible for the electric properties of particle. It can be easily shown from the equation (2) and equation for Dirac conjugated wave function that the variables $\rho_e(\vec{r}, t)$ and $\vec{j}_e(\vec{r}, t)$ obey the continuity equation

$$\frac{\partial \rho_e}{\partial t} + \vec{\nabla} \cdot \vec{j}_e = 0. \tag{56}$$

This is extremely important because it makes some proof of possibility of separation of the terms (45) from the total current density four-vector.

In accordance with the conventional definition the strength of the electric $\vec{E}$ and magnetic $\vec{B}$ fields are

$$\vec{E} = -\frac{1}{c} \frac{\partial \vec{A}}{\partial t} - \vec{\nabla} \varphi, \qquad \vec{B} = \vec{\nabla} \times \vec{A}. \tag{57}$$

To take these definitions as a matter of course we can rewrite the spatial and temporal components of the four-dimensional equation (3) as follows

$$\vec{\nabla} \times (\vec{B} + \vec{B}_0) = \frac{1}{c} \frac{\partial}{\partial t} (\vec{E} + \vec{E}_0) + \frac{4\pi}{c} \vec{j}_e, \tag{58}$$

$$\vec{\nabla} \cdot (\vec{E} + \vec{E}_0) = 4\pi \rho_e. \tag{59}$$

In Eqs. (48a-b), we have introduced the new variables

$$\vec{E}_0(\vec{r}, t) = -i \frac{2\pi q h}{m_0 c} \bar{\Psi} \vec{\alpha} \Psi, \qquad \vec{B}_0(\vec{r}, t) = -\frac{2\pi q h}{m_0 c} \bar{\Psi} \Sigma \Psi. \tag{60}$$

The complete set of equations for field includes the two additional equations

$$\vec{\nabla} \cdot \vec{B} = 0, \tag{61}$$

$$\vec{\nabla} \times \vec{E} = -\frac{1}{c}\frac{\partial \vec{B}}{\partial t}, \tag{62}$$

which follow from the definition (47).

It is seen from the equations (48a-b) and (49) that the fields $\vec{E}_0$ and $\vec{B}_0$ can be considered as the self-fields of a particle. On the one hand they are in similar positions with $\vec{E}$ and $\vec{B}$ in equations (48a-b). On the other hand they are solely determined by the intrinsic structure of particle. Let us introduce the integral fields

$$\vec{E}' = \vec{E} + \vec{E}_0, \qquad \vec{B}' = \vec{B} + \vec{B}_0, \tag{63}$$

which are the sum of the strength of the external field and self-field. Then the set of equations (48a-b) and (50a-b) become

$$\vec{\nabla} \cdot \vec{E}' = 4\pi\rho_e, \quad \vec{\nabla} \cdot \vec{B}' = 4\pi\rho_m,$$

$$\vec{\nabla} \times \vec{B}' = \frac{1}{c}\frac{\partial \vec{E}'}{\partial t} + \frac{4\pi}{c}\vec{j}_e, \quad \vec{\nabla} \times \vec{E}' = -\frac{1}{c}\frac{\partial \vec{B}'}{\partial t} - \frac{4\pi}{c}\vec{j}_m, \tag{64}$$

where

$$\rho_m(\vec{r}, t) = -\frac{qh}{2m_0 c}\vec{\nabla} \cdot (\bar{\Psi}\vec{\Sigma}\Psi), \quad \vec{j}_m(\vec{r}, t) = \frac{qh}{2m_0}\left[\frac{1}{c}\frac{\partial}{\partial t}(\bar{\Psi}\vec{\Sigma}\Psi) - i\vec{\nabla} \times (\bar{\Psi}\vec{\alpha}\Psi)\right]. \tag{65}$$

Thus we can see that invention of the idea of the particle self-fields enables us to transform the Maxwell set of equations to the symmetric form. In this case the variables ρ_e and $\vec{j}_e$ play the role of the electric charge and current density, and the variables ρ_m and $\vec{j}_m$ play the role of the magnetic charge and current density. The magnetic charge and current density obey the continuity equation as well

$$\frac{\partial \rho_m}{\partial t} + \vec{\nabla} \cdot \vec{j}_m = 0. \tag{66}$$

For electric charge and current density in the classic electrodynamics we have $\rho_e(\vec{r}, t) = -\vec{\nabla} \cdot \vec{P}$ and $\vec{j}_e = \partial \vec{P}/\partial t + \vec{\nabla} \times \vec{M}$, where the polarization polarization $\vec{P}$ and magnetization $\vec{M}$ vectors describe the inner fields of a medium. We can easily see the similarity between these equations and equations (53). The appearance of the polarization and magnetization of medium is due to the action of the external fields. The external fields break the central symmetry of initially disordered medium. If we introduce the magnetic charge and current density then in analogy with classic definitions we should assume $\rho_m(\vec{r}, t) = -\vec{\nabla} \cdot \vec{M}$ and $\vec{j}_m(\vec{r}, t) = \partial \vec{M}/\partial t + \vec{\nabla} \times \vec{P}$. Hence $\vec{M} = -\vec{B}_0/4\pi$ and $\vec{P} = \vec{E}_0/4\pi$.

Therefore the self-fields are simply the polarization and magnetization vectors originated from the interaction of a particle with the external fields and $\vec{E}' = \vec{D}\vec{E} + 4\pi\vec{P}$, $\vec{B}' = \vec{H} = \vec{B} - 4\pi\vec{M}$. Indeed the magnetic charge density is equal to zero for free particle, because the absolute square of the wave function in this case does not depend on coordinates. However when particle is in external field the spatial symmetry of its wave function is changed. As a result the non-zero magnetic charge density could appear. Notice that if particle is only in the magnetic field then the equations for two spinors constituent of the bispinor wave function become independent. It means that one of the spinors in the wave function (5) is equal to zero. In this case the polarization vector is equal to zero. However it is not equal zero when particle interacts with the electric field and should be described by the four-component wave function. It shows again that the polarization describes the electric properties of particle. The equations (53) demonstrate the fundamental difference between the electric and magnetic charges. The magnetic charge is equal to zero, because $\rho_m(\vec{r}, t)$ is integrated over the whole space is always equal to zero.

The variation principle of quantum mechanics tells us that the existence of the eigenstates for spinless particle is due to the availability of extremums for integral over atomic energy density. It means that the electron charge density has an optimal distribution when the kinetic energy and energy of electrostatic field of atom are minimal. From this point of view for particle with non-zero spin spin should simultaneously hold the conditions of the optimal distribution both electric and magnetic charge density. The vectorial nature of spin interaction with the external fields explains the two fold splitting of s states. Indeed for s states the radius vector is only vectorial characteristic of particle in Coulomb field. Therefore, the two $(\vec{r}\vec{s}$ sublevels should appear. The orbital momentum is another vectorial characteristic of the particle in Coulomb field. As a result for $\ell \neq 0$ each of $\vec{r}\vec{s}$ sublevels should be in its turn splitted in two $(\vec{\ell}\vec{s})$ sublevels, but in Coulomb field one of the $(\vec{r}\vec{s})$ sublevels is an eigenstate of orbital momentum at well. As a result we have triplet hyperfine structure for states with $\ell \neq 0$.

5. Conclusion

The analysis given above has shown that there is a fundamental difference between the Dirac equation and the second order differential equation (2). The equation (2) predicts the existence of hypefine structure even for electron that is in the Coulomb field of nucleus with zero magnetic moment. The origin of hyperfine splitting is in the interaction of electron spin with the electrostatic field of nucleus. As a result the calculated hydrogen spectrum (26) does not coincide with the well-known Sommerfeld equation (28) that follows from the Dirac theory. The expansion of equation (26) and Sommerfeld equation in series of $Z\alpha$ shows that the terms proportional to $(Z\alpha)^2$ are absolutely the same in both expansions and only the terms of expansions proportional to $(Z\alpha)^4$ are different. Thus for the low-charged hydrogen-like ions the predicted hyperfine splitting is of the order of α^4. However, the difference appears for highly charged ions. Indeed the energies of the two lowest states of hydrogenic ions become complex for $Z > 56$ and $Z > 69$, respectively. Therefore the doublet hyperfine structure of the lowest state should disappear for hydrogenic ions with $56 < Z < 68$ and the state with the quantum numbers $n = 0$ and $\ell = 0$ becomes unstable for ions with $Z > 69$. This specific feature of spectrum of hydrogenic ions could be verified experimentally if only the effects of the nucleus finite size will not start to play the decisive role for such highly ionized ions.

In contrast to Dirac equation, the equation (2) is absolutely symmetric with respect to particle and its antiparticle. This symmetry reflects the equivalence of the physical properties of particle and its antiparticle. By using this symmetry we can easily transform the positron solution into the electron solution with the help of substitution $\varphi \to \chi$ and $\chi \to \varphi$ in the wave function (5).

The predicted hyperfine structure, which is doublet for $\ell = 0$ and triplet for $\ell > 0$, is in agreement with hydrogen spectrum. However the non-zero value of hydrogen nucleus spin does not permit us to make a precise comparison without numerical solution of the problem allowing for the non-uniformity of the intra-atomic magnetic field. The decisive role for the verification of the proposed theory could play the experiments on the study of hyperfine structure of ions $_4He^{+1}$, $_{12}C^{+5}$, $_{16}O^{+7}$ with zero magnetic moment of nucleus. Unfortunately the absence of the reliable experimental data on the hyperfine structure of the above mentioned ions makes it impossible now to provide a precise comparison of theoretical and experimental spectra.

This work was partially supported by Russian Foundation for Basic Research (grant 02-02-17138) and program "Universities of Russia" (grant UR.01.03.001).

References

[1] W.E.Lamb and R.C.Retherford, Phys.Rev. 72, 241 (1957)

[2] W. Heitler, The quantum Theory of Radiation, Dover, New York, 1984

[3] H.A.Bethe, Phys.Rev. 72, 339 (1947)

[4] J.Schwinger, Phys.Rev. 75, 1912 (1949)

[5] M.D.Crisp and E.T.Jaynes, Phys.Rev. 179, 1253 (1969); C.R.Stroud, Jr. and E.T.Jaynes, Phys.Rev. A 1, 106 (1970)

[6] S.B.Lai, P.L.Knight, and J.H.Eberly, Phys.Rev.Lett. 32, 494 (1974); J.R.Ackerhalt and J.H.Eberly, Phys.Rev. D 10, 3350 (1974)

[7] A.O.Barut and J.F.Van Huele, Phys.Rev. A 32, 3187 (1985); A.O.Barut, J.P. Dowling and J.F. van Huele, Phys.Rev. A38, 4405 (1988)

[8] A.V.Andreev, PRL (submitted)

[9] V.B. Berestetskii, E.M. Lifshitz, and L.P. Pitaevskii, Quantum Electrodynamics, Pergamon Press, Oxford, 1982

[10] M.Abramowitz and I.A.Stegun "Handbook of Mathematical Functions", NBS (1964)

[11] NIST Atomic Spectra Database (http://physics.nist.gov/cgi-bin/AtData)

INTERACTION OF MESOSCOPIC DEVICES WITH NON-CLASSICAL ELECTROMAGNETIC FIELDS

A. Vourdas

Department of Computing,
University of Bradford,
Bradford BD7 1DP, United Kingdom

Abstract   AC Aharonov-Bohm phenomena in which electric charges interfere in the presence of non-classical microwaves, are studied. The relative phase factor between the two electron beams is a quantum mechanical operator, whose expectation value with respect to the density matrix ρ describing the microwaves, determines the interference. It is shown that the quantum noise of the microwaves destroys slightly the interference. The results are interpreted physically in terms of multiphoton exchange between the electrons and the microwaves. A similar effect is also studied in the context of Josephson devices interacting with non-classical microwaves, in the external field approximation. Dual phenomena with vortex condensates in Josephson array insulators, are also considered. ac Aharonov-Casher phenomena in which vortices interfere in the presence of non-classical microwaves, are studied. Here, the relative 'dual phase' factor between the two vortex beams is a quantum mechanical operator. Dual Josephson junctions for vortices, made from two insulators separated by a weak link through which the vortices tunnel, are also studied.

1.   Introduction

Mesoscopic devices with typical size $0.1\mu m$ operating at a typical temperature of $100mK$, exhibit interesting quantum phenomena [1]. Most of the experimental work involves the interaction of these devices with classical microwaves. In these lectures we review some of our recent work where we have studied the interaction of various mesoscopic devices with non-classical microwaves. In this case we have a fully quantum system, and we study how the quantum (and classical) noise of the microwaves affects the interference in the mesoscopic devices.

In section 2 we study the so-called ac Aharonov-Bohm phenomenon. The Aharonov-Bohm phenomenon [2] studies the interference of electric charges that follow two different paths in the presence of a magnetostatic flux threading

A.S. Shumovsky and V.I. Rupasov (eds.), Quantum Communication and Information Technologies, 195–210.

the surface between the two paths. Here we study the same phenomenon in the presence of ac electromagnetic fields [3]. An important aspect of our work is that we consider both classical and non-classical electromagnetic fields and we compare and contrast the results for these two cases. In the quantum case the relative phase phase factor between the two electron beams is a quantum mechanical operator, whose expectation value with respect to the density matrix density matrix ρ describing the microwaves, determines the interference. The results depend on the density matrix density matrix and are interpreted physically in terms of multiphoton exchange between the electrons and the microwaves. We show that the quantum noise of the non-classical electromagnetic fields causes a small destruction of the electron interference.

In section 3 we consider the interaction of mesoscopic Josephson junctions with non-classical microwaves. There is a lot of work in the literature [4] on mesoscopic Josephson junctions and the novel aspect here is that the microwaves are non-classical [5,6]. In section 3.1 we consider the external field approximation, in which only the external field is considered and the back-reaction (i.e., the electromagnetic field created by the the electron pairs in the superconductor) is neglected. Due to the non-classical nature of the microwaves, the phase difference appearing in the Josephson equations is a quantum mechanical operator, whose expectation value with respect to the density matrix density matrix ρ of the microwaves, determines the current. We consider various examples of non-classical microwaves (coherent states, squeezed states, etc) coherent state squeezed state and show that the quantum noise of the microwaves reduces the amplitude of the Shapiro steps. In section 3.2 we indicate briefly how we can go beyond the external field approximation if we consider explicitly the circuit that produces the microwaves[6].

In section 4 we consider dual phenomena[7] to the above based on vortex condensates in Josephson array insulators. In this context Aharonov-Casher phenomena where vortices which encircle an electric charge interfere have been studied in the literature [8,9]. We consider ac Aharonov-Casher phenomena[9,10] in which vortices interfere in the presence of classical and non-classical microwaves [10]. In the case of non-classical microwaves, the relative 'dual phase' factor between the two vortex beams is a quantum mechanical operator. We show that in this context also, the quantum noise of the microwaves destroys slightly the vortex interference.

Condensates lead to Josephson phenomena. For vortex condensates, we consider dual Josephson junctions[11], made from two insulators separated by a weak link through which the vortices tunnel. We study the 'dual Josephson equations' in this context and solve them for certain examples of electromagnetic fields.

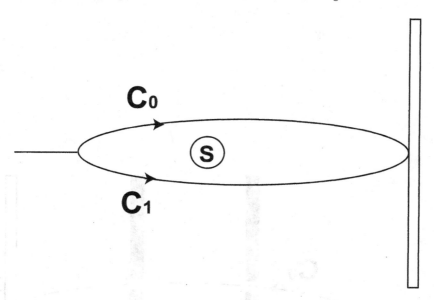

Figure 1.   Interference of electrons that follow the paths C_0 and C_1. The solenoid S is threading the surface between the two paths. In the dc Aharonov-Bohm phenomenon, a dc current in the solenoid S creates a magnetostatic flux perpendicular to the plane of the diagram. In the ac Aharonov-Bohm phenomenon, an ac current in the solenoid S creates an electromagnetic field whose magnetic field is perpendicular to the plane of the diagram and the electric field is parallel to it.

2.   AC Aharonov-Bohm Phenomena

2.1   AC Aharonov-Bohm phenomena with classical electromagnetic fields

The original Aharonov-Bohm phenomenon studies the interference of electric charges that follow two paths C_0 and C_1, when a magnetostatic flux is threading the surface between the two paths (fig 1). We refer to this as the dc Aharonov-Bohm phenomenon. In this case the electric charges feel only the vector potential but not the magnetostatic field, and the phenomenon demonstrates clearly the physical reality of the vector potential.

Here we study the ac Aharonov-Bohm phenomenon, where the magnetostatic flux is replaced by a time-dependent magnetic flux (electromagnetic field). Experimentally, this can be realised at low frequencies using a solenoid that threads the surface between the two paths and which has a suitable time-dependent current; and at high frequencies using a waveguide (Fig 2). In the ac Aharonov-Bohm experiment the electric charges do feel the electromagnetic field. The

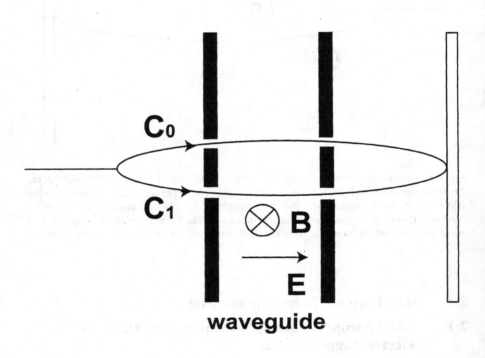

Figure 2.   High frequency ac Aharonov-Bohm phenomenon. The solenoid in fig.1, is replaced by a waveguide.

aim here is not to prove the physical reality of the vector potential, but to study how the electrons interact with the photons.

We consider the magnetic flux $\phi(t) = -Vt + A\sin(\omega t)$. Let ψ_0, ψ_1 be the wavefunctions corresponding to the two paths C_0 and C_1. The intensity at some point on the screen is:

$$I(t) = |\psi_0 + \psi_1 \exp[ie\phi(t)]|^2 = |\psi_0|^2 + |\psi_1|^2 + 2|\psi_0||\psi_1|\Re\{\exp[i(\sigma + e\phi)]\} \tag{1}$$

where $\sigma = arg(\psi_1) - arg(\psi_0)$. We consider the case of equal splitting in which $|\psi_0|^2 = |\psi_1|^2 = \frac{1}{2}$ and get:

$$I(t) = 1 + \Re\{\exp[i(\sigma + e\phi)]\} = 1 + \sum_{N=-\infty}^{\infty} J_N(eA)\cos[(-eV + N\omega)t] \tag{2}$$

where J_N are Bessel functions. We calculate the time-averaged intensity I_{dc} and see easily that $I_{dc} = 1$, with the exception of the values of eV for which $eV = K\omega$ (where K is an integer), in which case we get:

$$I_{dc} = 1 + J_K(eA)\cos(\sigma) \tag{3}$$

Therefore a plot of I_{dc} versus eV will have peaks at the values $eV = K\omega$. We call these peaks Shapiro steps (a terminology taken from Josephson junctions).

2.2 AC Aharonov-Bohm phenomena with non-classical microwaves

We consider now the case of microwaves at low temperatures ($k_B T \ll \hbar\omega$) which have been carefully prepared in a particular quantum state. A good source (e.g., maser) and very low dissipation are prerequisites for such experiments.

In quantized electromagnetic fields the vector potential and the electric field are dual quantum variables. Assuming that the area between the paths C_0 and C_1 is small in comparison to the square of the wavelength of the microwaves, we can integrate the vector potential and electric field and treat the ϕ, V_{EMF} as dual quantum variables. We can express them in terms of creation and annihilation operators as

$$\phi(t) = 2^{-\frac{1}{2}}\xi[\exp(i\omega t)a^\dagger + \exp(-i\omega t)a]$$
$$V_{EMF}(t) = 2^{-\frac{1}{2}}i\omega\xi[\exp(i\omega t)a^\dagger - \exp(-i\omega t)a] \tag{4}$$

ξ is a constant that depends on the dimensions of the device. We note that these relations are in the external field approximation where the back-reaction from the electrons on the electromagnetic field is neglected. The phase factor is the displacement operator

$$\exp(ie\phi) = D\left[i2^{-\frac{1}{2}}e\xi\exp(i\omega t)\right]; \quad D(A) \equiv \exp[Aa^\dagger - A^*a]. \quad (5)$$

The intensity of particles is given by:

$$I(t) = \text{Tr}\left\{\rho|\psi_0 + \psi_1\exp[i\phi(t)]|^2\right\} = 1 + \Re\{e^{i\sigma}\text{Tr}[\rho\exp[i\phi(t)]]\} \quad (6)$$

where ρ is the density matrix density matrix describing the non-classical electromagnetic field.

In addition to the non-classical ac electromagnetic field we also impose the classical flux $-Vt$. In this case we get:

$$I(t) = 1 + \Re\{\exp[i(\sigma - eVt)]W(A)\} \quad (7)$$

$$W(A) = \text{Tr}[\rho D(A)]; \quad A = i2^{-1/2}e\xi\exp(i\omega t) \quad (8)$$

$W(A)$ is known as the Weyl or characteristic function and is related to the Wigner function Wigner function through a two-dimensional Fourier transform [12]. We rewrite Eq(7) as

$$I(t) = 1 + |W(A)|\cos(\sigma - eVt + \lambda(t)); \quad \lambda(t) \equiv \arg[W(A)] \quad (9)$$

It is known that $|W(A)| \leq 1$ and this shows that the visibility is less than 1. We interpret this as partial destruction of the interference due to the quantum fluctuations in the electromagnetic field. In order to see this more clearly we use the following expansion of $|W(A)|$ around the origin :

$$\begin{aligned}|W(A)| &= [1 - 2(\Delta x)^2 A_R^2 - 2(\Delta p)^2 A_I^2 + 4K^2 A_R A_I + ...]^{1/2} \\ &= \{1 - \frac{(e\xi)^2}{2}[(\Delta x)^2 + (\Delta p)^2] + \frac{(e\xi)^2}{2}[(\Delta x)^2 - (\Delta p)^2]\cos(2\omega t) \\ &\quad - (e\xi)^2 K^2\sin(2\omega t) + ...\}^{1/2}\end{aligned} \quad (10)$$

where A_R, A_I are the real and imaginary parts of A (given in Eq(8)). $(\Delta x)^2$, $(\Delta p)^2$ and K^2 are the quantum uncertainties of the electromagnetic field defined as $\Delta x = [\langle x^2\rangle - \langle x\rangle^2]^{1/2}$; $\Delta p = [\langle p^2\rangle - \langle p\rangle^2]^{1/2}$; $K^2 = \frac{1}{2}\langle xp + px\rangle - \langle x\rangle\langle p\rangle$. The operators x and p are here the flux and the $\omega^{-1}V_{EMF}$ correspondingly. Eq(10) shows how the quantum noise of the electromagnetic field reduces $|W(A)|$ below 1 and therefore destroys slightly the interference.

>From a physical point of view, the Weyl function $\text{Tr}[\rho D(A)]$ describes the exchange of photons between the electrons and the external electromagnetic field. Expansion of the exponential gives an infinite sum of terms of the type $\text{Tr}\left[\rho(ae^{-i\omega t})^N(a^\dagger e^{i\omega t})^M\right]$ which describe processes in which the electrons emmit M photons to the external electromagnetic field and at the same time

absorb N photons from the external electromagnetic field. Therefore, in a Fourier series expansion of the Weyl function the $\exp[iK\omega t]$ term describes multiphoton processes where the electrons emmit $N + K$ photons and at the same time absorb N photons (for all N).

As an example, we consider coherent states coherent state with thermal noise. They are described by the density matrix density matrix

$$\rho(\alpha, \beta) = D(\alpha)\rho_{th}(\beta)[D(\alpha)]^\dagger$$
$$\rho_{th}(\beta) = (1 - e^{-\beta\omega}) \sum_N e^{-\beta\omega N}|N\rangle\langle N| \qquad (11)$$

In the zero temperature limit this becomes the coherent state coherent state $|\alpha\rangle$. We evaluate first eq(7) and then the time-averaged intensity:

$$eV = K\omega \rightarrow I_{dc} = 1 + \exp\left[-\frac{1}{2}e^2\xi^2\left(\frac{1}{2} + <N_T>\right)\right] J_K\left(2^{\frac{1}{2}}e\xi|\alpha|\right)$$
$$\times \cos\left(\sigma - K\theta_\alpha + \frac{K\pi}{2}\right)$$
$$eV \neq K\omega \rightarrow I_{dc} = 1 \qquad (12)$$

where $<N_T> = (e^{\beta\omega} - 1)^{-1}$ is the average number of thermal photons. This shows that the peaks are here smaller than in the classical case by a factor

$$f = f_{qua}f_{clas}; \quad f_{qua} = \exp\left[-\frac{1}{4}e^2\xi^2\right]; \quad f_{clas} = \exp\left[-\frac{1}{2}e^2\xi^2 <N_T>\right] \qquad (13)$$

The factor f_{qua} is due to quantum noise and mathematically is related to the non-commutativity of the operators a, $a^\dagger$. The factor f_{clas} is due to classical (thermal) noise. Physically this result shows that we have partial destruction of the interference due to both quantum and classical noise. In the zero temperature limit $f_{clas} = 1$, but we still have destruction of the interference due to quantum noise.

In the case of squeezed vacua we get peaks in the time-averaged intensity I_{dc}, only for even Shapiro steps $eV = 2\omega K$. This result is different from the classical one, in the sense that only the even peaks appear. This is due to the fact that squeezed vacua are superpositions of even number eigenstates and the electrons can only absorb an even number of photons. Similar result can be obtained for even coherent states coherent state $(\mathcal{N}(|A\rangle + |-A\rangle))$ which are also superpositions of even number eigenstates. We recall here that we use the external field approximation. If we go beyond the external field approximation, the back-reaction is expected to modify slightly these results. In other words the back-reaction will create some odd number eigenstates which will create some odd peaks the magnitude of which will depend on the magnitude of the back-reaction. But for negligible back-reaction, we only get the even peaks.

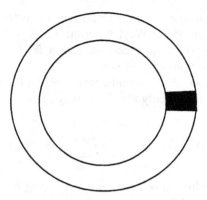

Figure 3.   Superconducting ring with a Josephson junction. The ring is irradiated with non-classical microwaves.

3.   Interaction of Josephson Devices with non-classical Microwaves

In this section we study the interaction of mesoscopic Josephson junctions with non-classical microwaves. In section 3.1 we consider the external field approximation, and study the Josephson equations. Due to the non-classical nature of the microwaves, the phase phase difference is a quantum mechanical operator, whose expectation value with respect to the density matrix density matrix ρ of the microwaves, determines the current. We show that the quantum noise of the microwaves reduces the amplitude of the Shapiro steps, in this context. In section 3.2 we indicate briefly how a Hamiltonian Hamiltonian approach can go beyond the external field approximation if we include explicitly the circuit that produces the microwaves[6].

3.1   The external field approximation

We consider a superconducting ring with a Josephson junction (Fig.3). This is known as SQUID (superconducting quantum interference device). The ring is coupled with an external source of microwaves. In addition to that, a coil creates an extra flux which increases linearly with time ($V_0 t + \phi_0$) and which is also threading the ring. So the total external flux threading the ring is

$$\phi(t) = A\sin(\omega t) + V_0 t + \phi_0 \tag{14}$$

Apart from the external flux, there is also a flux created by the Cooper pairs in the superconductor (back-reaction) and which we neglect in this section.

Let ψ be the wavefunction of the Cooper pair condensate in one end of the Josephson junction. Then the wavefunction at the other end is $\psi \exp[i\theta]$ where the phase difference θ across the junction is $\theta = 2e\phi(t)$. Using this we can prove the Josephson equations:

$$I = I_{cr} \sin\theta; \quad \partial_t\theta = 2eV \tag{15}$$

where I is the current through the junction, I_{cr} is the 'critical current' (the value of which depends on the junction) and V is the voltage across the junction. It is seen that there is a non-linear (sinusoidal) relationship between current and voltage. In the case of the flux of Eq.(12), we insert $\theta = 2e\phi(t)$ in Eq.(13) and get

$$
\begin{aligned}
I &= I_{cr} \sin[2eA\sin(\omega t) + 2eV_0 t + 2e\phi_0] \\
&= I_{cr} \sum_{N=-\infty}^{\infty} J_N(2eA) \sin[(2eV_0 + N\omega)t + 2e\phi_0].
\end{aligned} \tag{16}
$$

We now calculate the time-averaged value I_{dc} of this current and get

$$
\begin{aligned}
2eV_0 = M\omega &\rightarrow I_{dc} = I_{cr} J_{-M}(2eA)\sin(2e\phi_0) \\
2eV_0 \neq M\omega &\rightarrow I_{dc} = 0
\end{aligned} \tag{17}
$$

where M is an integer. It is seen that I_{dc} is non-zero only at integral steps of the potential V_0 (Shapiro steps).

We now consider the case of microwaves which are carefully prepared in a particular quantum state described by a density matrix density matrix ρ. The system operates at low temperatures so that the thermal noise is less than the quantum noise in the microwaves and the device. In this case $\theta = 2e\phi(t)$ is an operator (given in Eq(4)), and consequently the current is the operator

$$\hat{I} = I_{cr} \sin[2^{\frac{1}{2}}e\xi(e^{i\omega t}a^\dagger + e^{-i\omega t}a) + 2eV_0 t + 2e\phi_0] \tag{18}$$

Expectation values of the current are given by $< I > = Tr[\rho\hat{I}]$. We calculate the time-averaged component of $< I >$ (which we denote as $< I >_{dc}$) for various density matrices of the non-classical microwaves; and we compare and contrast the results with the classical result.

For microwaves in a coherent state coherent state we get Shapiro steps as in the classical case. But the quantum dc currents $< I >_{dc}$ are smaller than the classical dc currents I_{dc}:

$$< I >_{dc} = I_{dc} \exp(-e^2\xi^2) \tag{19}$$

For microwaves in a squeezed vacuum we get only even Shapiro steps.

It is seen that there is a similarity between the results derived in this section and the results for the ac Aharonov-Bohm phenomena discussed earlier. This is because in both cases we have electrons circulating ac magnetic fluxes (electromagnetic fields). In the quantum case the quantum noise destroys slightly the interference and we get smaller peaks.

3.2   Hamiltonian approach: Inclusion of the back-reaction

In order to go beyond the external field approximation we need to consider explicitly the quantum circuit that produces the non-classical microwaves. Such a system consists of two coupled oscillators. The first is the circuit that produces the microwaves and it is assumed to have inductance L_{mw} and capacitance C_{mw}. The frequency of this harmonic oscillator is $\omega_{mw} = (L_{mw}C_{mw})^{-1/2}$. The second oscillator is non-linear and it consists of the SQUID ring. It is assumed to have inductance L and capacitance C. The frequency of the linear part of this oscillator is $\omega = (LC)^{-1/2}$.

The Hamiltonian Hamiltonian describing this system contains an inductive term, a capacitive term and a Josephson sinusoidal term. Assuming that at $t = 0$ the system is described by the density matrix density matrix $\rho(0)$ we calculate the density matrix $\rho(t) = \exp[iHt]\rho(0)\exp[-iHt]$ and the reduced density matrices $\rho_1(t) = Tr_2\rho(t)$ and $\rho_2(t) = Tr_1\rho(t)$. Using the reduced density matrices we can calculate the average number of quanta in each mode and the second order correlations:

$$\langle N_i \rangle = Tr[a^\dagger a \rho_i]; \quad g_i^{(2)} = \frac{\langle N_i^2 \rangle - \langle N_i \rangle}{\langle N_i \rangle^2} \tag{20}$$

The $g_i^{(2)}$ describe electron pair and photon bunching bunching or antibunching. antibunching Such calculations show explicitly how the quantum statistics of the photons affects the quantum statistics of the tunneling electron pairs. Several calculations of this type have been presented in refs[6].

4.   Dual Phenomena with Vortex Condensates in Josephson Array Insulating Rings

Duality between electricity and magnetism has been studied in many contexts. Above we have studied several phenomena where electrons circulate around magnetic fluxes. The important quantity was the phase of the electron wavefunction associated with the vector potential. In this section we will study phenomena where vortices (magnetic flux tubes) circulate around electric charges. The important quantity will be the 'dual' phase phase of the vortex wavefunction associated with the dual vector potential. This will be used in the context of vortex interference in ac Aharonov-Casher phenomena in section 4.1.

In section 3 we have considered superconductors, with electron pair condensates. In this section we will consider Josephson array insulators, with vortex condensates. All types of condensates exhibit in principle Josephson phenomena. In this spirit we study in section 4.2 'dual Josephson junctions' for vortices[11].

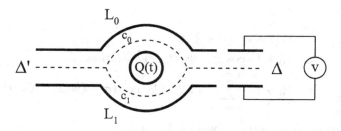

Figure 4.   A Josephson array insulator. Vortices travel ballistically from the point Δ' to the point Δ. The center contains a time-dependent electric charge $Q(t)$.

4.1   Vortex interference in ac Aharonov-Casher phenomena

Following refs[9] we consider the experiment of Fig.4 where a two-dimensional array of superconducting islands coupled through Josephson junctions is enclosed by the superconducting boundaries L_0 and L_1. The Josephson coupling energy E_J is smaller than the Coulomb charging energy E_C, so that the system is in the insulating phase, where the vortices are quantum objects with low mass, moving with a high mobility. The dissipation is assumed to be negligible.

The vortices travel ballistically from the point Δ' to the point Δ. The center contains electric charge $Q(t) = It + Q_0 \sin(\omega t)$. At low frequencies this charge can be generated by an ac source. For higher frequencies, the ac source is a waveguide carrying microwaves.

The vortex condensate is described by a wavefunction $S(x)$. Let $S_0(x)$ and $S_1(x)$ be the wavefunctions corresponding to vortices that follow the paths c_0 and c_1 correspondingly (in the absense of the electric charge Q). The relative 'dual' phase factor between the two wavefunctions is $\exp[i\Phi_0 \int_c a_i dx_i] = \exp[i\Phi_0 Q]$ (where a_i is the 'dual potential').

The total wavefunction at the point Δ is:

$$S(\Delta) = S_0 + \exp[i\Phi_0 Q]S_1 \qquad (21)$$

and the intensity of the vortices is given by

$$I(\Delta) = |S_0 + S_1 \exp(i\Phi_0 Q)|^2 = |S_0|^2 + |S_1|^2 + 2|S_0||S_1|\cos(\sigma + \Phi_0 Q) \qquad (22)$$

where $\sigma = arg(S_1) - arg(S_0)$. We consider the case of equal splitting in which $|S_0|^2 = |S_1|^2 = \frac{1}{2}$ and taking into account that the voltage V between the boundaries L_0 and L_1 is proportional to the intensity of vortices, we conclude that

$$
\begin{aligned}
V(t) &= V_0[1 + \cos(\sigma + \Phi_0 Q)] \\
&= V_0 + V_0 \cos[\sigma + \Phi_0 It + \Phi_0 Q_0 \sin(\omega t)] \\
&= V_0 + V_0 \sum_{K=-\infty}^{\infty} J_K(\Phi_0 Q_0) \cos[(\sigma + (\Phi_0 I + K\omega)t]. \quad (23)
\end{aligned}
$$

We calculate the time-averaged voltage V_{dc} and get:

$$
\begin{aligned}
\Phi_0 I = N\omega &\quad \rightarrow \quad V_{dc} = V_0 + V_0 J_{-N}(\Phi_0 Q_0) \cos \sigma \\
\Phi_0 I \neq N\omega &\quad \rightarrow \quad V_{dc} = V_0 \quad\quad\quad\quad\quad\quad\quad (24)
\end{aligned}
$$

where N is an integer. Therefore if we plot V_{dc} against $\Phi_0 I$ we get a constant, with peaks (or antipeaks) at the values $\Phi_0 I = N\omega$ (Bloch steps). Their physical interpretation is that the vortices absorb an integer number of photons to compensate the effect of the magnetomotive force I on the vortex.

We next consider an experiment similar to that described above, but here the waveguide is carrying non-classical microwaves at low temperatures. In a quantum treatment of the electromagnetic field the A_i and E_i are usually taken as dual quantum variables. In this section we are interested in the dual potential and we use the a_i and B_i as dual quantum variables, in the gauge $a_0 = 0$ where $B_i = \partial_0 a_i$. Since the dimensions of the device are much smaller than the wavelength of the microwaves, we integrate these variables over a small loop around the origin and get the charge $Q = \int a_i dx_i$ at the center and the magnetomotive force $S_{MMF} = \int B_i dx_i$, which obey the commutation relation $[Q, S_{MMF}] = i\xi$ where ξ is a constant depending on the dimensions and geometry of the device.

The charge operator $Q(t)$ can be written in terms of creation and annihilation operators (in the external field approximation) as:

$$
Q(t) = (2\omega)^{-1/2} \left[\exp(i\omega t) a^\dagger + \exp(-i\omega t) a \right] \quad (25)
$$

The relative phase factor $\exp[i\Phi_0 Q(t)]$ between the two vortex beams, is now the quantum mechanical operator

$$
\exp[i\Phi_0 Q(t)] = D(A); \quad A = i\Phi_0 (2\omega)^{-1/2} \exp(i\omega t) \quad (26)
$$

where $D(A)$ is the displacement operator. The expectation value of this operator with respect to the density matrix density matrix ρ describing the non-classical microwaves, determines the interference.

Using the above formalism we easily see that

$$
\begin{aligned}
V(t) &= V_0 + V_0 \Re \{\exp[i(\sigma + \Phi_0 It)] W(A)\} \\
&= V_0 + V_0 |W(A)| \cos[\sigma + \Phi_0 It + \arg(W(A))] \quad (27)
\end{aligned}
$$

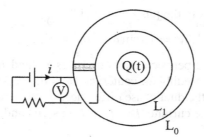

Figure 5. A ring made from a Josephson array insulator. The shaded region contains a weak link and plays the role of a dual Josephson junctions for vortices. The center contains a time-dependent electric charge $Q(t)$ fed by a current $I = \partial_t Q$. The voltmeter V measures the vortex current.

where $W(A) = \text{Tr}[\rho D(A)]$ is the Weyl function.

For coherent states $|\alpha\rangle$ we find that for $\Phi_0 I \neq N\omega$ where N is any integer, we get $\frac{V_{dc}}{V_0} = 1$; and for $\Phi_0 I = N\omega$ we get:

$$\frac{V_{dc}}{V_0} = 1 + \exp\left(-\frac{\Phi_0^2}{4\omega}\right) J_{-N}\left[\Phi_0\left(\frac{2}{\omega}\right)^{1/2}|\alpha|\right] \cos\left[\sigma - K\left(\theta_\alpha - \frac{\pi}{2}\right)\right]$$

$$(28)$$

where $\theta_\alpha = Arg(\alpha)$. It is seen that in comparison to the classical case the values of the peaks have been reduced by a factor $\exp\left(-\frac{\Phi_0^2}{4\omega}\right)$. This is similar ('dual') to the result for the ac Aharonov-Bohm phenomena. For squeezed vacua we can show that we have only even Bloch steps.

4.2 Dual Josephson Junctions

We consider the experiment depicted in Fig.5, where we have a ring made from a Josephson array insulator. A weak magnetic field perpendicular to the plane of the diagram creates magnetic vortices which are forced by the current i (which is very weak in the absence of dissipation) to circulate the ring. The ring contains a 'weak link' which obstructs the movement of vortices and plays the role of a dual Josephson junction for vortices. The link couples weakly the two insulators where vortices move with a high mobility and provides an obstruction to vortices which go through it. The center of the ring contains a time-dependent electric charge $Q(t)$ fed by a current $I = \partial_t Q$.

Let $S_0(x)$ and $S_1(x)$ be the wavefunctions corresponding to vortices in the two sides of the weak link in Fig.2. The 'dual Josephson equations' which describe macroscopically the junction are [11]:

$$V = V_0 \sin \delta; \quad \partial_t \delta = \Phi_0 I \qquad (29)$$

where V is the voltage between the boundaries L_0 and L_1; V_0 is a constant; $I = \partial_t Q$ is the current feeding the center; and $\delta = \theta_1 - \theta_0$ (where $\theta_1 = \arg(S_1)$ and $\theta_0 = \arg(S_0)$) is the dual phase difference.

We next consider the current: $I = I_0 + I_1 \cos(\omega t)$ and solving Eqs.(27) we get

$$
\begin{aligned}
V(t) &= V_0 \sin[\delta_0 + \Phi_0 I_0 t + \Phi_0 \frac{I_1}{\omega} \sin(\omega t)] \\
&= V_0 \sum_{K=-\infty}^{\infty} J_K(\Phi_0 \frac{I_1}{\omega}) \sin[\delta_0 + (\Phi_0 I_0 + K\omega)t] \qquad (30)
\end{aligned}
$$

We can now calculate the time-averaged value of $V(t)$ which we denote as V_{dc}:

$$
\begin{aligned}
\Phi_0 I_0 = N\omega &\rightarrow V_{dc} = V_0 J_{-N}(\Phi_0 \frac{I_1}{\omega}) \sin \delta_0 \\
\Phi_0 I_0 \neq N\omega &\rightarrow V_{dc} = 0 \qquad (31)
\end{aligned}
$$

Therefore if we plot V_{dc} against $\Phi_0 I_0$ we get zero, with peaks (or antipeaks) at the Bloch steps $\Phi_0 I_0 = N\omega$. If we now consider non-classical electromagnetic fields we can show a reduction in the size of the peaks that depends on the density matrix density matrix of the field.

It is seen that there is a similarity between the results derived in this section and the results for the ac Aharonov-Casher phenomena discussed earlier. This is because in both cases we have vortices circulating time dependent charges. In the quantum case the quantum noise destroys slightly the interference and we get smaller peaks.

5. Discussion

We have considered ac Aharonov-Bohm phenomena, i.e., electron interference in the presence of a time-dependent magnetic flux (electromagnetic field). We have also considered non-classical electromagnetic fields in which case the relative phase factor between the two electron beams is a quantum mechanical operator, whose expectation value with respect to the density matrix density matrix ρ describing the microwaves, determines the interference. The results have been interpreted physically in terms of multiphoton exchange between the electrons and the microwaves. We have shown how the quantum (and also classical) noise in the electromagnetic field destroys partly the interference.

Similar ideas have also been applied in the context of Josephson devices interacting with non-classical microwaves. We have shown how the quantum noise in the microwaves affects the current and in particular its dc component.

Dual phenomena that involve vortex condensates in Josephson array insulators, have also been considered. ac Aharonov-Casher phenomena in which vortices circulate around time-dependent electric charges (electromagnetic fields) have been studied. Dual Josephson junctions that involve Josephson array insulators separated by a weak link through which vortices tunnel, and the corresponding dual Josephson equations have been discussed.

All the devices considered exhibit non-linear (sinusoidal) behaviour. and might be used as amplifiers and frequency converters at microwave frequencies.

6. References

1 Y. Imry 'Introduction to Mesoscopic Physics' (Oxford Univ. Press, Oxford, 1997)
   S. Datta 'Electronic Transport in Mesoscopic Systems' (Cambridge Univ. Press, Cambridge, 1995)
   D.K. Ferry, S.M. Goodnick 'Transport in Nanostructures' (Cambridge Univ. Press, Cambridge, 1997)

2 Y. Aharonov, D. Bohm, Phys. Rev. 115, 485 (1959);
   W.H. Furry, N.F. Ramsey Phys. Rev. 110, 629 (1960)
   S.Olariu, I.I.Popescu, Rev. Mod. Phys. 57, 339 (1985)
   M. Peshkin, A. Tonomura, "The Aharonov-Bohm effect", Lecture notes in Physics, Vol. 340, (Springer, Berlin, 1989).

3 A.Vourdas, Europhys. Lett. 32, 289 (1995); Phys. Rev. B54, 13175(1996); Phys. Rev. A65, 042321 (2002)
   A.Vourdas, B.C.Sanders Europhys. Lett. 43, 659 (1998)

4 G.Schon and A.D. Zaikin Phys. Rep., 198, 237 (1990)
   M.A. Kastner, Rev. Mod. Phys., 64, 849, (1992)
   Y. Makhlin, G. Schon, A. Shnirman, Rev. Mod. Phys. 73, 357 (2001)
   R.Fazio, H. van der Zant, Phys. Rep. 355, 235 (2001).

5 A. Vourdas, Phys. Rev. B49, 12040 (1994)
   A. Vourdas, Z.Phys. B100, 455 (1996)
   A.A. Odintsov and A. Vourdas, Europhys. Lett. 34, 385 (1996)

6 A. Vourdas, T.P. Spiller, Z. Physik B102, 43 (1997)
   M.J. Everitt, et al Phys. Rev. B63, 144530 (2001); Phys. Rev. B64, 184517 (2001)

7 S.M. Girvin Science, 274, 524 (1996)

8 Y. Aharonov and A.Casher Phys. Rev. Lett. 53, 319 (1984)
   B. Reznik and Y. Aharonov Phys. Rev. D40, 4178 (1989)
   A.S. Goldhaber Phys. Rev. Lett. 62, 482 (1989)

9 B.J. van Wees Phys. Rev. Lett. 65, 255 (1990); Phys. Rev. B44, 2264
   (1991)
   T.P. Orlando and K.A. Delin Phys. Rev. B43, 8717 (1991)
   W.J. Elion, J.J. Wachters, L.L. Sohn, J.E. Mooij, Phys. Rev. Lett. 71,
   2311 (1993)

10 A. Vourdas, Phys. Rev. BB60, 620 (1999)

11 A. Vourdas, Europhys. Lett. 48, 201 (1999)
    A. Vourdas, T.P. Spiller, T.D. Clark, D. Poulton, Phys. Rev. B63, 104501
    (2001)
    A. Konstadopoulou et al IEE Proc. Sci. Meas. Tech. 148, 229 (2001)

12 S. Chountasis and A. Vourdas, Phys. Rev. A, **58**, 848 (1998); Phys. Rev.
    A, **58**, 1794 (1998).

RESONANCE QUANTUM OPTICS OF PHOTONIC CRYSTALS

Valery I. Rupasov

Landau Institute for Theoretical Physics, Moscow, Russia

Abstract   Quantum effects of light propagation through an extended absorbing system of two-level atoms placed within a frequency gap media are studied in the framework of the generalized quantum Maxwell-Bloch model. The model is proven to exhibit hidden integrability and is diagonalized exactly by the Bethe ansatz method, provided that the atomic system is described in the gas approximation. Apart from polaritons and their bound complexes - "ordinary" solitons, the multiparticle spectrum of the radiation plus medium plus atoms system is shown to contain also massive pairs of confined gap excitations, which do not exist out of pairs, and bound complexes of these pairs - "gap" solitons. Furthermore, "composite" solitons are predicted as bound states of "deformed" ordinary and gap solitons. Both gap and composite solitons containing even a few excitations could be quite short to propagate through the absorbing atomic system without dissipation and should be associated with the self-induced transparency pulses in doped frequency gap media.

1.   Introduction

The self-induced transparency (SIT) of short light pulses propagating in a resonance absorbing medium has been predicted and observed in the pioneering work of McCall and Hahn [1]. The SIT pulses may be regarded [2] as solitons of the classical Maxwell-Bloch (CMB) model, describing classical radiation propagating in a single direction and coupled to an extended system of two-level atoms two-level atom. The CMB model is completely integrable and the time evolution of an arbitrary light pulse incident on the atomic system is described [3, 4, 5] by means of the inverse scattering method [6]. A comprehensive review of the present status of the SIT theory and corresponding mathematical methods has recently been done in Ref. [7].

In the standard SIT theory [1, 2, 3, 4, 5, 7], due to the existence of strong radiation in an initial condition of the problem, quantum corrections are a priori expected to be negligibly small. Therefore, the quantum version of the CMB model - the quantum Maxwell-Bloch (QMB) model - has been studied [8, 9]

A.S. Shumovsky and V.I. Rupasov (eds.), Quantum Communication and Information Technologies, 211–249.

by means of the Bethe ansatz technique [10] only in the context of superfluorescence phenomenon where quantum effects play a crucial role [11]. But the situation is drastically changed for frequency gap media (FGM), such as a frequency dispersive medium (DM) [12], a photonic bandgap (PBG) material [13, 14], and a one-dimensional Bragg reflector [15]. In the single-atom problem [16, 17], "spherical" quantum solitons containing an even number of gap excitations propagate within a frequency gap of a medium, while an "odd" gap soliton is pinned to an atom and forms a multiparticle-atom bound state, in which the radiation and the medium polarization are localized in the vicinity of the atom. This essential difference in the behavior of even and odd solitons is preserved even in the limit of a macroscopically large number of particles. Thus, we deal with a quantum macroscopic phenomenon, therefore electrodynamics of a FGM doped with resonance atoms cannot be described in terms of classical field theory and quantum nature of phenomena under consideration has to be taken into account.

In the present paper [18], we generalize the McCall-Hahn theory to study quantum effects of light propagation through an extended homogeneous system of two-level atoms [19] placed within a FGM. We study the generalized quantum Maxwell-Bloch model which describes polaritons ("photons in a medium") propagating in a single direction and coupled to an extended system of identical two-level atoms. As in the standard SIT theory, we treat the atomic system in the gas approximation assuming that the characteristic times of interatomic resonance dipole-dipole interaction (RDDI) and other collisional dephasing effects are much longer than the light pulse duration. In empty space (or in a nondispersive medium), an atom-atom interaction is automatically eliminated from consideration due to a local (pointlike) model of photon-atom coupling. In a FGM, a polariton-atom coupling is *nonlocal*. Therefore, in contrast to the case of empty space, an extension of the Bethe ansatz method from the case of a single atom embedded in a FGM [16, 17] to the case of an extended atomic system requires a thorough analysis.

In the case of a Bragg reflector with a Kerr nonlinearity, numerous analytical and numerical studies [15] show the existence of a nonlinear solitary wave propagating within a frequency gap where propagating linear modes are completely forbidden. Unfortunately, nonlinear equations describing light propagation in these materials are not integrable that does not allow one to construct many-pulse solutions to describe the time evolution of an arbitrary incident light. In a linear FGM doped with two-level atoms, a nonlinearity is generated by a photon-atom scattering. Unlike the Bragg-Kerr reflectors, the radiation plus medium plus atoms system exhibits hidden integrability [16] that allows us to construct many soliton solutions and to describe thus the time evolution of an arbitrary incident field.

The outline of this paper is as follows. In Sec. II we derive the generalized QMB model for the case of a dispersive medium, which is treated as a continuous set of charge harmonic oscillators coupled to radiation. To clarify our approach to the generalized model, we first discuss in Sec. III the ordinary QMB model to develop a quantum version of the McCall-Hahn theory. We show that a one-dimensional free radiation can be described in terms of either the ordinary basis of free photons or the Bethe basis scattering particles. While the basis of free photons is completely destroyed by the photon-atom scattering, the Bethe basis is preserved and the Bethe wave functions acquire only additional phase phase factors to account for scattering on the atoms. The spectrum of photons plus atoms system contains the Bethe strings (bound many-photon complexes or quantum solitons), whose velocity of propagation within the atomic systems depends on a number of photons in a string. An arbitrary state of the radiation field incident on the atomic system can be represented as a linear superposition of Bethe strings. A propagation of strings through the atomic system generates a spatial separation in their initial superposition. Moreover, the Bethe strings containing a macroscopically large number of photons are shorter than the characteristic relaxation times in the atomic system, and hence propagate without dissipation. Namely these strings should be associated with the classical SIT pulses or with solitons of CMB model.

Further, we apply the string conception to the generalized QMB model. In Sec. IV we first solve the problem of a single polariton to find a polariton-atom scattering matrix. In contrast to the case of empty space, the polariton spectrum is *nonlinear* and cannot be linearized in the vicinity of the gap. Moreover, the polariton atom-coupling is *nonlocal* that leads to an effective *nonlocal* polariton-polariton coupling. In other words, polariton wave functions are *continuous* and no ordinary Bethe construction, which describes a scattering process in terms of discontinuous functions, can be applied to the case of FGM. To diagonalize the generalized QMB model, we introduce *auxiliary* particles in an *auxiliary* space. In Sec. V we construct an auxiliary Bethe basis for a one-dimensional free polariton field in a gap medium. The continuous polariton wave functions in the real space are then found as a result of an integral "dressing" procedure of the discontinuous Bethe wave functions of auxiliary particles. We show then in Sec. VI that the polariton-atom scattering does not destroy the Bethe basis that allow us to construct many-particle eigenstates of the polaritons plus atoms system. Thus, as well as in the one-atom problem, the integrability of the generalized QMB model is *hidden* and manifested only in the limit of large interpolariton separations.

Although the Bethe ansatz equations, determining the spectrum of the system, are valid in all three cases of physical interest, their possible solutions are controlled by the spectrum of elementary electromagnetic excitations of a medium and could be different in different FGM. In Sec. VII, we study solutions of the Bethe ansatz equations only in the particular case of a DM, where simple and explicit expressions for the polariton spectrum are well-known. We first show that, unlike an attractive effective photon-photon coupling in empty space, an effective polariton-polariton coupling is attractive for polaritons of the lower branch and repulsive for polaritons of the upper branch, provided the atomic transition frequency ω_{12} lies below or inside the frequency gap of the medium. Therefore bound many-polariton complexes - "ordinary" solitons can be constructed only from polaritons of the lower branch. Apart from polaritons and ordinary solitons, the spectrum of the system is then shown to contain massive pairs of confined gap excitations (or gap particles) which do not exist out of a pair. Pairs of confined gap particles can form bound many-pair complexes - "gap" solitons, which are highly correlated quantum many-body states and are distinct from single photon hopping conductivity [20] through the photonic impurity gap created by the atoms. We derive the energy-momentum dispersion relations for gap solitons and evaluate the widths of soliton bands and the group velocities of their propagation inside the atomic system as functions of the number of pairs. Furthermore, we predict the existence of "composite" solitons as bound states of "deformed" ordinary and gap solitons. Both gap or composite solitons containing even a few particles could be quite short to propagate through the atomic system without dissipation. Thus, they realize the self-induced transparency effect in FGM.

If ω_{12} lies above the gap, the gap excitations vanish, but now both intrabranch and interbranch polariton-polariton couplings become attractive, that allows one to construct both ordinary solitons in each branch and composite solitons as bound states of ordinary solitons from different branches.

Finally, in Sec. VIII the theory developed is extended to the case of a degenerate atomic transition that allows one to study effects of an alternative polarization of an incident light.

2.   Generalized Quantum Maxwell-Bloch Model

The Hamiltonian Hamiltonian of the system field plus medium plus identical two-level atoms two-level atom has the form

$$H = H_F + H_M + H_A + V_{MF} + V_{AF}, \tag{1}$$

where

$$H_F = \frac{1}{8\pi} \int d\mathbf{r} \left[\mathbf{E}^2(\mathbf{r}) + \mathbf{H}^2(\mathbf{r}) \right], \tag{2}$$

$$H_M = \frac{1}{2m} \int d\mathbf{r} \left[\mathbf{P}^2(\mathbf{r}) + m^2 \Omega^2 \mathbf{Q}^2(\mathbf{r}) \right], \tag{3}$$

$$H_A = \omega_{12} \sum_{a=1}^{M} \left(\sigma_a^z + \frac{1}{2} \right) \tag{4}$$

are the Hamiltonians of a free radiation field, a dispersive medium, and M two-level atoms, respectively. A dispersive medium is treated here as a continuous set of charge harmonic oscillators, each with frequency Ω, charge e, and mass m. The operators of displacement, $\mathbf{Q}(\mathbf{r})$, and momentum, $\mathbf{P}(\mathbf{r})$, obey the commutation relations

$$[Q^i(\mathbf{r}), P^j(\mathbf{r}')] = i\delta_{ij}\delta(\mathbf{r} - \mathbf{r}'), \quad i, j = x, y, z, \tag{5}$$

and the medium-field coupling operator is written as

$$V_{MF} = e\sqrt{n_0} \int d\mathbf{r} \, \mathbf{Q}(\mathbf{r}) \cdot \mathbf{E}(\mathbf{r}), \tag{6}$$

where n_0 is the density of the number of oscillators. In the dipole approximation, the operator of atom-field coupling has obviously the form

$$V_{AF} = -\sum_{a=1}^{M} \mathbf{d}_a \cdot \mathbf{E}(\mathbf{r}_a), \tag{7}$$

where $\mathbf{d}_a$ is the atomic dipole operator and $\mathbf{E}(\mathbf{r}_a)$ is the electric field operator at the point of atom's position. An atom is treated as a quantum two-level system with the transition frequency ω_{12} and is described by the spin operators $\sigma_a^i = (\sigma_a^x, \sigma_a^y, \sigma_a^z)$, $\sigma^{\pm} = \sigma^x \pm i\sigma^y$ with the commutation rules

$$[\sigma_a^i, \sigma_b^j] = \delta_{ab}e^{ijk}\sigma_a^k, \tag{8}$$

where e^{ijk} is the unit antisymmetric tensor. In terms of spin operators the atomic dipole operator takes the form

$$\mathbf{d}_a = \mathbf{e}_z d(\sigma_a^+ + \sigma_a^-), \tag{9}$$

where $\mathbf{e}_z$ is the unit vector and d is the matrix element of the dipole operator.

2.1 One-dimensional approximation

As it has been mentioned above, the standard SIT theory assumes that the effect can be described within the framework of a one-dimensional model where the field and medium operators depend only on one spatial variable x – the coordinate along a propagation axis of a light incident on the medium. The operator of the vector potential of the radiation field can be expended in terms of one-dimensional transverse plane waves,

$$\mathbf{A}(x) = -i\mathbf{e}_z \int_{-\infty}^{\infty} \frac{dk}{2\pi} \left(\frac{2\pi}{S_0|k|}\right)^{1/2} [c(k)e^{ikx} - \text{h.c.}]. \tag{10}$$

The operators of the electric and magnetic components are then defined as

$$\mathbf{E}(x) \quad = \quad \mathbf{e}_z \int_{-\infty}^{\infty} \frac{dk}{2\pi} E_k [c(k)e^{ikx} + \text{h.c.}] \tag{11}$$

$$\mathbf{H}(x) \quad = - \quad \mathbf{e}_y \int_{-\infty}^{\infty} \frac{dk}{2\pi} H_k [c(k)e^{ikx} + \text{h.c.}], \tag{12}$$

where $E_k = \left(\frac{2\pi |k|}{S_0} \right)^{1/2}$, $H_k = k \left(\frac{2\pi}{S_0 |k|} \right)^{1/2}$, while the medium operators are represented in the form

$$\mathbf{Q}(x) \quad = \quad \mathbf{e}_z Q \int_{-\infty}^{\infty} \frac{dk}{2\pi} [b(k)e^{ikx} + \text{h.c.}], \tag{13}$$

$$\mathbf{P}(x) \quad = -i \quad \mathbf{e}_z P \int_{-\infty}^{\infty} \frac{dk}{2\pi} [b(k)e^{ikx} - \text{h.c.}]. \tag{14}$$

Here $Q = \left(\frac{1}{2m\Omega S_0} \right)^{1/2}$, $P = \left(\frac{m\Omega}{2S_0} \right)^{1/2}$, $\mathbf{e}_{x,y,z}$ are orthonormal vectors of the Cartesian coordinates, S_0 is the cross-section area of the light beam, and the operators $c(k)$ and $b(k)$ with the commutation relations

$$[c(k), c^\dagger(k')] = 2\pi\delta(k - k'), \quad [b(k), b^\dagger(k')] = 2\pi\delta(k - k')$$

describe field and medium excitations, respectively.

Substituting now Eqs. (2.9) and (2.10) into the model Hamiltonian, one obtains

$$H_F \quad = \quad \int_{-\infty}^{\infty} \frac{dk}{2\pi} |k| \, c^\dagger(k)c(k), \tag{15}$$

$$H_M \quad = \quad \Omega \int_{-\infty}^{\infty} \frac{dk}{2\pi} b^\dagger(k)b(k), \tag{16}$$

$$V_{MF} \quad = \quad \int_{-\infty}^{\infty} \frac{dk}{2\pi} \left(\frac{|k|\Delta}{2} \right)^{1/2} [c(k) + c^\dagger(-k)]$$
$$\times [b(-k) + b^\dagger(k)] \tag{17}$$

$$V_{AF} \quad = \quad -\sum_{a=1}^{M} (\sigma_a^- + \sigma_a^+) \int_{-\infty}^{\infty} \frac{dk}{2\pi} \sqrt{\gamma(k)}$$
$$\times [c(k) + c^\dagger(-k)] \, e^{ikx_a}, \tag{18}$$

where x_a are the atomic coordinates along the x-axis and parameters

$$\Delta = \frac{2\pi e^2 n_0}{m\Omega}, \quad \gamma(k) = \frac{2\pi}{S_0} |k| d^2 \tag{19}$$

represent the medium-field and atom-field couplings, respectively.

In contrast to the previous studies [16], we do not use here the Heitler-London approximation and conserve in the medium-field coupling operator the terms $b^\dagger c^\dagger$ (bc) which create (annihilate) both the medium and field excitations simultaneously. It will be seen below that these terms play an extremely important role for a correct description of the structure and parameters of gap excitations of the system.

2.2   Polariton variables

The field-medium part of the model Hamiltonian,

$$H_{FM} = H_F + H_M + V_{MF}, \tag{20}$$

can easily be diagonalized in terms of polariton variables [21, 22]. Introducing the polariton operators $p_\alpha(k)$ for the lower ($\alpha = -$) and upper ($\alpha = +$) branches of medium-field (polariton) excitations, we obtain

$$H_{FM} = \sum_{\alpha=\pm} \int_{-\infty}^{\infty} \frac{dk}{2\pi} \, \omega_\alpha(k) p_\alpha^\dagger(k) p_\alpha(k), \tag{21}$$

where the polariton spectra

$$\omega_\pm^2(k) = \frac{1}{2} \left\{ (\Omega^2 + k^2) \pm \sqrt{(\Omega^2 - k^2)^2 + 8k^2 \Omega \Delta} \right\} \tag{22}$$

are found from the dispersion relation

$$k^2(\omega) = \omega^2 \varepsilon(\omega), \quad \varepsilon(\omega) = \frac{\omega^2 - \Omega_\parallel^2}{\omega^2 - \Omega_\perp^2}, \tag{23}$$

where $\varepsilon(\omega)$ is the dielectric permeability of the medium. The polariton frequency varies from zero to $\Omega_\perp = \sqrt{\Omega^2 - 2\Omega\Delta} \approx \Omega - \Delta$ within the lower branch, and from $\Omega_\parallel = \Omega$ to $+\infty$ within the upper branch, while the frequency interval of the width Δ between $\Omega_\perp$ and $\Omega_\parallel$ is forbidden for propagating polariton modes. It is convenient to introduced also the refractive index of the medium, $n(\omega) = \sqrt{\varepsilon(\omega)}$, and express the polariton wave vector as $k(\omega) = \omega n(\omega)$.

In terms of polariton variables the atom-field coupling operator takes the form:

$$
\begin{aligned}
V_{AF} = \; & -\sum_{a=1}^{M} (\sigma_a^- + \sigma_a^+) \sum_{\alpha=\pm} \int_{-\infty}^{\infty} \frac{dk}{2\pi} \sqrt{\Gamma_\alpha(k)} \\
& \times [p_\alpha(k) + p_\alpha^\dagger(-k)] e^{ikx_a},
\end{aligned} \tag{24}
$$

where

$$\Gamma_\alpha(k) = \gamma(k) \frac{|k|}{\omega_\alpha(k)} \frac{|\Omega^2 - \omega_\alpha^2(k)|}{\omega_+^2(k) - \omega_-^2(k)}.$$

As well as in the standard SIT theory, we now can make use of the single-direction approximation and restrict thus our consideration to polaritons propagating in a single, say positive, direction of the x-axis. In other words, in the single direction approximation we can omit in Eqs. (2.14) and (2.17) the integrals over negative polariton wave vectors. It is convenient then to introduce the polariton operators on the "energy scale",

$$
p(\omega) = \begin{cases} (d\omega_-/dk)^{-1/2}p_-(k), & 0 \le \omega < \Omega_\perp \\ (d\omega_+/dk)^{-1/2}p_+(k), & \Omega_\| \le \omega < \infty \end{cases} \tag{25}
$$

$$
[p(\omega), p^\dagger(\omega')] = 2\pi\delta(\omega - \omega'). \tag{26}
$$

and to turn from the integration over k to the integration over the frequency ω. Then, we finally find for the Hamiltonian of the QMB model generalized to the case of a frequency gap medium:

$$
H = H_0 + V, \tag{27}
$$

$$
H_0 = \omega_{12} \sum_{a=1}^{M} \left(\sigma_a^z + \frac{1}{2} \right) + \int_C \frac{d\omega}{2\pi} \omega\, p^\dagger(\omega)p(\omega), \tag{28}
$$

$$
V = -\sqrt{\gamma} \sum_{a=1}^{M} \int_C \frac{d\omega}{2\pi} z(\omega)[\sigma_a^+ p(\omega)e^{ik(\omega)x_a} + \text{h.c.}], \tag{29}
$$

where the first term describes free atoms and polaritons propagating in the positive direction of the x-axis, while the second one represents their coupling. The coupling constant $\gamma = 2\pi\omega_{12}d^2/S_0$, represents the atom-field coupling in empty space, while the atomic form factor

$$
z(\omega) = (\omega/\omega_{12})^{1/2}n^{3/2}(\omega) \tag{30}
$$

reflects the behavior of the polariton spectra of the medium. The integration contour C consists of two intervals, $C = (0, \Omega_\perp) \bigcup (\Omega_\|, \infty)$. Accordingly to the rotating wave approximation [23], we omitted in Eq. (2.19c) the terms $p^\dagger\sigma^+$ ($p\sigma^-$) which create (annihilate) both the polariton and atomic excitations simultaneously. In the absence of dispersion [$n(\omega) = 1$, $z(\omega) = 1$, $C = (-\infty, \infty)$, where the integration in the lower limit is extended to $-\infty$ due to the resonance approximation [23]], the equations of motion for dynamical variables of the system coincide with the standard MB equations.

The model Hamiltonian Hamiltonian derived here for the case of a dispersive medium has quite a general structure and can easily be derived in the cases of 1D periodic structures and PBG materials. The information about the medium spectrum is contained into the atomic form factor $z(\omega)$, the integration contour C, and the dispersion relation $k = k(\omega)$. In what follows, we analyze the SIT problem within the framework of the model Hamiltonian (2.19) with arbitrary $z(\omega)$, C and $k(\omega)$. Therefore, all the qualitative results obtained below are also valid for both PBG materials and 1D Bragg reflectors doped with resonance atoms.

2.3 Radiation field in dispersive medium

The Heisenberg operators of the electric and magnetic fields obey the Maxwell equations in vacuum,

$$\mathbf{curl}\,\mathbf{E}_v(x,t) = -\frac{\partial}{\partial t}\mathbf{H}_v(x,t), \tag{31}$$

$$\mathbf{curl}\,\mathbf{H}_v(x,t) = \frac{\partial}{\partial t}\mathbf{E}_v(x,t), \tag{32}$$

provided the time evolution of the Schrödinger operators (2.9) is determined by the free radiation Hamiltonian H_F, e. g. $\mathbf{E}_v(x,t) = \exp(iH_Ft)\mathbf{E}(x)\exp(-iH_Ft)$. Moreover, in deriving Eqs. (2.9) from Eq. (2.8) for the vector potential,

$$\mathbf{E}_v(x,t) = -\frac{\partial}{\partial t}\mathbf{A}_v(x,t) \tag{33}$$

$$\mathbf{H}_v(x,t) = \mathbf{curl}\,\mathbf{A}_v(x,t) \tag{34}$$

we already assumed that the time evolution is determined by H_F. Our purpose now is to find the field operators in the polariton variables which obey the Maxwell equations in the medium,

$$\mathbf{curl}\,\mathbf{E}_m(x,t) = -\frac{\partial}{\partial t}\mathbf{H}_m(x,t), \tag{35}$$

$$\mathbf{curl}\,\mathbf{H}_m(x,t) = \frac{\partial}{\partial t}\mathbf{D}_m(x,t), \tag{36}$$

where $\mathbf{D}_m(x,t)$ is the operator of the electric induction. In other words, we give here an alternative derivation of the model Hamiltonian (2.19) not making use of too tedious mathematical manipulations with the unitary transform to polariton operators [21, 22].

Let us first rewrite the Schrödinger operator of the vector potential in terms of polariton operators,

$$\mathbf{A}(x) = -i\mathbf{e}_z \int_C \frac{d\omega}{2\pi}\left(\frac{2\pi}{S_0 k(\omega)}\right)^{1/2}$$
$$\times [p(\omega)e^{ik(\omega)x} - p^\dagger(\omega)e^{-ik(\omega)x}], \tag{37}$$

where we again restricted our consideration to waves propagating in the positive direction of the x-axis. The operator of the magnetic field is then given by

$$
\mathbf{H}(x) = \mathbf{curl}\,\mathbf{A}(x) = -\mathbf{e}_y \int_C \frac{d\omega}{2\pi} \left(\frac{2\pi k(\omega)}{S_0} \right)^{1/2}
$$
$$
\times [p(\omega)e^{ik(\omega)x} + p^\dagger(\omega)e^{-ik(\omega)x}], \tag{38}
$$

and coincides obviously with the operator (2.9b) expressed in polariton variables.

To find the operator of the electric field in the medium, one has to introduce the Heisenberg operator of the vector potential whose time evolution is determined by the Hamiltonian H_{FM},

$$
\mathbf{A}_m(x, t) = \exp\left(iH_{FM}t\right)\mathbf{A}(x)\exp\left(-iH_{FM}t\right). \tag{39}
$$

Evaluating the operator $\mathbf{E}_m(x, t) = -\partial\mathbf{A}_m(x, t)/\partial t$, and setting then $t = 0$, we find the following representation for the Schrödinger operator of the electric field in the medium:

$$
\mathbf{E}_m(x) = \mathbf{e}_z \int_C \frac{d\omega}{2\pi} \left(\frac{2\pi\omega}{S_0} \right)^{1/2} n^{-1/2}(\omega)
$$
$$
\times [p(\omega)e^{ik(\omega)x} + p^\dagger(\omega)e^{-ik(\omega)x}]
$$
$$
\equiv \int_C \frac{d\omega}{2\pi}\mathbf{E}_m(x, \omega), \tag{40}
$$

which naturally differs from the operator (2.9a) expressed in polariton variables. The induction operator is found from the second Maxwell equation as

$$
\mathbf{D}_m(x) = \mathbf{e}_z \int_C \frac{d\omega}{2\pi} \left(\frac{2\pi\omega}{S_0} \right)^{1/2} n^{3/2}(\omega)
$$
$$
\times [p(\omega)e^{ik(\omega)x} + p^\dagger(\omega)e^{-ik(\omega)x}]
$$
$$
\equiv \int_C \frac{d\omega}{2\pi}\mathbf{D}_m(x, \omega). \tag{41}
$$

This expression coincides with Eq. (2.9a) expressed in polariton operators. The integrands in Eqs. (2.26) and (2.27), $\mathbf{E}_m(x, \omega)$ and $\mathbf{D}_m(x, \omega)$, are linked by

$$
\mathbf{D}_m(x, \omega) = \varepsilon(\omega)\mathbf{E}_m(x, \omega). \tag{42}
$$

It is easy to see now that the operator of the polariton-atom coupling (2.17) is represented in the conventional form

$$
V_{AF} = -\sum_a \mathbf{d}_a \cdot \mathbf{D}_m(x_a), \tag{43}
$$

describing an interaction of an external dipole with the electromagnetic field in a medium [12].

3.   Quantum Version of McCall-Hahn Theory

Before to analyze the SIT effect within the framework of the model Hamiltonian Hamiltonian (2.19), let us study the limiting case of empty space where $n(\omega) = 1$, $z(\omega) = 1$, and $C = (-\infty, \infty)$. It is convenient then to introduce the Fourier images of the photon operators,

$$c(x) = \int_{-\infty}^{\infty} \frac{dk}{2\pi} c(k) e^{ikx} \tag{44}$$

$$[c(x), c^{\dagger}(x')] = \delta(x - x'), \tag{45}$$

and to rewrite the model Hamiltonian in the x-space,

$$H_0 = \omega_{12} \sum_{a=1}^{M} \left(\sigma_a^z + \frac{1}{2} \right) - i \int_{-\infty}^{\infty} dx c^{\dagger}(x) \partial_x c(x) \tag{46}$$

$$V = -\sqrt{\gamma} \sum_{a=1}^{M} \int_{-\infty}^{\infty} dx \delta(x - x_a)[\sigma_a^+ c(x) + \text{h.c.}]. \tag{47}$$

Eqs. (3.2) define the QMB model, which has earlier been studied by means of the quantum inverse scattering method [8] and the Bethe ansatz technique [9] in connection with the superfluorescence phenomenon, where the quantum effects play a crucial role. In the case of SIT effect, the initial condition of the problem contains a strong radiation field incident on the atomic system, therefore all quantum corrections to the classical McCall-Hahn theory are expected to be negligibly small. The existence of the frequency gap in the medium spectrum will be shown to lead to essential quantum corrections to the classical SIT theory. Therefore in this Section, we first analyze a physical background of the McCall-Hahn theory from a quantum point of view making use of some basic results of Refs. [8, 9].

3.1   Exact diagonalization of QMB model

The model (3.2) has one obvious integral of motion,

$$N = \sum_{a=1}^{M} \left(\sigma_a^z + \frac{1}{2} \right) + \int_{-\infty}^{\infty} dx c^{\dagger}(x) c(x), \tag{48}$$

which is naturally treated as the particle number operator of the model. We look for one-particle eigenstates,

$$N|\Psi_1\rangle = 1|\Psi_1\rangle, \quad (H - \lambda)|\Psi_1\rangle = 0,$$

where λ is the eigenenergy, in the form

$$|\Psi_1\rangle = \left[\sum_a g_a \sigma_a^+ + \int_{-\infty}^{\infty} dx \psi(x) c^{\dagger}(x) \right] |0\rangle,$$

where the vacuum state is defined by $\sigma_a^-|0\rangle = c(x)|0\rangle = 0$. The Schrödinger equation then reads

$$(-i\partial_x - \lambda)\psi(x|\lambda) = \sqrt{\gamma}\sum_a g_a(\lambda)\delta(x - x_a), \tag{49}$$

$$(\omega_{12} - \lambda)g_a(\lambda) = \sqrt{\gamma}\int_{-\infty}^{\infty} dx\delta(x - x_a)\psi(x|\lambda). \tag{50}$$

Its solution

$$\psi(x|\lambda) = e^{i\lambda x}\prod_{a=1}^{M}\frac{h(\lambda) - (i/2)\mathrm{sgn}(x - x_a)}{h(\lambda) + i/2}, \tag{51}$$

$$g_a(\lambda) = -\frac{\gamma^{-1/2}}{h(\lambda) + i/2}\left(\frac{h(\lambda) - i/2}{h(\lambda) + i/2}\right)^{a-1}e^{i\lambda x_a}, \tag{52}$$

where $\mathrm{sgn}(x) = \{-1, x < 0; 0, x = 0; 1, x > 0\}$ and $h(\lambda) = (\lambda - \omega_{12})/\gamma$, describes a photon propagating in the positive direction of the x-axis and *subsequently* scattering on atoms. As a result of scattering on an atom, the photon wave function acquires the phase factor

$$S_a(\lambda) = \frac{h(\lambda) - i/2}{h(\lambda) + i/2} = \frac{\lambda - \omega_{12} - i\gamma/2}{\lambda - \omega_{12} + i\gamma/2}, \tag{53}$$

which should be treated as the photon-atom scattering matrix.

The discontinuous expression for the one-photon wave function stems from the pointlike model of the photon-atom coupling in Eq. (3.2b). Strictly speaking, the pointlike model should be considered only as a convenient mathematical trick to eliminate an atom-atom interaction already from the model Hamiltonian Hamiltonian, and to describe thus the atomic system in the gas approximation. Indeed, in a real physical system an average interatomic "separation" along x-axis is much less than even the nuclear size. For example, for $S_0 \sim 10^{-2}\,cm^2$ and the density of the number of atoms $D \sim 10^{18}\,cm^{-3}$, the number of atoms having the x-coordinate within an interval $\Delta x \sim 10^{-8}\,cm$ is estimated as $n_{\Delta x} \sim 10^8$. Therefore, the pointlike model is justified only by our understanding that the dependence of atom-atom coupling only on the difference of atomic coordinates along x-axis is an obvious artefact of the one-dimensional approximation. In a real physical system an atom-atom coupling depends on a real interatomic separation and leads to a transfer of an atomic excitation from one atom to another. But if a characteristic time of this transfer is much longer than a duration of light pulses, one can neglect an atom-atom coupling and describe the atomic system within the framework of the gas approximation.

The QMB model is completely integrable and allows us to construct N-particle eigenstates by means of the Bethe ansatz technique, the N-photon wave eigenfunction being found as [9]

$$\Psi(x_1, \ldots, x_N) = A(x_1, \ldots, x_N) \prod_{j=1}^{N} \psi(x_j|\lambda_j). \qquad (54)$$

Here $\psi(x_j|\lambda_j)$ are the one-photon wave functions (3.5a), while the function A describing an effective photon-photon scattering is factorized into the product of all the possible two-photon scattering functions,

$$A(x_1, \ldots, x_N|\lambda_1, \ldots, \lambda_N) = \prod_{j<l} A(x_j, x_l|\lambda_j, \lambda_l), \qquad (55)$$

where

$$
\begin{aligned}
A(x_1, x_2|\lambda_1, \lambda_2) &= 1 + \frac{i}{h(\lambda_1) - h(\lambda_2)} \mathrm{sgn}(x_1 - x_2) \\
&= 1 + \frac{i\gamma}{\lambda_1 - \lambda_2} \mathrm{sgn}(x_1 - x_2). \qquad (56)
\end{aligned}
$$

To find the spectrum of the photons plus atoms system, one can put the system in a finite box of size L and impose the periodic boundary conditions (PBC) on the photon wave function (3.7),

$$\Psi(\ldots, x_j = -L/2, \ldots) = \Psi(\ldots, x_j = L/2, \ldots). \qquad (57)$$

That leads to the Bethe ansatz equations for "rapidities" λ_j,

$$e^{i\lambda_j L} \left(\frac{\lambda_j - \omega_{12} - i\gamma/2}{\lambda_j - \omega_{12} + i\gamma/2} \right)^M = -\prod_{l=1}^{N} \frac{\lambda_j - \lambda_l - i\gamma}{\lambda_j - \lambda_l + i\gamma}, \qquad (58)$$

where $E = \sum \lambda_j$ is the eigenenergy of the state. As $L \to \infty$, Eqs. (3.9), apart from real solutions, admit complex ones, in which rapidities λ_j are grouped into "strings",

$$\lambda_j = \Lambda + \frac{i\gamma}{2}(n + 1 - 2j), \qquad (59)$$

where Λ is a common real part, and n is the order of a string.

3.2   Quantum nature of SIT effect

The N-particle eigenstates of system are thus represented as a set of the Bethe strings of different orders. For example, the two-particle eigenstates are represented by two possible different configurations of strings. The first configuration contains two one-particle strings with two real rapidities λ_1 and

λ_2, while the second one contains only one two-particle string, in which complex rapidities are grouped into string structure, $\lambda_1 = \Lambda + i\gamma/2$, $\lambda_2 = \Lambda - i\gamma/2$. For large N, the number of all the possible configurations is very large and is given by the number of ways in which the number N is represented as a sum of positive integers. For example, the four-particle states ($N = 4$) has five different configurations: $4 = 1 + 1 + 1 + 1$, $4 = 1 + 1 + 2$, $4 = 2 + 2$, $4 = 1 + 3$, $4 = 4$, where the first one contains four one-particle strings, while the last contains only one four-particle string.

Note now that the Bethe ansatz equations (3.9) still have string solutions even if the number of atoms in the atomic subsystem is taken to be zero, $M = 0$. This surprising fact should be discussed specially, because in the absence of atoms the model Hamiltonian

$$H_0 = -i \int_{-\infty}^{\infty} dx c^{\dagger}(x) \partial_x c(x) \tag{60}$$

describes the *free* radiation field, and N-photon eigenfunctions can immediately be written as

$$\Psi_F(x_1, \dots, x_N | k_1, \dots, k_N) = \prod_{j=1}^{N} \exp(ik_j x_j), \tag{61}$$

where the photon momenta k_j are real, and $E = \sum k_j$ is the eigenenergy.

To clarify this point, let us consider the two-particle Schrödinger equation corresponding to the Hamiltonian (3.11),

$$(i\partial_{x_1} + i\partial_{x_2} + E)[\Psi(x_1, x_2) + \Psi(x_2, x_1)] = 0. \tag{62}$$

It is easy to see that Eq. (3.13a), apart from the solution

$$\Psi_F(x_1, x_2 | k_1, k_2) = \exp[i(k_1 x_1 + k_2 x_2)] \tag{63}$$

describing free photons with real momenta, admit also solutions

$$\Psi_B(x_1, x_2 | \lambda_1, \lambda_2) = A(x_1 - x_2) \exp[i(\lambda_1 x_1 + \lambda_2 x_2)], \tag{64}$$

$E = \lambda_1 + \lambda_2$, with an arbitrary function A depending only on the difference of the particle coordinates. One can choose, for example, the function A in the form given in Eq. (3.7c). Then the Bethe function (3.13c) describe the free radiation field in terms of particles scattering on each other. Moreover, one can construct a bound state of the particles grouping the rapidities λ_1 and λ_2 into the two-particle string, whose wave function has the form

$$\Psi_B(x_1, x_2 | \Lambda) = \exp\{i\Lambda(x_1 + x_2) - (\gamma/2)|x_1 - x_2|\}, \tag{65}$$

where the common real part of the rapidities Λ describes the propagation of the center of gravity of the particles, while the parameter γ determines the string's size.

Thus, the free field with the Hamiltonian Hamiltonian (3.11) can be described in terms of two different bases: (i) the basis of free photons with real momenta, or (ii) the Bethe basis in which particles are scattered on each other. The wave functions of the Bethe basis are written as

$$\Psi_B(\{x_j\}|\{\lambda_j\}) = \prod_{j<l} A(x_j, x_l|\lambda_j, \lambda_l) \prod_{j=1}^N \exp(i\lambda_j x_j), \qquad (66)$$

where the multiparticle scattering process is factorized into two-particle ones. The N-particle Bethe function is characterized by a set of N rapidities λ_j, which can be grouped into different strings. The wave function of the n-particle string,

$$\Psi_B(\{x_j\}|\Lambda) = \exp\left\{i\Lambda \sum_{j=1}^n x_j - \frac{\gamma}{2} \sum_{j<l} |x_j - x_l|\right\}, \qquad (67)$$

describes an n-particle bound state of size $L_s = (n\gamma)^{-1}$ propagating with the momentum per one particle Λ.

Both of bases can be used to describe the free field. But the situation is drastically changed if the field-atom coupling is taken into account. The strong field-atom scattering destroys completely the basis of free photons and all the Bethe bases, in which the two-particle function A differs from that given in Eq. (3.7c). The Bethe basis (3.15), (3.7c) is preserved and wave functions of the Bethe particles acquire only the additional phase factors as result of scattering on atoms [24]. In other words, to obtain the N-particle eigenfunctions of the photons plus atoms system (3.7), one needs only to replace the functions $\exp(i\lambda_j x_j)$ in the expression (3.15) for the free radiation field by the one-photon wave functions $\psi(x_j|\lambda_j)$ (3.5a) accounting for scattering on the atomic system.

After passing through M atoms, the wave function of n-particle string takes the form

$$\Psi_B(\{x_j\}|\Lambda) = \exp\left\{i\Lambda \sum_{j=1}^n x_j - 2iM \arctan \frac{n\gamma/2}{\Lambda - \omega_{12}}\right\}$$

$$\times \exp\left\{-\frac{\gamma}{2} \sum_{j<l} |x_j - x_l|\right\}. \qquad (68)$$

Introducing now the string's center of gravity $x_c = (1/n)\sum x_j$ and representing the number of atoms as $M = \rho x_c$, where ρ is the linear density of the number of atoms, one can define the string wave vector per one particle,

$$K(\Lambda) = \Lambda - \frac{2\rho}{n} \arctan \frac{n\gamma}{2(\Lambda - \omega_{12})}. \tag{69}$$

The group velocity of the string propagation inside the atomic system $V_s = d\Lambda/dK$ is then given by the expression

$$\frac{1}{V_s} = \frac{1}{c} + \frac{\gamma\rho}{(\Lambda - \omega_{12})^2 + (n\gamma/2)^2}, \tag{70}$$

which coincides with the corresponding expressions in the classical SIT theory. For the sake of convenience, we restored here the speed of light in empty space c.

An arbitrary state of the radiation field incident on the atomic system is represented as a linear superposition of the Bethe strings. Because the velocity of the string propagation through the atomic system depends on the order of a string, this propagation generates a *spatial separation* in an initial system of strings. Moreover, "short" stings containing a small number of particles disappear due to relaxation process in the atomic system, while "long" strings containing quite a large number of particles propagate without dissipation, provided that their size is shorter than the characteristic relaxation times. Namely these long strings containing macroscopically large number of particles should be associated with the SIT pulses of the classical McCall-Hahn theory or with soliton solutions of the CMB model.

4.   One-particle Problem in Generalized QMB Model

The generalized QMB model (2.19) also has one obvious integral of motion,

$$N = \sum_{a=1}^{M} \left(\sigma_a^z + \frac{1}{2} \right) + \int_C \frac{d\omega}{2\pi} p^\dagger(\omega) p(\omega), \tag{71}$$

which is naturally treated as the particle number operator. We look for one-particle eigenstates of the model in the form

$$|\Psi_1\rangle = \left[\sum_{a=1}^{M} g_a \sigma_a^+ + \int_C \frac{d\omega}{2\pi} z(\omega) f(\omega) p^\dagger(\omega) \right] |0\rangle, \tag{72}$$

where the polariton wave function $\psi(\omega)$ is written as the product of the atomic form factor $z(\omega)$ and the auxiliary wave function $f(\omega)$, $\psi(\omega) = z(\omega) f(\omega)$. The vacuum state is defined by $\sigma_a^- |0\rangle = p(\omega)|0\rangle = 0$. The Schrödinger equation then reads

$$(\omega - \lambda) f(\omega|\lambda) = \sqrt{\gamma} \sum_a g_a(\lambda) e^{-ik(\omega)x_a}, \tag{73}$$

$$(\omega_{12} - \lambda) g_a(\lambda) = \sqrt{\gamma} \int_C \frac{d\omega}{2\pi} z^2(\omega) f(\omega|\lambda) e^{ik(\omega)x_a}, \tag{74}$$

where λ is the eigenenergy.

Substituting the general solution of Eq. (4.3a)

$$f(\omega|\lambda) = 2\pi\delta(\omega - \lambda) + \frac{\sqrt{\gamma}}{\omega - \lambda - i0} \sum_a g_a(\lambda) e^{-ik(\omega)x_a} \tag{75}$$

into Eq. (4.3b), one obtains the closed set of equations for the atomic wave functions,

$$[\omega_{12} - \lambda - \Sigma(\lambda)] g_a(\lambda) = \sqrt{\gamma} z^2(\lambda) e^{ik(\lambda)x_a} + \gamma \sum_{b \neq a} I_{ab}(\lambda) g_b(\lambda), \tag{76}$$

where

$$\Sigma(\lambda) = \gamma \int_C \frac{d\omega}{2\pi} \frac{z^2(\omega)}{\omega - \lambda - i0} \tag{77}$$

is the self-energy, while the function

$$I_{ab}(\lambda) = \gamma \int_C \frac{d\omega}{2\pi} \frac{z^2(\omega)}{\omega - \lambda - i0} e^{ik(\omega)(x_a - x_b)} \tag{78}$$

describes an effective atom-atom coupling.

4.1 Gas approximation

In empty space, the function $I_{ab}(\lambda)$, due to the pointlike model of the atom-field coupling, is given by $I_{ab}^{(\text{emp})}(\lambda) = i\gamma\theta(x_a - x_b) \exp[i\lambda(x_a - x_b)]$, and thus the gas approximation is automatically taking into account. In the generalized model, to account for the gas approximation, one has to select appropriate solutions of the Schrödinger equation (4.3) not containing effects of the interatomic influence. To eliminate these effects from Eq. (4.5a), one has to evaluate the integral in Eq. (4.5c) taking the contribution only in the pole $\omega = \lambda + i0$. Then, one obtains

$$I_{ab}(\lambda) \simeq i\gamma z^2(\lambda)\theta(x_a - x_b) \exp[ik(\lambda)(x_a - x_b)], \tag{79}$$

and hence

$$g_a(\lambda) = \frac{\sqrt{\gamma}}{\omega_{12} - \lambda - \Sigma(\lambda)} \left(\frac{h(\lambda) - i/2}{h(\lambda) + i/2}\right)^{a-1} e^{ik(\lambda)x_a}. \tag{80}$$

Now only atoms having coordinates $x_b < x_a$ contribute to the wave function of the a-th atom as it must occur in a model with a single direction of a particle propagation. In Eq. (4.7) the function $h(\lambda)$ is defined as

$$h(\lambda) = \frac{\lambda + \Sigma'(\lambda) - \omega_{12}}{2\Sigma''(\lambda)}, \tag{81}$$

where

$$\Sigma'(\lambda) \equiv \mathrm{Re}\,\Sigma(\lambda) = \gamma \int_C \frac{d\omega}{2\pi} \frac{z^2(\omega)}{\omega - \lambda}. \tag{82}$$

To renormalize this integral, we assume that the integration over k is cut off at the point $k_{\max} = \pi/\mathcal{L}$, where $\mathcal{L}$ is a lattice constant. Because the atom-field coupling constant γ is very small, one can neglect λ-dependence of Σ' replacing λ by ω_{12}. Then the value $\bar{\omega}_{12} = \omega_{12} - \Sigma'(\omega_{12})$ can be treated as the transition frequency of the atom placed within the medium, and Eq. (4.8a) takes a more simple form

$$h(\lambda) = \frac{\lambda - \bar{\omega}_{12}}{2\Sigma''(\lambda)}. \tag{83}$$

The imaginary part of the self-energy is evaluated explicitly as

$$\Sigma''(\lambda) \equiv \mathrm{Im}\,\Sigma(\lambda) = \begin{cases} \gamma z^2(\lambda)/2 & , \lambda \quad \text{outside the gap} \\ 0 & , \lambda \in (\Omega_\perp, \Omega_\parallel) \end{cases} \tag{84}$$

For what follows, it is convenient to introduce the Fourier transform of the auxiliary wave function $f(\omega|\lambda)$,

$$f(\tau|\lambda) = \int_{-\infty}^{\infty} \frac{d\omega}{2\pi} f(\omega|\lambda) e^{i\omega\tau}, \tag{85}$$

and to rewrite Eq. (4.3a) in the auxiliary τ-space,

$$(-i\partial_\tau - \lambda)f(\tau|\lambda) = \sqrt{\gamma} \sum_a g_a(\lambda)\varphi_a(\tau), \tag{86}$$

where

$$\varphi_a(\tau) = \int_{-\infty}^{\infty} \frac{d\omega}{2\pi} \exp\left[i(\omega\tau - k(\omega)x_a)\right]. \tag{87}$$

As $\tau \to -\infty$, the asymptotic solution of Eq. (4.10a)

$$f(\tau|\lambda) \sim \exp(i\lambda\tau) \tag{88}$$

describes an incident particle, while as $\tau \to \infty$, the asymptotic solution

$$f(\tau|\lambda) \sim e^{i\lambda\tau}\left[1 + i\sum_a g_a(\lambda)e^{-ik(\lambda)x_a}\right] = e^{i\lambda\tau}S(\lambda), \tag{89}$$

where

$$S(\lambda) = \left(\frac{h(\lambda) - i/2}{h(\lambda) + i/2}\right)^M, \tag{90}$$

incorporates the result of scattering on the atomic system.

It is easy to see that the phase factors $\exp[ik(\omega)x_a]$ in the model Hamiltonian effect only on the behavior of the polariton wave function in the vicinity of an atom, but do not contribute to its asymptotics. Therefore, one can approximate the smooth function $f(\tau|\lambda)$ by its discontinuous form

$$f(\tau|\lambda) \simeq f_D(\tau|\lambda) \equiv e^{i\lambda\tau}\prod_{a=1}^{M}\frac{h(\lambda) - (i/2)\mathrm{sgn}(\tau - x_a)}{h(\lambda) + i/2}, \tag{91}$$

and modify simultaneously the polariton-atom coupling operator (2.19c) replacing the phase factors $\exp[ik(\omega)x_a]$ by $\exp(i\omega x_a)$,

$$V \Rightarrow V_m = -\sqrt{\gamma}\sum_{a=1}^{M}\int_C \frac{d\omega}{2\pi}z(\omega)[\sigma_a^+ p(\omega)e^{i\omega x_a} + \mathrm{h.c.}]. \tag{92}$$

The expression for the atomic wave functions then takes the form

$$g_a(\lambda) = \frac{\sqrt{\gamma}}{\omega_{12} - \lambda - \Sigma(\lambda)}\left(\frac{h(\lambda) - i/2}{h(\lambda) + i/2}\right)^{a-1}e^{i\lambda x_a}. \tag{93}$$

It should be emphasized once more that the approximations made for the one-particle wave functions are not specific for a frequency gap medium, but they are also used implicitly in the case of empty space due to the pointlike model of photon-atom coupling. As well as in the case of empty space, Eqs. (4.13) and (4.15) describe a *subsequent* scattering of a polariton on the atoms, the exact asymptotics of the polariton wave function (4.11) being preserved.

4.2 Real x-space

In contrast to the case of empty space, the discontinuous form (4.13) occurs only for the auxiliary function in the auxiliary τ-space, while the real polariton wave function in the real x-space, $\psi(x|\lambda)$, is continuous and expressed in terms of auxiliary function in quite a complicated way. To find the one-polariton wave function in the real x-space, we first rewrite the one-particle eigenstates (4.2) in terms of the auxiliary τ-space,

$$|\lambda\rangle = \left[\sum_{a=1}^{M} g_a(\lambda)\sigma_a^+ + \int_{-\infty}^{\infty} \psi(\tau|\lambda)p^\dagger(\tau) \right] |0\rangle, \tag{94}$$

where

$$\psi(\tau|\lambda) = \int_{-\infty}^{\infty} dt f(t|\lambda) \int_C \frac{d\omega}{2\pi} z(\omega)e^{i\omega(\tau-t)}. \tag{95}$$

Making use now of the definition

$$\psi(x|\lambda) \equiv \langle 0|p(x)|\lambda\rangle, \tag{96}$$

where

$$\begin{aligned} p(x) &= \frac{1}{\sqrt{2}} \sum_{\alpha=\pm} \int_{-\infty}^{\infty} \frac{dk}{2\pi} p_\alpha(k)e^{ikx} \\ &= \frac{1}{\sqrt{2}} \int_C \frac{d\omega}{2\pi} \left(\frac{dk(\omega)}{d\omega} \right)^{1/2} e^{ik(\omega)x} p(\omega) \end{aligned} \tag{97}$$

is the operator of annihilation of a polariton at the point x, we arrive at the following integral relationship between the polariton wave function and the auxiliary function:

$$\psi(x|\lambda) = \int_{-\infty}^{\infty} d\tau u(x,\tau)f(\tau|\lambda) \tag{98}$$

where

$$u(x,\tau) = \int_C \frac{d\omega}{2\pi} \left(\frac{dk(\omega)}{d\omega} \right)^{1/2} z(\omega) \exp\{i[k(\omega)x - \omega\tau]\}. \tag{99}$$

In empty space, the "dressing" function $u(x,\tau)$ becomes the Dirac δ-function, $u(x,\tau) \to \delta(x-\tau)$, and Eqs. (4.20) lead immediately to the expression (3.5a) for the photon wave function. In the gap medium, one can easily find only the asymptotics of the polariton wave function as $x \to \pm\infty$. Substituting into Eqs. (4.20) the asymptotic expressions (4.11) for the auxiliary function, one gets

$$\psi(x|\lambda) = z(\lambda) \left(\frac{dk(\lambda)}{d\lambda} \right)^{1/2} e^{ik(\lambda)x} \begin{cases} 1 & , x \to -\infty \\ S(\lambda) & , x \to \infty \end{cases} \tag{100}$$

where the scattering matrix $S(\lambda)$ describes now the result of polariton scattering on the atomic system.

The information about the polariton spectrum of the medium is contained also into the relationship between the eigenenergy λ and the function $h(\lambda)$, which is convenient now to call "rapidity". It is easy to see that the one-particle wave functions vanish for λ lying within the gap that play an extremely important role in a construction of many-particle eigenstates of the generalized QMB model.

5. Hidden Bethe Basis for Free Polariton Field

The exact diagonalization of the QMB model is possible because one can construct the Bethe basis for the free radiation field, which is not destroyed by the photon-atom scattering. The photon's spectrum is linear, therefore a solution of the first-order Schrödinger equation (3.13a) can be written in the form (3.13c) with an arbitrary scattering function $A(x_1 - x_2)$ depending only on the difference of the particle coordinates. The polariton's spectrum in a gap medium is significantly *nonlinear* and cannot be linearized in the vicinity of the gap. Therefore, it seems obviously that no Bethe basis can be constructed for a free polariton field with a nonlinear spectrum.

In this Section, we show that a Bethe basis can be constructed for the 1D free polariton field with the Hamiltonian (2.14) for *auxiliary* particles in the *auxiliary* τ-space (see also [16]). The auxiliary wave functions are written in a discontinuous Bethe ansatz form, while the continuous polariton wave functions in the real x-space are found as a result of the dressing procedure defined in Eqs. (4.20). Thus, the Bethe ansatz construction for the polariton field is *hidden* because of the nonlinear spectrum of the particles and manifested only in the limit of large interpolariton separations.

5.1   Two-particle eigenstates

In the single-direction approximation the Hamiltonian Hamiltonian of the free polariton field has the form

$$H_{FM} = \int_C \frac{d\omega}{2\pi} \, \omega \, p^\dagger(\omega) p(\omega). \tag{101}$$

We start our analysis from the two-particle eigenstates

$$|\Psi_2\rangle = \int_C \frac{d\omega_1}{2\pi} \frac{d\omega_2}{2\pi} z(\omega_1) z(\omega_2) F(\omega_1, \omega_2) p^\dagger(\omega_1) p^\dagger(\omega_2)|0\rangle, \tag{102}$$

where the auxiliary two-particle wave function obeys the Schrödinger equation,

$$(\omega_1 + \omega_2 - E)[F(\omega_1, \omega_2) + F(\omega_2, \omega_1)] = 0, \tag{103}$$

with the eigenenergy E. Apart from the ordinary solution

$$F_F(\omega_1, \omega_2 | \epsilon_1, \epsilon_2) = (2\pi)^2 \delta(\omega_1 - \epsilon_1)\delta(\omega_2 - \epsilon_2) \qquad (104)$$

$E = \epsilon_1 + \epsilon_2$, describing free polaritons with real frequencies ϵ_1 and ϵ_2 lying outside the gap, Eq. (5.3a) admits also the Bethe solution, which, for example, can be written in the form

$$
\begin{aligned}
F_B(\omega_1, \omega_2 | \lambda_1, \lambda_2) &= (2\pi)^2 \delta(\omega_1 - \lambda_1)\delta(\omega_2 - \lambda_2) \\
&+ \frac{1}{h_1 - h_2}\left(\frac{1}{\omega_1 - \lambda_1} - \frac{1}{\omega_2 - \lambda_2}\right) \\
&\times 2\pi\delta(\omega_1 + \omega_2 - \lambda_1 - \lambda_2),
\end{aligned} \qquad (105)
$$

where $\lambda_1 + \lambda_2 = E$ and the rapidities $h_j \equiv h(\lambda_j)$ are assumed to be determined by Eq. (4.8).

To avoid too tedious expressions for many-particle Bethe functions in the ω-space, it is convenient to rewrite Eqs. (5.3) in the auxiliary τ-space,

$$(i\partial_{\tau_1} + i\partial_{\tau_2} + E)F(\tau_1, \tau_2) = 0 \qquad (106)$$

$$F_F(\tau_1, \tau_2 | \epsilon_1, \epsilon_2) = e^{i(\epsilon_1 \tau_1 + \epsilon_2 \tau_2)} \qquad (107)$$

$$F_B(\tau_1, \tau_2 | \lambda_1, \lambda_2) = A(\tau_1 - \tau_2)e^{i(\lambda_1 \tau_1 + \lambda_2 \tau_2)}, \qquad (108)$$

where the two-particle scattering function is given by

$$A(\tau_1 - \tau_2) = 1 + \frac{i}{h_1 - h_2}\mathrm{sgn}(\tau_1 - \tau_2). \qquad (109)$$

The Bethe basis (5.4c) describes the free polariton field in terms of particles scattering on each other. While the basis of free polaritons (5.4b) does not admit an analytical continuation to the complex ϵ-plane, because this leads to states which cannot be normalized, one can construct a normalizable state of the Bethe particles grouping the complex rapidities h_1 and h_2 into a two-particle string,

$$h_1 = H + i/2, \quad h_2 = H - i/2, \qquad (110)$$

where H is a common real part, which we will call a "carrying rapidity" of a string. The wave function of the two-particle string describes a bound state of Bethe particles and has the form

$$F_B(\tau_1, \tau_2 | H) = \exp\left\{i\Lambda(H)(\tau_1 + \tau_2) - \frac{\Gamma(H)}{2}|\tau_1 - \tau_2|\right\}, \qquad (111)$$

where the parameters $\Lambda(H)$ and $\Gamma(H)$ are found from the analytical continuation of the function $h(\lambda)$ in the complex λ-plane.

For what follows, it is convenient to derive the Bethe ansatz equations which determine the spectrum of the system. Making use of the dressing equations (4.20a), one can rewrite the Bethe wave function (5.4c) in the real x-space,

$$\Psi(x_1, x_2|\lambda_1, \lambda_2) = \int_{-\infty}^{\infty} d\tau_1 d\tau_2 u(x_1, \tau_1) u(x_2, \tau_2)$$
$$\times F(\tau_1, \tau_2|\lambda_1, \lambda_2), \qquad (112)$$

where the dressing function $u(x, \tau)$ is defined in Eq. (4.20b). If now we put the system in a box of large size L, the periodic boundary conditions (3.8) lead to the following Bethe ansatz equations:

$$e^{ik_j L} = -\prod_{l=1,2} \frac{h_j - h_l - i}{h_j - h_l + i}, \qquad (113)$$

where $k_j \equiv k(\lambda_j)$. To get Eqs. (5.8), we have assumed that the box size L is so large that we can use the asymptotic expressions for the wave function at the points $x = \pm L/2$. Possible solutions of the Bethe ansatz equations of the generalized QMB model will be studied in Sec. VII.

5.2   Many-particle eigenstates

Now we are able to generalized the results obtained to the case of N-particle eigenstates of the free polariton field. The N-particle Bethe wave functions are written in the form

$$F_B(\{\tau_j\}|\{h_j\}) = A(\{\tau_j\}|\{h_j\}) \prod_{j=1}^{N} e^{i\lambda_j \tau_j}, \qquad (114)$$

where $E = \sum \lambda(h_j)$ is the eigenenergy, and the many-particle scattering function is factorized into two-particle ones,

$$A(\{\tau_j\}|\{h_j\}) = \prod_{j<l} A(\tau_j, \tau_l|h_j, h_l). \qquad (115)$$

The N-particle states are characterized by a set of rapidities $\{h_j, j = 1, \ldots, N\}$, which are found from the Bethe ansatz equations

$$e^{ik_j L} = -\prod_{l=1}^{N} \frac{h_j - h_l - i}{h_j - h_l + i}. \qquad (116)$$

The spectrum of the system is described in terms of the Bethe strings

$$h_l^{(\alpha,n)} = H^{(\alpha,n)} + \frac{i}{2}(n+1-2l). \qquad (117)$$

of different orders n and with different carrying rapidities $H^{(\alpha, n)}$. It will be shown below that namely this Bethe basis is not destroyed by the polariton-atom scattering.

6.   Exact Diagonalization of Generalized QMB Model

6.1   Two-particle scattering matrix

We look for two-particle eigenstates of the generalized QMB model in the form

$$
\begin{aligned}
|\Psi_N\rangle &= \left[\int_C \frac{d\omega_1}{2\pi} \frac{d\omega_2}{2\pi} z(\omega_1) z(\omega_2) F(\omega_1, \omega_2) p^\dagger(\omega_1) p^\dagger(\omega_2) \right. \\
&\quad + \sum_{a=1}^M \int_C \frac{d\omega}{2\pi} z(\omega) G_a(\omega) p^\dagger(\omega) \\
&\quad + \left. \sum_{a_1=1}^M \sum_{a_2=1}^M H_{a_1 a_2} \sigma_{a_1}^+ \sigma_{a_2}^+ + \right] |0\rangle.
\end{aligned}
\tag{118}
$$

The Schrödinger equation in the auxiliary space then reads

$$
\begin{aligned}
(-i\partial_{\tau_1} - i\partial_{\tau_2} - E)[F(\tau_1, \tau_2) + F(\tau_2, \tau_1)] &= \sqrt{\gamma} \\
\times \sum_{a=1}^M [G_a(\tau_1)\delta(\tau_2 - x_a) + G_a(\tau_2)\delta(\tau_1 - x_a)]
\end{aligned}
\tag{119}
$$

$$
\begin{aligned}
(-i\partial_\tau + \omega_{12} - E) G_a(\tau) &= \sqrt{\gamma} \\
\times \left[\int_{-\infty}^\infty d\tau' \zeta(\tau' - x_a)[F(\tau, \tau') + F(\tau', \tau)] \right. \\
\left. + \sum_{b \neq a} (H_{ab} + H_{ba}) \delta(\tau - x_b) \right]
\end{aligned}
\tag{120}
$$

$$
\begin{aligned}
(2\omega_{12} - E)(H_{a_1 a_2} + H_{a_2 a_1}) &= \sqrt{\gamma} \\
\times \int_{-\infty}^\infty d\tau [G_{a_1}(\tau)\zeta(\tau - x_{a_2}) + G_{a_2}(\tau)\zeta(\tau - x_{a_1})],
\end{aligned}
\tag{121}
$$

where E is the eigenenergy and

$$
\zeta(\tau - x_a) = \int_C \frac{d\omega}{2\pi} z^2(\omega) e^{i\omega(x_a - \tau)}.
\tag{122}
$$

To get Eqs. (6.2) and (6.3) we have accounted for the gas approximation and used the modified operator of the polariton-atom coupling (4.14). The second term in r.h.s. of Eq. (6.2b) does not contain the term H_{aa}, because the two-level atom cannot be excited twice. The absence of this term generates an effective particle-particle scattering and leads to nontrivial solutions of Eqs. (6.2). It is clear that if the atomic spin operators are replaced by some Bose operators, a solution of the corresponding two-particle Schrödinger equation is immediately found as the product of two one-particle wave functions, as it has to take place for the Hamiltonians of the second order in Bose operators.

In our case of the two-level model of the atom, the constrain $(\sigma_a^\pm)^2 = 0$ leads to an effective polariton-polariton coupling, therefore we look for the two-particle wave function in the form

$$F(\tau_1, \tau_2 | \lambda_1, \lambda_2) = A(\tau_1 - \tau_2) f(\tau_1 | \lambda_1) f(\tau_2 | \lambda_2), \tag{123}$$

where $A(\tau_1 - \tau_2)$ is an unknown scattering function depending only on the difference of the particle "coordinates", and the one-particle functions $f(\tau_j | \lambda_j)$ are determined in Eq. (4.13). The substitution of Eq. (6.4a) into Eq. (6.2a) yields

$$\begin{aligned} G_a(\tau | \lambda_1, \lambda_2) = \; & A(\tau - x_a) f(\tau | \lambda_1) g_a(\lambda_2) \\ & + A(x_a - \tau) g_a(\lambda_1) f(\tau | \lambda_2), \end{aligned} \tag{124}$$

where the atomic wave functions $g_a(\lambda)$ are determined in Eq. (4.15). Assuming now that the function $H_{a_1 a_2}$ is given by

$$\begin{aligned} H_{a_1 a_2} + H_{a_2 a_1} = \; & A(x_{a_1} - x_{a_2}) g_{a_1}(\lambda_1) g_{a_2}(\lambda_2) \\ & + A(x_{a_2} - x_{a_1}) g_{a_1}(\lambda_2) g_{a_2}(\lambda_1) \end{aligned} \tag{125}$$

we find from Eq. (6.4b) the following equation for the scattering function:

$$\begin{aligned} & -i[f(\tau | \lambda_1) g_a(\lambda_2) A'(\tau - x_a) + g_a(\lambda_1) f(\tau | \lambda_2) A'(x_a - \tau)] \\ & + 2\sqrt{\gamma} A(0) g_a(\lambda_1) g_a(\lambda_2) \delta(\tau - x_a) \\ & = \int_{-\infty}^{\infty} d\tau \zeta(\tau' - x_a) \\ & \times \{[A(\tau - \tau') - A(\tau - x_a)] f(\tau | \lambda_1) f(\tau' | \lambda_2) \\ & + [A(\tau' - \tau) - A(x_a - \tau)] f(\tau | \lambda_2) f(\tau' | \lambda_1)\}. \end{aligned} \tag{126}$$

In the quantum version of the McCall-Hahn theory, the function $\zeta(\tau - x_a)$ is replaced by $\delta(\tau - x_a)$, and the r.h.s. of Eq. (6.5) vanishes. That immediately leads to the discontinuous solution (3.7c). As it has been discussed in Sec. III, the pointlike model of the photon-atom coupling is only a convenient mathematical trick to eliminate the effects of atom-atom interaction from the model Hamiltonian Hamiltonian and the Schrödinger equation. In our case of the nonlocal polariton-atom coupling, Eqs. (6.4) and (6.5) incorporate effects of the atom-atom interaction. Therefore, as well as in the one-polariton problem, we have to account for the gas approximation selecting appropriate solutions of these equations.

In the one-atom problem, $M = 1$, despite the nonlocal behavior of the function $\zeta(\tau - x_a)$, the solution of Eq. (6.5) is given by [16]

$$A(\tau_1 - \tau_2 | h(\lambda_1), h(\lambda_2)) = 1 + \frac{i}{h(\lambda_1) - h(\lambda_2)} \text{sgn}(\tau_1 - \tau_2) \qquad (127)$$

with the rapidity $h(\lambda)$ defined in Eqs. (4.8). It is easy to see also that the function (6.6) is the exact solution of Eq. (6.5) if the interatomic separations are much greater than the vicinity in which the function $\zeta(\tau - x_a)$ sufficiently differs from zero. In the opposite case, the expression (6.6) contains some additional function $\delta A(\tau_1 - \tau_2)$ which incorporates effects of atom-atom coupling and has to be omitted in the gas approximation. In a similar manner, one can show that the functions (6.4b) and (6.4c) obey Eq. (6.2c).

It is interesting to note also that the function δA vanishes for large interparticle distances $|\tau_1 - \tau_2|$, and thus does not contribute to the asymptotic behavior of the exact solution of Eq. (6.5),

$$A(\tau | h_1, h_2) = \frac{1}{h_1 - h_2} \begin{cases} h_1 - h_2 - i & , \tau \to -\infty \\ h_1 - h_2 + i & , \tau \to \infty. \end{cases} \qquad (128)$$

This allows us to assume that the generalized QMB model may be *asymptotically* integrable even beyond the gas approximation.

The two-particle scattering matrix

$$S(h_1, h_2) \equiv \frac{A(\tau_1 > \tau_2)}{A(\tau_1 < \tau_2)} = \frac{h_1 - h_2 + i}{h_1 - h_2 - i} \qquad (129)$$

is an obvious solution of the Yang-Baxter equations [25, 26, 27], and hence a many-particle scattering process of the auxiliary particles is factorized into two-particle ones. Thus, we have shown that the polariton atom scattering does not destroy the Bethe basis of the free polariton field defined in the previous Section. The auxiliary two-particle wave function for the polariton-atom system (6.4a) is obtained from the auxiliary Bethe wave function for the free polariton field (5.4c) by the replacement of the exponential factors $\exp(i\lambda_j\tau_j)$ by the functions $f(\tau_j | \lambda_j)$ accounting for the particle-atom scattering, which determines also the expression for the rapidity $h(\lambda)$.

6.2   Bethe ansatz equations

The auxiliary many-particle wave functions of the generalized QMB model are thus written in the form

$$F(\{\tau_j\} | \{h_j\}) = \prod_{j<l} A(\tau_j, \tau_l | h_j, h_l) \prod_{j=1}^{N} f(\tau_j, \lambda_j), \qquad (130)$$

while the many-polariton wave functions in the real x-space are determined by the dressing equation

$$\Psi(\{x_j\}|\{\lambda_j\}) = \int_{-\infty}^{\infty} d\tau_1 \ldots d\tau_N F(\{\tau_j\}|\{\lambda_j\})$$
$$\times \prod_{j=1}^{N} u(x_j, \tau_j). \tag{131}$$

The Bethe ansatz equations take now the form

$$e^{ik_j L} \left(\frac{h_j - i/2}{h_j + i/2} \right)^M = -\prod_{l=1}^{N} \frac{h_j - h_l - i}{h_j - h_l + i}, \tag{132}$$

$$E = \sum_{j=1}^{N} \lambda_j \tag{133}$$

where the scattering matrix $S(h_j)$ (4.12) accounts for the polariton scattering on the atomic system, while the scattering matrix $S(h_j - h_l)$ (6.8) accounts for the polariton-polariton scattering. As $L \to \infty$, the full spectrum of the polaritons plus atoms system is described in terms of Bethe strings

$$h_l^{(\alpha,n)} = H^{(\alpha,n)} + \frac{i}{2}(n + 1 - 2l). \tag{134}$$

of different orders n and with different carrying rapidities $H^{(\alpha,n)}$. The eigenenergy of a state with a given set of rapidities h_j is determined by Eq. (6.11b), where the frequencies λ_j are found from the analytical continuation of Eqs. (4.8) in the complex plane. Correspondingly, the momenta k_j are determined by the analytical continuation of the dispersion relation (2.16).

At last, it should be emphasized that the string (6.12) is a solution of the Bethe ansatz equations if and only if the imaginary parts of the rapidity h_j and the corresponding momentum k_j have the same sign,

$$\text{sgn}(\text{Im}\, k_j) = \text{sgn}(\text{Im}\, h_j). \tag{135}$$

This necessary condition imposes, as it will be shown in the next Section, strong restrictions on possible values of the carrying rapidity, eigenenergy and momentum of the strings.

7. SIT Pulses in Dispersive Medium

Hitherto we have studied the effective model (2.19) not specifying the atomic form factor $z(\omega)$ and the dispersion relation $k(\omega)$ which, of course, to be different for different gap media. Therefore, our main result given in Eqs. (6.11) is valid for arbitrary $z(\omega)$ and $k(\omega)$ and has a correct transition to the case of empty space.

For a further analysis of the problem, we need to specify the functions $z(\omega)$ and $k(\omega)$. In this paper we confine ourselves to the case of a dispersive medium where these functions are determined in Eqs. (2.16) and (2.19d) in terms of the medium's refractive index. Correspondingly, the relationship between the rapidity and the frequency is given by

$$h(\omega) = \frac{\omega - \omega_{12}}{\omega n^3(\omega)} \qquad (136)$$

For the sake of convenience, we redefined here the function $h(\omega)$ multiplying Eq. (4.8c) by $\beta = \gamma/\omega_{12}$. In what follows we also neglect a small difference between the atomic transition frequencies in empty space and in the medium, $\bar{\omega}_{12} \approx \omega_{12}$. A string solution (6.12) containing N particles takes now the form

$$h_j = H + \frac{i\beta}{2}(N + 1 - 2j), \qquad (137)$$

where the carrying rapidity H is arbitrary.

7.1   Ordinary solitons

To avoid a possible confusing, in what follows we use the term "string" for a solution of the Bethe ansatz equations for rapidities h_j and the term "soliton" for corresponding constructions in terms of particle frequencies ω_j and momenta k_j. We start our analysis from the simplest case of a soliton in which the real parts of frequencies lie outside the gap, $\mathrm{Re}\,\omega_j \in C = C_- + C_+$, where $C_- = (0, \Omega_\perp)$ and $C_+ = (\Omega_\|, \infty)$. Then, the analytical continuations of the functions $k(\omega)$ and $h(\omega)$ to the complex plane are given by

$$k(\omega) \simeq k(\xi) + i\eta k'(\xi), \quad h(\omega) \simeq h(\xi) + i\eta h'(\xi), \qquad (138)$$

where ξ and η are respectively the real and imaginary parts of ω, $\omega = \xi + i\eta$.

Since $k'(\xi) \equiv dk/d\xi > 0$, the imaginary parts of the momentum and the rapidity have the same sign only if $h'(\xi) \equiv dh/d\xi > 0$ as well. Under this condition the string in the h-space (7.2) has the following soliton image in the ω- and k-spaces:

$$\omega_j = \xi + \frac{i\beta}{2h'(\xi)}(N + 1 - 2j) \qquad (139)$$

$$k_j = k(\xi) + \frac{i\beta}{2}\frac{k'(\xi)}{h'(\xi)}(N + 1 - 2j), \qquad (140)$$

where a common real part of particle frequencies ξ is found from two conditions

$$h(\xi) = H, \quad h'(\xi) > 0. \qquad (141)$$

If the atomic transition frequency ω_{12} lies below the gap, $\omega_{12} < \Omega_{\perp}$, the function $h(\xi)$, as it is easy to see from its plot, has the maximum $h = h_{\max}$ at the point $\xi = \xi_{\max} < \Omega_{\perp}$. On the intervals $C = (\xi_{\max}, \Omega_{\perp})$ and C_+, $h'(\xi)$ is negative, and hence these intervals are forbidden for all solitons except one-particle ones ($N = 1$), when $\operatorname{Im} k(\xi) = \operatorname{Im} h(\xi) = 0$. If ξ lies far from the gap, the solution (7.4) is smoothly transformed into the ordinary quantum soliton of QMB model discussed in Sec. III. But in a sharp contrast to the case of empty space, the soliton exist only if its carrying frequency ξ lies below the point $\xi_{\max}$, $\xi < \xi_{\max}$. Moreover, in real physical systems the parameter β is so small that only solitons containing a macroscopically large number of particles N could be shorter the characteristic relaxation times in the atomic system, $\tau_s \sim (\beta N)^{-1} < \tau_{\text{rel}}$, to propagate without dissipation. As $\xi \to h_{\max} - 0$, the soliton is squeezed because of $h'(\xi) \to 0$, and can become shorter relaxation times even for a few particles N. In the case of too short solitons, we cannot keep in Eqs. (7.3) only the terms of the first order in η, therefore the vicinity of the point $\xi_{\max}$ requires additional studies.

If ω_{12} lies inside the gap, $\Omega_{\perp} < \omega_{12} < \Omega_{\parallel}$, $h'(\xi)$ is positive for $\xi \in C_-$ and still negative for $\xi \in C_+$. Correspondingly, the soliton exist now only below the gap, while the frequency interval above the gap is still forbidden.

At last, if ω_{12} lies above the gap, $h'(\xi)$ is positive for all $\xi \in C$, and there is no forbidden intervals. But in this case, the string (7.1) with a negative H has two different soliton images (7.4), because the equation $h(\xi) = H$ for $H < 0$ has now two roots $\xi_1 \in C_-$ and $\xi_2 \in C_+$. While the string with $H \in (0, 1)$ is mapped into a single soliton with the carrying frequency $\xi \in C_+$.

To estimate the velocity of a soliton propagation, one can multiply Eqs. (6.11a) to obtain

$$\exp\left\{ iL \sum_{j=1}^{N} k_j - 2iM \tan^{-1} \frac{\beta N}{2H} \right\} = 1. \tag{142}$$

Representing now the total number of atoms in the atomic system as $M = \rho L$, where ρ is the linear density of the number of atoms, one can define the carrying momentum of the soliton inside the atomic system from the condition $\exp(iKNL) = 1$, where

$$K(\xi) = k(\xi) - \frac{2\rho}{N} \arctan \frac{\beta N}{2h(\xi)}, \tag{143}$$

while the carrying momentum of the soliton outside the atomic system $k(\xi)$ is given by $k(\xi) = N^{-1} \sum k_j$. If the velocity of the soliton propagation is defined as the group velocity of a wave packet, $V_s = d\xi/dK(\xi)$, we obtain

$$\frac{1}{V_s} = \frac{1}{v_s} + \frac{\rho\beta}{h^2(\xi) + (\beta N/2)^2} \frac{dh(\xi)}{d\xi}, \tag{144}$$

where the soliton's velocity outside the atomic system is determined as $v_s = d\xi/dk(\xi)$. Eq. (7.7) naturally generalizes the expression (3.19) of the McCall-Hahn theory to the case of a dispersive medium. Because the solitons exist only when $h'(\xi) > 0$, the second term is positive to describe a soliton's delay due to the polariton-atom scattering.

The highly unusual behavior of the ordinary solitons of the generalized QMB model occurs for an arbitrary small gap and has a clear physical meaning. It is easy to understand that the necessary condition (6.13) for string solutions of the Bethe ansatz equations is fulfilled if and only if the two-particle scattering matrix (6.8) corresponds to an attractive effective particle-particle coupling. In empty space, where $k(\omega) = \omega$, the two-particle scattering matrix with the rapidity

$$h_{\text{emp}}(\omega) = \frac{\xi^2 + \eta^2 - \omega_{12}}{\xi^2 + \eta^2} + \frac{i\eta\omega_{12}}{\xi^2 + \eta^2} \tag{145}$$

corresponds to an attractive coupling at all $\xi \in (-\infty, \infty)$, therefore solitons of the QMB model exist with an arbitrary carrying frequency ξ. In the case of a dispersive medium, we deal with a physical system which exhibits an alternative character of an effective particle-particle coupling. For example, if ω_{12} lies above the gap, particle-particle coupling is attractive for all particles, and solitons exist with any $\xi \in C$. If ω_{12} lies inside the gap, particles below the gap still exhibit an attractive coupling, while the interaction between particles lying above the gap becomes now repulsive, and hence they do not compose any bound complexes.

7.2   Heavy gap solitons

Let us now try to map a string (7.2) to a soliton in which the real parts of all the frequencies lie inside the gap, $\xi \in \mathcal{G} = (\Omega_\perp, \Omega_\parallel)$. We need first to chose an appropriate branch of the function $n(\omega)$. Let $n(\xi \pm i0) = \pm i\nu(\xi)$, where $\nu(\xi) = \sqrt{|\varepsilon(\xi)|}$. The analytical continuations of the functions $k(\omega)$ and $h(\omega)$ are then given by

$$k(\omega) \simeq \begin{cases} -\eta\kappa'(\xi) + i\kappa(\xi), & \eta > 0 \\ \eta\kappa'(\xi) - i\kappa(\xi), & \eta < 0 \end{cases} \tag{146}$$

$$h(\omega) \simeq \begin{cases} -\eta\phi'(\xi) + i\phi(\xi), & \eta > 0 \\ \eta\phi'(\xi) - i\phi(\xi), & \eta < 0 \end{cases} \tag{147}$$

where $\kappa(\xi) = \xi\nu(\xi)$, $\phi(\xi) = (\xi - \omega_{12})\chi(\xi)$, and $\chi(\xi) = [\xi\nu^3(\xi)]^{-1}$. Since $\kappa(\xi) > 0$, the necessary condition (6.13) is equivalent now to the condition $\phi(\xi) > 0$.

If ω_{12} lies below the gap, the function $\phi(\xi)$ is positive for all $\xi \in \mathcal{G}$, while at ω_{12} lying within the gap, $\phi(\xi) > 0$ only if $\xi \in (\omega_{12}, \Omega_\parallel)$. If ω_{12} lies above the gap, the function $\phi(\xi)$ is negative for all $\xi \in \mathcal{G}$. Thus, a frequency interval, in which "gap excitations" exhibit an attractive coupling, exists only under condition $\omega_{12} < \Omega_{\|}$, and one can look for a construction in ω- and k-spaces corresponding to a string solution (7.2).

Let us start with a string containing an even number of rapidities ("even" string), $N = 2l$. A pair of the complex conjugated rapidities $h_j = H + i\beta(l + 1/2 - j)$ and $h_j^* = H - i\beta(l + 1/2 - j)$, where $j = 1, \ldots, l$, is mapped to a pair of complex conjugated frequencies

$$\omega_j = \xi_j + i\eta_j, \quad \omega_j^* = \xi_j - i\eta_j, \tag{148}$$

where ξ_j and $\eta_j > 0$ are determined by

$$\phi(\xi_j) = \beta(l + 1/2 - j), \quad \eta_j = -H/\phi'(\xi_j). \tag{149}$$

A correspondent pair of momenta is given by

$$k_j = \eta_j |\kappa'(\xi_j)| + i\kappa(\xi_j), \quad k_j^* = \eta_j |\kappa'(\xi_j)| - i\kappa(\xi_j), \tag{150}$$

where we took into account that $\kappa'(\xi) < 0$. Since the function $\phi'(\xi)$ is positive on the allowed interval where $\phi > 0$ and the imaginary parts of the frequencies η_j are chosen to be positive, the mapping (7.10) occurs only for a string with a negative carrying rapidity H.

Thus, an even string from the space of rapidities with $H < 0$ is mapped into a set of l pairs in the spaces of frequencies and momenta. We will use the phrase "gap soliton" to refer to the solution (7.10). Unlike an ordinary soliton, the imaginary part of a rapidity, $\operatorname{Im} h_j = \beta(l + 1/2 - j)$, determines now a common real part (carrying frequency), ξ_j, of a corresponding pair of frequencies, while the imaginary part of pair's frequencies, η_j, is determined by the carrying rapidity of a string H. Since the function $\phi(\xi)$ monotonically grows, carrying frequencies are positioned on the ξ-axis in the inverse order with respect to the growth of a rapidity's number in a string, $\omega_{12} < \xi_{\frac{N}{2}} < \ldots < \xi_1 < \Omega_{\|}$.

In particular case of a two-particle string, $N = 2$, the solution (7.10) describes a confined pair of "gap excitations" ("gap particles"). We use here the term "confinement" to emphasize that gap particles, unlike polaritons bounded in an ordinary soliton, do not exist separately from each other. Moreover, a pair of gap particles cannot be treated as a bound state of polaritons from different polariton branches (like a Wannier-Mott exciton in semiconductors), because under the gap excitation existence condition $\omega_{12} < \Omega_{\|}$ an interbranch polariton-polariton coupling is repulsive. The imaginary part of particle momenta $\kappa'(\xi)$ is nothing but the penetration length of the radiation with the frequency $\omega = \xi \in \mathcal{G}$ incident on the medium's interface [12]. In the one-particle problem, the wave function of a single gap particle, $f_g(x) \sim \exp[\kappa(\xi)x]$ or

$f_g(x) \sim \exp[-\kappa(\xi)x]$, cannot be normalized in bulk medium, and hence a single gap excitation does not exist. But already in the two-particle problem, the effective particle-particle coupling allows one to construct the normalizable wave function of a pair from two unnormalizable one-particle wave functions. Indeed, at large interparticle separations, the wave function of a pair is given by

$$\Psi(x_1, x_2|\xi) \sim \exp\{iq(\xi)(x_1 + x_2) - \kappa(\xi)|x_1 - x_2|\}, \qquad (151)$$

where the real part of particle momenta $q(\xi) = \eta|\kappa'(\xi)|$ describes propagation of the pair's center of gravity, while the penetration length $\kappa(\xi)$ determines the spatial size of a pair.

Despite a nonstring structure of gap soliton's frequencies and momenta, all pairs compose a many-pair bound state. Indeed, let us consider, for example, a four-particle gap soliton whose wave function at large interparticle separations is given by

$$\begin{aligned}
\Psi(x_1, \ldots, x_4) \sim \ & \theta(x_1 > \ldots > x_4) \\
& \times \exp[iq_1(x_1 + x_4) + iq_2(x_2 + x_3)] \\
& \times \exp[-\kappa_1 x_1 - \kappa_2 x_2 + \kappa_2 x_3 + \kappa_1 x_4] \\
& + \text{permutations of coordinates,} \qquad (152)
\end{aligned}$$

where $q_j = \eta|\kappa'(\xi_j)|$ are the real parts of momenta. It is clear that neither a single particle nor a pair of particles can be removed far from the rest particles, because the soliton wave function vanishes.

In contrast to the case of ordinary solitons, where only quite a "long" string, $N \gg 1$, corresponds to quite a "short" soliton, $\delta_{\mathrm{ord}} \sim (\beta k' N/h')^{-1}$, even a two-particle gap soliton ($l = 1$) has a very small size, $\delta_1 \sim \kappa^{-1}(\xi)$, which varies from ∞ to 0 when ω_{12} varies from $\Omega_\perp$ to $\Omega_\parallel$. Hence even few-pair gap solitons propagate without dissipation through absorbing atomic system. Moreover, the size of a soliton containing l pairs is determined by the value $\kappa(\xi_1)$, $\delta_l \sim \kappa^{-1}(\xi_1)$. Since $\kappa(\xi)$ falls when ξ grows, the gap soliton's size grows with the number of soliton's pairs. Therefore, if ω_{12} lies close to $\Omega_\parallel$, many-pair gap solitons, $l \gg 1$, become quite long to be destroyed by the relaxation processes in the atomic system.

Because of a smallness of the parameter β, all the carrying frequencies lie very close to ω_{12}. It means that for not too long strings the function $\phi(\xi)$ can be linearized at the point $\xi = \omega_{12}$,

$$\phi(\xi) \simeq a(\omega - \omega_{12}), \quad a = \chi(\omega_{12}). \qquad (153)$$

Then, all the imaginary parts of frequencies become equal to each other, $\eta_j = \eta = |H|/a$, while the real parts of frequencies are given by

$$\xi_j^{(0)} = \omega_{12} + (\beta/a)(l + 1/2 - j), \qquad (154)$$

where the upper index indicates that the frequencies are evaluated making use of the linear approximations (7.9) and (7.13), which we will call now the zero approximation. The eigenenergy of a gap soliton per one particle is then easily evaluated as

$$\epsilon_l^{(0)} = l^{-1} \sum_{j=1}^{l} \xi_j^{(0)} = \omega_{12} + \frac{\beta}{a} l. \tag{155}$$

The energy is independent of the soliton momentum per one particle,

$$q = l^{-1} \sum_{j=1}^{l} q_j \simeq \frac{\eta}{l} \sum_{j=1}^{l} |\kappa(\xi_j)|, \tag{156}$$

and hence gap soliton states are infinitely degenerate in the zero approximation.

To estimate the first order corrections to Eqs. (7.14), we need to keep the next term of the Taylor series (7.9b) for the function $h(\omega)$, $-i(\eta^2/2)\phi''(\xi)$, and the next term of the expansion of the function $\phi(\xi)$ at the point $\xi = \omega_{12}$,

$$\phi(\xi) \simeq a(\xi - \omega_{12}) + b(\xi - \omega_{12})^2, \tag{157}$$

where $b = \chi'(\omega_{12}) > 0$. Then we find

$$\xi_j - \xi_j^{(0)} = -(b/a)(\xi_j^{(0)} - \omega_{12})^2 + (b/a)\eta^2, \tag{158}$$

while the soliton momentum per one particle is still given by Eq. (7.15) and $\eta = -H/a$. The first term in r.h.s. of Eq. (7.17) determines obviously the width of a soliton zone, while the second one represents a contribution of the kinetic energy to the total soliton energy

$$\epsilon_l = \epsilon_l^{(0)} - \Delta_l + \frac{q^2}{2m_l}, \tag{159}$$

where

$$\Delta_l = \frac{b\beta^2}{12a^3}(4l^2 - 1), \quad m_l = \frac{a}{2b}\left(\frac{1}{l}\sum_{j=1}^{l}|\kappa'(\xi_j^{(0)})|\right)^2 \tag{160}$$

are the zone width and the mass of a gap soliton containing l pairs. As the number of soliton's pairs grows, the width of the soliton zone also grows, while the soliton's mass falls. But, for quite small momenta, gap solitons are heavy and even motionless at $q = 0$, because the group velocity of the soliton propagation, $v_l = d\epsilon_l/dq = q/m_l$, vanishes at $q = 0$.

Let us consider now an odd string, $N = 2l + 1$, in which one of rapidities has to lie on the real axis. The extra real rapidity can be mapped into the frequency

$\xi = \omega_{12} \in \mathcal{G}$. In the case of a one-particle string, $l = 0$, this condition determines the eigenenergy of a polariton-atom bound state predicted in Ref. [28] for the case of 3D PBG materials doped with resonance atoms. In this state the polariton field is localized in the vicinity of the atom. It is easy to understand that an odd gap soliton should be pinned to the atom to form many-polariton–atom bound state [16]. In this paper we do not discuss the soliton pinning, because its correct treatment lies obviously beyond our 1D model and requires special studies [17].

For a gap soliton, Eq. (7.6), describing a motion of the soliton's center of gravity, takes the form

$$\exp(iQNL) = 1, \quad Q(\epsilon_l) = q(\epsilon_l) - \frac{\rho}{l} \tan^{-1} \frac{\beta l}{H}, \tag{161}$$

where the soliton momentum outside the atomic system, $q(\epsilon_l)$, is defined in Eq. (7.15). Then, we find that the velocities of the gap soliton propagation inside the atomic system, $V_l = d\epsilon_l/dQ$, and outside the atomic system are related by

$$\frac{1}{V_l} = \frac{1}{v_l}\left(1 - \frac{al}{\sum_{j=1}^{l}|\kappa(\xi_j^{(0)})|} \frac{\rho\beta}{H^2 + \beta^2 l^2}\right), \tag{162}$$

where $dH/d\epsilon_l$ is found from Eq. (7.17)

$$\frac{dH}{d\epsilon_l} = \frac{dH}{dq}\frac{dq}{d\epsilon_l} = -\frac{1}{v_l}\left(\frac{1}{al}\sum_{j=1}^{l}|\kappa(\xi_j^{(0)})|\right)^{-1}. \tag{163}$$

In contrast to the case of ordinary solitons, the velocity of the gap soliton propagation inside the atomic system is greater than its velocity outside the atomic system, $V_l > v_l$. In other words, the particle-atom scattering speeds up a gap soliton.

But, it should be emphasized that the results obtained are valid only for quite small soliton momenta q when the terms in r.h.s. of Eq. (7.17) can be treated as small corrections to the zero approximation. At arbitrary q or for quite big solitons ($l \gg 1$), we cannot use only the first terms of the Taylor expansions (7.9) and (7.16) and need to solve the exact equations for soliton parameters ξ_j and η_j,

$$\mathrm{Re}\,h(\xi_j, \eta_j) = H, \quad \mathrm{Im}\,h(\xi_j, \eta_j) = \beta(l + 1/2 - j), \tag{164}$$

which require simple but numerical calculations.

If ω_{12} lies below the gap, frequencies ξ_j lie close to the frequency $\Omega_\perp$, and one needs to account for strong relaxation processes in the medium in the vicinity of the transverse frequency $\Omega_\perp$, which should lead to a quick dissipation of a gap soliton. When ω_{12} goes to $\Omega_\parallel$, all the pairs of complex conjugated frequencies of a gap soliton are squeezed to each other on the narrow interval between ω_{12} and $\Omega_\parallel$, and a soliton is deformed accordingly to Eqs. (7.22) The gap solitons vanish when ω_{12} goes out of the gap, but simultaneously ordinary solitons with carrying frequency on the interval C_+ appear. The moving of ω_{12} push out the gap solitons from the gap to the upper polariton branch.

7.3   Composite solitons

Hitherto we have looked for a soliton image of a string assuming that all the string's rapidities are mapped to the soliton's frequencies whose real parts lie either outside the gap (ordinary soliton) or inside the gap (gap soliton). In other words, we have used either Eqs. (7.3) or Eqs. (7.9) for all the rapidities of a string. It is easy to see that this assumption is not necessary, and one can try to look for a soliton structure mapping one part of string's rapidities to frequencies lying outside the gap, while the rest string's rapidities are mapped to frequencies lying inside the gap.

To clarify this idea, let us consider the simplest example of a three-particle string with a negative carrying rapidity H, assuming first that ω_{12} lies inside the gap. Then, the pair of the complex conjugated rapidities $h_1 = H + i\beta$ and $h_3 = h_1^* = H - i\beta$ is mapped to the pair of the complex conjugated frequencies $\omega_g = \xi_g + i\eta$ and $\omega_g^* = \xi_g - i\eta$, where $\xi_g = \omega_{12} + \beta/a \in \mathcal{G}$ and $\eta = |H|/a$, and momenta $k_g = q + i\kappa$ and $k_g^* = q - i\kappa$; where $q = \eta|\kappa'(\omega_{12})|$ and $\kappa = \kappa(\omega_{12})$. Under the condition $H < 0$, the rest real rapidity $h_2 = H$ is mapped to the real frequency $\xi_- \in C_-$ and the corresponding real momentum $k_- = k(\xi_-)$, where ξ_- is determined by $h(\xi_-) = H$. If ω_{12} lies above the gap, the gap part of the construction under consideration vanishes, but simultaneously the second possible solution, $\xi_+ \in C_+$, of the equation $h(\xi_+) = H$ appears. Therefore now, one can map the rapidities h_1 and h_3 to the pair of frequencies $\omega_+ = \xi_+ + i\beta/h'(\xi_+)$ and $\omega_+^* = \xi_+ - i\beta/h'(\xi_+)$.

In both cases all the three particles compose a bound state, that can be shown as well as in the case of gap solitons. For example, if $\omega_{12} \in \mathcal{G}$ the asymptotic wave function,

$$\Psi(x_1, x_2, x_3) \sim \theta(x_1 > x_2 > x_3) \exp\left[-\kappa x_1 + \kappa x_3\right]$$
$$\times \exp\left[iq(x_1 + x_3) + ik_- x_2\right]$$
$$+\text{permutations of coordinates}, \tag{165}$$

vanishes if one of the particles is removed far from the rest two ones. It is easy to see also that the eigenenergy of the states constructed, $E_{-g} = \xi_- + 2\xi_g$ or $E_{-+} = \xi_- + 2\xi_+$, is obviously a real value. In what follows we use the phrase "composite solitons" to refer to bound many-particle complexes containing particles from different interval, either C_- and $\mathcal{G}$ or C_- and C_+.

Now we are able to generalize the composite soliton construction to the many-particle case. Let us consider again an N-particle string with an negative H assuming that $\omega_{12} \in \mathcal{G}$. A pair of the complex conjugated rapidities or any number of such pairs could be mapped to corresponding pairs of frequencies whose real parts lie inside the gap, while the rest rapidities of the string are mapped to frequencies whose real parts lie below the gap. In other words, one part of the string rapidities is used to construct a "gap soliton", while the rest rapidities of the same string are used to construct an "ordinary soliton". All the particles together compose a bound many-particle complex, so that an composite soliton can be treated as a bound state of an "ordinary" and "gap soliton", but these "solitons" are different from corresponding ordinary and gap solitons discussed above and do not exist separately from each other.

If ω_{12} lies above the gap, all gap states vanish, but now an composite soliton can be constructed in the same manner as a bound state of two "ordinary solitons" from intervals C_- and C_+. It is interesting to note that a three-particle composite soliton can also be treat like the simplest "Wannier-Mott polariton exciton", which describes a bound state of two polaritons of the lower branch and one polariton of the upper branch, or vise versa, one polariton of the lower branch and two polariton of the upper one.

8.   Polarization Effects

The theory developed in the previous sections of the paper is valid provided the polarization polarization of radiation field incident on a medium is fixed, i. e. it does not depend on the time and spatial variables. Only under this condition one can represent the field and medium operators and the atomic dipole operator in the forms given in Eqs. (2.7), (2.9) and (2.10). If the polarization of an incident light is not fixed, we have to modify our theory to account for a degeneration of the excited level of atoms.

In this paper we consider the simplest model of a two-level atom two-level atom, in which the ground and excited level are assumed to have the angular momentum Angular momentum $J_1 = 0$ and $J_2 = 1$, respectively. The excited level is triply degenerate with respect to three possible angular momentum Angular momentum projections, $M_2 = -1, 0, 1$, on the x-axis, which is regarded as a quantization axis of atomic states. The transverse radiation field propagating along the x-axis is then coupled only to the atomic transitions $M_1 = 0 \leftrightarrow M_2 = \pm 1$ and the atomic dipole operator can be represented as

$$\mathbf{d} = d \sum_{\sigma = \pm} \left(\frac{1+i}{2} \mathbf{e}_- X_{\sigma 0} + \frac{1-i}{2} \mathbf{e}_+ X_{0\sigma} \right), \qquad (166)$$

where $e_\pm = \mathbf{e}_y \pm i\mathbf{e}_z$ and the Hubbard operators X_{ab} obey the commutation rule

$$[X_{ab}, X_{cd}] = \delta_{cb}X_{ad} - \delta_{ad}X_{cb}. \tag{167}$$

Here the index $a = 0, \sigma$ enumerates both the ground state ground state $(a = 0)$ and two excited states of interest $(\sigma = \pm)$.

In terms of the Hubbard operators the model Hamiltonian Hamiltonian takes the form

$$H = H_0 + V, \tag{168}$$

$$H_0 = \omega_{12} \sum_{a=1}^{M} X_{\sigma\sigma}^{(a)} + \int_C \frac{d\omega}{2\pi} \, \omega \, p_\sigma^\dagger(\omega)p_\sigma(\omega), \tag{169}$$

$$V = -\sqrt{\gamma} \sum_{a=1}^{M} \int_C \frac{d\omega}{2\pi} \, z(\omega) \, [p_\sigma(\omega)X_{\sigma 0}^{(a)}e^{ik(\omega)x_a}$$
$$+ X_{0\sigma}^{(a)}p_\sigma^\dagger(\omega)e^{-ik(\omega)x_a}], \tag{170}$$

where the summation over repeated polarization index $\sigma = \pm$ is assumed. The Schrödinger equation leads now to the following two-particle scattering matrix:

$$S_{\sigma_1\sigma_2}^{\sigma_1'\sigma_2'}(\lambda_1, \lambda_2) = a(h_1, h_2)I_{\sigma_1\sigma_2}^{\sigma_1'\sigma_2'} + b(h_1, h_2)P_{\sigma_1\sigma_2}^{\sigma_1'\sigma_2'}, \tag{171}$$

where

$$I_{\sigma_1\sigma_2}^{\sigma_1'\sigma_2'} = \delta_{\sigma_1\sigma_1'}\delta_{\sigma_2\sigma_2'}, \qquad P_{\sigma_1\sigma_2}^{\sigma_1'\sigma_2'} = \delta_{\sigma_1\sigma_2'}\delta_{\sigma_2\sigma_1'} \tag{172}$$

are the unit and permutation operators, and the scattering amplitudes

$$a(h_1, h_2) = \frac{h_1 - h_2}{h_1 - h_2 - i\beta}, \quad b(h_1, h_2) = \frac{i\beta}{h_1 - h_2 - i\beta} \tag{173}$$

correspond to processes in which particles either preserve their polarizations or exchange them with each other, respectively. Note also that Eqs. (8.3) is quite obvious due to $SU(2)$ symmetry of the problem.

The two-particle scattering matrix is obviously a solution of the Yang-Baxter equations [25, 26, 27], and hence the generalized QMB model is completely integrable in the case of different polarizations of the particles as well. The periodic boundary conditions (3.8) lead now to the hierarchy of the Bethe ansatz equations,

$$e^{ik_j L} \left(\frac{h_j - i\beta/2}{h_j + i\beta/2} \right)^M \prod_{\mu=1}^{N_+} \frac{h_j - r_\mu - i\beta/2}{h_j - r_\mu + i\beta/2} =$$

$$- \prod_{l=1}^{N} \frac{h_j - h_l - i\beta}{h_j - h_l + i\beta} \tag{174}$$

$$\prod_{j=1}^{N} \frac{r_\mu - h_j - i\beta/2}{r_\mu - h_j + i\beta/2} = - \prod_{\nu=1}^{N_+} \frac{r_\mu - r_\nu - i\beta}{r_\mu - r_\nu + i\beta} \tag{175}$$

$$E = \sum_{j=1}^{N} \omega_j \tag{176}$$

Since the model (8.2) has the additional discrete degree of freedom, the additional set of rapidities, describing the motion in the space of particle polarizations, is introduced, r_μ, $\mu = 1, \ldots, N_+$, where N_+ is the number of particles with the polarization $\sigma = +$. If $N_+ = 0$, i. e. all N particles have the same polarization $\sigma = -$, Eq. (8.4a) coincides with Eq. (6.11). The existence of the particles with $\sigma = +$ is described in terms of N_+ waves of polarization. The l.h.s. of Eq. (8.4a) acquires the additional factor accounting for scattering of particles on the waves of polarization, while Eq. (8.4b) describes scattering of the waves of polarization on each other.

In the limit of empty space, Eqs. (8.4) coincide with the hierarchy of the Bethe ansatz equations in the color QMB model [29]. It can be shown that the polarization of a soliton in empty space is independent of time and spatial variables. In the case of a gap medium, due to the nontrivial behavior of the soliton solutions discussed above, we have the right to expect an appearance also nontrivial polarization effects, which we hope to study elsewhere.

References

[1] S. L. McCall and E. L. Hahn, Phys. Rev. Lett. **18**, 908 (1967); Phys. Rev. **183**, 457 (1969).

[2] G. L. Lamb, Jr., **43**, 99 (1971); *Elements of Soliton Theory* (Wiley, New York, 1980).

[3] M. J. Ablowitz, D. J. Kaup, A. C. Newell, and H. Segur, Studies Appl. Math. **53**, 249 (1974).

[4] M. J. Ablowitz, D. J. Kaup, and A. C. Newell, J. Math. Phys. **15**, 1852 (1974).

[5] D. J. Kaup, **16**, 704 (1977).

[6] V. E. Zakharov, S. V. Manakov, S. P. Novikov, and L. P. Pitaevskii, *Theory of Solitons: the Inverse Scattering Method* (Consultants Bureau, New York, 1984).

[7] A. I. Maimistov, A. M. Basharov, S. O. Elyutin, and Yu. M. Sklyarov, Phys. Rep. **191**, 1 (1990).

[8] V. I. Rupasov, Pis'ma Zh. Eksp. Teor. Fiz. **36**, 115 (1982) [JETP Lett. **36**, 142 (1982)]; Zh. Eksp. Teor. Fiz. **83**, 1711 (1982) [Sov. Phys. JETP **56**, 989 (1982)].

[9] V. I. Rupasov and V. I. Yudson, Zh. Eksp. Teor. Fiz. **86**, 819 (1984) [Sov. Phys. JETP **59**, 478 (1984)].

[10] H. B. Thacker, **53**, 253 (1981).

[11] Quantum effects in the context of SIT theory were discussed by A. LeClair, Nucl. Phys. B **450**, 753 (1995).

[12] L. D. Landau and E. M. Lifshitz, *Electrodynamics of Continuous Media* (Pergamon Press, Oxford, 1984).

[13] E. Yablonovitch, **58**, 2059 (1987).

[14] S. John, **58**, 2486 (1987); for a recent review, see the articles in *Photonic Bandgaps and Localization*, edited by C. Soukoulis (Plenum, New York, 1993).

[15] See, for a review, S. M. de Sterke and J. E. Sipe, in *Progress in Optics*, edited by E. Wolf, Vol. XXXIII, p. 203 (North-Holland, Amsterdam, 1994).

[16] V. I. Rupasov and M. Singh, **77**, 338 (1996); J. Phys. A **29**, L205 (1996); **54**, 3614 (1996).

[17] S. John and V. I. Rupasov, **79**, 821 (1997).

[18] See also S. John and V. I. Rupasov, Europhys. Lett. **46**, 326 (1999).

[19] A propagation of light pulses through a periodic array of thin resonance films has been discussed by B. I. Mantsyzov, Phys. Rev. A **51**, 4939 (1995), and A. Kozhekin and G. Kurizki, Phys. Rev. Lett. **74**, 5020 (1995).

[20] S. John and T. Quang, **52**, 4083 (1995).

[21] V. M. Agranovich, *Theory of Excitons* (Nauka, Moscow, 1968).

[22] A. S. Davydov, *Theory of Molecular Excitons* (Plenum, New York, 1971).

[23] L. Allen and J. H. Eberly, *Optical Resonance and Two-Level Atoms* (Wiley, New York, 1975).

[24] A similar treatment is applicable to all "impurity" problems solved by the Bethe ansatz [25].

[25] A. M. Tsvelick and P. B. Wiegmann, Adv. Phys. **32**, 453 (1983).

[26] N. Andrei, K. Furuya, and J. H. Lowenstein, Rev. Mod. Phys. **55**, 331 (1983).

[27] R. J. Baxter, *Exactly Solved Models in Statistical Mechanics* (Academic Press, London, 1982).

[28] S. John and J. Wang, bf 64, 2418 (1990); **43**, 12 772 (1991).

[29] V. Ya. Chernyak and V. I. Rupasov, Phys. Lett. **114A**, 77 (1986).

PHOTONIC CRYSTALS FOR COMMUNICATIONS

Miniaturised waveguide devices for novel integrated optics applications

Thomas F. Krauss

School of Physics and Astronomy, University of St. Andrews, St. Andrews, UK, KY16 9SS

Abstract   Photonic crystals promise to provide a new platform for advanced miniaturised communications systems. Photonic crystals interact with light on a wavelength scale, which allows the design of novel miniaturised components. The concepts enabling this technology are presented as well as the fabrication technology for semiconductor-based planar structures. Propagation losses and their origins and possible methods for reduction are discussed. One of the key property of photonic crystals that can be exploited for communication devices is their strong dispersion, which allows applications in pulse compression and wavelength selective devices. Examples for dispersive devices include coupled cavity and defect waveguides as well as the photonic crystal superprism.

Introduction

Photonic crystals provide a fascinating platform for a new generation of integrated optical devices and components. Circuits of similar integration density as hitherto only known from electronic VLSI can be envisaged, finally bringing the dream of true photonic integration to fruition. What sets photonic crystals apart from conventional integrated optical circuits is their ability to interact with light on a wavelength scale, thus allowing the creation of devices, components and circuits that are several orders of magnitude smaller than currently possible. Apart from miniaturisation., a key property of photonic crystal components is their "designer dispersion", i.e. their ability to implement desired dispersion characteristics into the circuit. Functionalities include optical delay lines ("slow light"), optical pulse compression/dilation and wavelength splitters. In order to realise these functions, we need to understand the fundamental properties of photonic crystals and be able to fully master the technology of making them. Another key aspect is the understanding of the origin of propagation losses and their reduction to acceptable levels. While competitive values in terms of "loss per device" have now been demonstrated, the full potential of photonic crystal

A.S. Shumovsky and V.I. Rupasov (eds.), Quantum Communication and Information Technologies, 251–271.
© 2003 *Kluwer Academic Publishers.*

circuits will only be realised once many components can be cascaded together in complex circuits.

In the following, the underlying concepts, technology and device realisations of planar photonic crystals will be discussed. Another theme of the paper is the identification of areas where photonic crystals have particularly favourable properties. The discussion is restricted to semiconductor-based structures, mainly GaAs/AlGaAs and silicon on insulator (SOI), which provide the required high refractive index contrast (n>2) and have a mature fabrication technology. Among the many possible realisations of photonic crystals, the triangular lattice of holes is chosen because of its high degree of symmetry that leads to large and robust bandgaps. All experimental work is conducted in the "holes in dielectric" instead of the reverse "pillars in air" geometry because of its connected nature and the fact that the dielectric provides the required vertical confinement.

1.   Fabrication

1.1   Lithography

Photonic crystals consist of features well below 1 µm in size, and need to be fabricated with a precision of tens of nanometres. In photonic crystal research, the most commonly used pattern definition technique is e-beam lithography, where a focussed beam of electrons is scanned across the surface to generate the pattern. The resolution of this process tends to be limited by the resists and substrates used. Although the beam can be focussed down to less than a 2 nm spot, minimum feature sizes are typically of order 10-20 nm, because of electron scattering in the resist and electron backscattering from the substrate. Despite this limitation, the achievable resolution is very impressive, as illustrated in Fig. 1.

State-of-the-Art e-beam pattern generators use 16 bit D-A converters, which means that patterns with a resolution of 65k by 65k pixels can be written; in other words, for a resolution of 10 nm, the maximum writefield that can be exposed is a 650 µm square, although this size may also be limited by writefield distortions that are machine-dependent. For larger writing areas, writefields are "stitched" together, using high-resolution laser laser interferometry stages. These stages operate with high precision, and have remaining stitching errors of typically 30-50 nm, which lead to unacceptable phase-jumps in the photonic crystal pattern. The critical parts of a photonic crystal circuits are therefore always placed in a single writefield.

E-beam lithography, being a serial process, presents inherent limitations to high-volume mass-manufacture, so it is worthwhile to explore other routes to generating photonic crystal patterns. Classical photolithography, which is used for the current generation of planar lightwave circuits, lacks the resolution

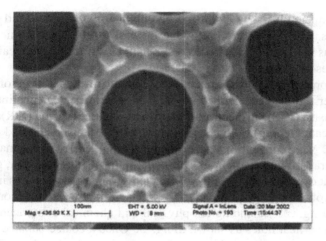

Figure 1.   Photonic crystal lattice exposed by electron-beam lithography after transfer into an SiO_2 mask layer. The central hole (200 nm diameter) highlights the high resolution achievable with the process, as it was generated as a dodekanon (12-sided polygon) and most of the 12 corners are still visible. The hole at the top right is distorted, possibly due to some resist-related fluctuations. The uneven layer on the ridges between the holes is partially eroded resist and the smooth layer underneath is patterned SiO_2.

to fabricate these submicron structures with sufficient precision. Extending photolithography into the deep UV, however, where excimer lasers laser are used as light sources, changes the picture. With illumination wavengths of 248 nm, moving on to 193 and eventually 157 nm, DUV lithography offers both the resolution and speed required for the high volume mass-manufacture of photonic circuits. Recent results obtained using 8" SOI wafers are very promising indeed and highlight the viability of the process [1].

1.2   Pattern transfer

Once written, the pattern is transferred into the semiconductor by dry etching techniques. Silicon-on-insulator based photonic crystals can be etched using the resist directly as a mask, because the required etch-depth is relatively low (200-300 nm). The etch chemistry is typically fluorine-based, e.g. SF6, but chlorine-based chemistries have also been used [1]. III-V based waveguides require a two-stage etching process because of the greater etching depth required [2]. A 200nm-thick SiO2 layer is typically used as an intermediate mask and is etched with fluorine chemistry (e.g. CHF3). The second, deep etching step is carried out with chlorine chemistry, which forms volatile compounds with the respective constituent atoms, e.g. GaCl3 (GaAs) and InCl3 (InP). Since InCl3 is not volatile until about 130-150°C [3], etching of InP-compounds requires elevated temperatures. Dry etching is a combination of physical and chemical components. The physical component arises from the bombardment of the reactive ions whereas the chemical component describes the chemical reaction between these ions and the etched material.

Etching processes where the plasma is generated in a separate part of the etching chamber and then accelerated towards the specimen, e.g. in an inductively coupled plasma (ICP), an electron-cyclotron resonance plasma (ECR) or an ion gun (Ion beam etching), allow separate control of the physical and chemical parameters. More conventional reactive ion etching, where the plasma is generated in a parallel-plate configuration directly in the etching chamber, gives less independent control, as most of the parameters are interlinked, but is inherently stable and robust. An illustration of the effects of the two etching components is shown in fig. 2. Balancing the physical and chemical components allows one to etch vertical holes as shown in fig. 3. Once a coarse balance has been achieved, the parameters such as gas flow, pressure, acceleration voltage and substrate temperature need to be optimised individually to optimise the process. Some results arising from such an optimisation are shown in fig. 3b) and c).

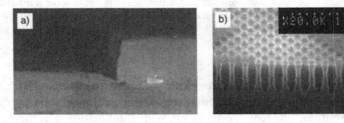

Figure 2.   Dry etching. a) Excessive physical component. The ion bombardment dominates the etching process which leads to an overcut profile and a notch in the etch floor ("Ricoche" effect caused by ions deflecting off the sidewall). b) Excessive chemical component leading to undercutting and layer selectivity in a GaAs/AlGaAs heterostructure. The chemical etching has preferentially removed layers of high Al-content that form the waveguide cladding, whereas the waveguide core half-way down the hole bulges out due to a lower lateral etch rate.

2.   Waveguide Propagation Losses

While two-dimensional photonic crystals have the advantage of being amenable to fabrication via the sophisticated planar technology developed by the silicon industry, their disadvantage is the lack of periodicity in the third dimension, so some other form of confinement is required. The most obvious approach is also the most commonly used, whereby the photonic crystal is formed in a slab waveguide structure, so photonic crystal effects in the plane are combined with total internal reflection (TIR) guiding in the vertical dimension. This seemingly simple "hybrid" combination of PhC + TIR has two major problems. Firstly, it

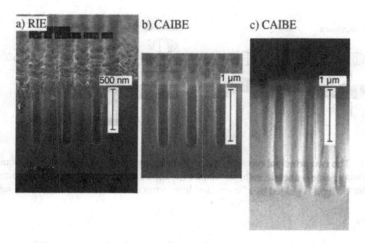

Figure 3. Cross-sections of photonic crystal lattices etched using different techniques. The respective lattices are all of different size (260-340 nm range) and have been scaled to allow a direct comparison of the aspect ratio. a) Reactive Ion Etching (RIE). Having optimised all parameters, an aspect ratio of 1:6 (hole diameter: etch depth) is the best we were able to achieve while maintaining verticality of the holes. b) and c) Chemically assisted ion beam etching (CAIBE). By allowing independent control of the relevant parameters, aspect ratios of up to 10 have been achieved, with further improvement possible. Parameters are beam voltage 300V, beam current 17mA, substrate temperature $105_{o}C$ and chlorine flow 10sccm. Etching time in b) is 15 min, whereas it is 30 min in c), enabled by the use of a more durable mask.

assumes that the periodic structure extends infinitely deep into the waveguide and secondly, it suffers from the fact that there is no waveguiding in the low-index region, i.e. the holes. Both of these effects lead to out-of-plane leakage and diffraction, which provide the major source of propagation loss. Additionally, there is roughness scattering arising from fabrication imperfections, lattice positioning and hole size variations. Finally, the finite size of waveguide components leads to losses, as every change in waveguide geometry constitutes a discontinuity that acts like a point scatterer. Which of these is dominant depends on the particular case and is difficult to assess in general, but some principal trends can be observed.

Before discussing the different loss mechanisms, we need to introduce the two types of waveguides commonly used, i.e. membrane and heterostructure. Membranes are typically made by patterning the waveguide core and completely removing the cladding, e.g. by selective wet etching, thus producing a suspended structure with air on both sides. Silicon on insulator (SOI) waveguides can also be classified as membranes, because the silicon waveguide core sits on a silica cladding that offers equally high vertical confinement. In con-

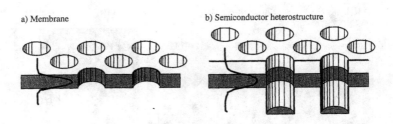

Figure 4.   The two principal types of planar photonic crystal waveguides. a) Membrane-type, where the waveguide is a semiconductor membrane suspended in air or silica. b) Heterostructure-type, where the light is guided by a "laser-like" heterostructure and sits on a high index substrate.

trast, the "laser-like" laser semiconductor heterostructures only offer moderate vertical confinement (typically between n=3.4 (core) and n=3.2 (cladding)). This leads to leakage losses into the high-index substrate, but offers all the possibilities inherent to semiconductor optoelectronics, such as gain, carrier effects and electro-optic tuning as well as better heatsinking. Channel waveguides are formed in these geometries by removing a single or multiple line of holes from the otherwise regular lattice.

2.1   Out-of plane leakage

Out-of plane leakage is the dominant loss mechanism for heterostructure waveguides. As all waveguide modes are situated above the light line, there is always the possibility of coupling to radiation modes. In simple terms, the origin of these losses is a) the lack of waveguiding in the holes, with resulting diffraction loss and b) the fact that the holes are not etched sufficiently deeply, so the tail of the mode radiates into the substrate. Both of these effects have been used to describe the losses in recent numerical treatments. A model by Hadley based on solving the 3D Helmoltz equation numerically [4] observes good agreement between the frequency ranges of high waveguide propagation losses and those where coupling to dominant radiation modes takes place. Correspondingly, there are also regions where coupling to radiation modes is weak and propagation loss is low. An alternative semi-analytical model proposed by Benisty [5] describes the losses of the structure by placing a material with complex refractive index into the holes, which combines the effects of diffractive scattering at the holes and of radiation into the substrate. The model achieves reasonable agreement with experimental data and concludes that losses increase parabolically with vertical index contrast. In other words, it suggests that the losses are lower for heterostructure waveguides than for membranes, which appears surprising - we would expect that higher vertical index contrast leads

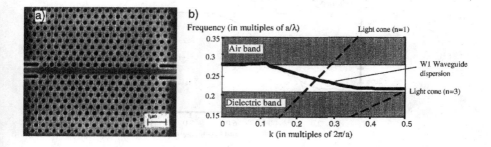

Figure 5. a) SEM micrograph of a W1 waveguide. The term "W1" refers to a waveguide consisting of a single line of missing holes. Here, the light enters from the left and exits to the right via ridge waveguides that have been tapered down to the size of the W1. The dark lines are deeply-etched trenches that laterally confine the ridge waveguides. b) Dispersion and the light cone for a W1 waveguide. For membrane-type waveguides, there is a region below the light cone (Here: frequency range between $0.21 < a/\lambda < 0.23$) where propagation is principally loss-less. Heterostructure waveguides tend to operate inside (or above) the light cone, i.e. there is the possibility of coupling to radiation modes at all operating frequencies.

to better confinement because it offers operation below the light line (fig. 5b). This apparent contradiction is addressed by Bogaerts [6] who has studied losses with an eigenmode expansion code and concluded that the losses, for small index contrast, indeed have the parabolic dependence proposed by Benisty, but then drop again once the index contrast is high enough for the waveguide to operate below the light line. Therefore, the model by Benisty appears correct for small index index contrast, but the inherent assumptions do not hold for high index contrast and do not account for loss-less Bloch modes. All of the models conclude that low loss propagation above the light line is indeed possible and that current waveguide designs are not yet optimised. The key parameters to improve are diffractive loss at the holes (reduced by smaller hole size), scattering into the substrate (reduced by deeper holes) and improvement of the waveguide geometry, away from the current asymmetric surface waveguides. As can be seen from the origins of these losses, most of the of the possible improvements are technology-dependent, which explains the large amount of effort currently devoted to dry etch technology with the aim of etching holes with higher aspect ratio. Another method involves embedding the waveguide in a 3D photonic crystal material, as recently proposed by [7], which would

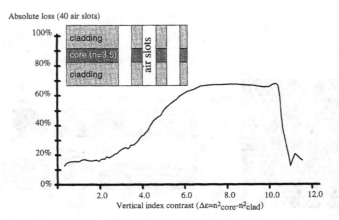

Figure 6.   Propagation loss of a waveguide mode after transversing 40 air slots as a function of vertical index contrast calculated numerically using an eigenmode expansion code. Please note that this is a 2D model, i.e. it assumes that the air features are slots rather than holes, so only gives a qualitative indication of losses in real photonic crystal waveguides. The losses increase quadratically from an initial low value, in agreement with [5], then saturate and eventually drop with the onset of Bloch modes around $\Delta\varepsilon \approx 11$.

eliminate the possibility of substrate leakage altogether. Tentative projections suggest that values around 10dB/cm or lower will indeed be possible, initially in W3 waveguides [8], but should eventually also be achieved in W1 waveguides with these improvements in both design and technology, as well as sidewall roughness, which is considered next.

2.2   Roughness scattering

Waveguides that operate below the light line and that are in principle loss-less, such as air-clad membranes and silicon on insulator (SOI) types mainly suffer from roughness scattering. Roughness scattering is a well-known challenge in high-contrast waveguides, since the loss scales with (n)3. Typical values of roughness are of order 10 nm rms, which is well sub-wavelength, but still significant because of the (n)3- dependence. In silicon-based waveguides, the surface roughness can be reduced significantly by wet oxidation and subsequent wet etching, as demonstrated recently on narrow ridge SOI waveguides [9] where a loss-figure of 0.8dB/cm was achieved. While this result provides a useful benchmark, it can not, however, be compared directly with photonic-crystal based waveguides. The 0.8dB/cm result was obtained with ridges that had been oxidised down a 50 nm Si-core. The effective index of such a guide is essentially that of the cladding, and it behaves mainly like a low-index waveguide, so the sharp bends and other attractive features of high-contrast waveguides are not possible. Nevertheless, this figure is clearly encouraging and points to the fact that low losses are ultimately achievable, even with high-index materials.

In photonic crystal waveguides, oxidation-smoothing has also shown promising results, as exemplified by W1 waveguides fabricated in SOI (Arentoft et al. 2002)[10]. The losses achieved in this system are 4dB/mm or lower, whereas the best results achieved with similar waveguides without oxidation are 6dB/mm [11]. While it is difficult to compare results from different groups that use different measurement techniques, the lower loss obtained with oxidation-smoothed guides supports the evidence that oxidation indeed reduces the sidewall roughness and leads to lower propagation losses.

2.3   Finite structures

The assumption of loss-less guiding breaks down as soon as discontinuities are introduced into the structure (fig. 7). A bend, for example, can be understood as the superposition of a an infinte guide and a point defect. Propagation is then only possible where there is an allowed state in both the waveguide and the point defect, i.e. where the two dispersion curves overlap. This limits the bandwidth of the bend (here: to the Q of the defect, which is indicated by the width of the horizontal dispersion line) and opens an exit route for radiation modes, because the dispersion of the point defect extends into the light cone.

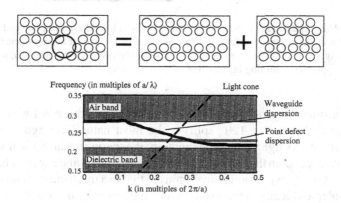

Figure 7.   Real structures are finite. A bend can be understood as the superposition of a line defect and a point defect. The point defect leads to a reduction of the bandwidth and a projection of the waveguide dispersion into the light cone causing radiation losses.

It is difficult to assess the magnitude of this effect, but according to [6], the out-of-plane scattering at discontinuities is larger for higher vertical index contrast structures, so such losses are expected to be higher in membranes than in heterostructures. Currently, there is little experimental evidence or 3D modelling available to draw a more quantitative conclusion, and we can also expect that the effect will be minimised by improved design.

3.   Y-junctions

While straight waveguides and bends have now been studied extensively, the very important problem of bends and junctions that is essential for the operation of more complex circuits has only recently received attention [12-14]. Photonic crystals offer a real advantage in this case, as they allow, in principle, to construct 60° or even 90° bends at junctions, unlike in conventional integrated optics, where splitting angles are restricted to values around 2°.

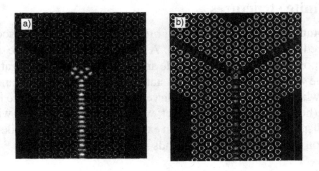

Figure 8.   Mode expansion at a Y-junction. Two additional holes have been reduced in order to highlight the mode expansion effect. Mode expansion still occurs at a simple junction, but not as obviously. b) By adding a single, smaller hole at the centre of the junction, the mode expansion is suppressed, resulting in clean and efficient splitting.

Let us consider the problem in more detail. If we join three W1 waveguides together, the 60° bend (or 120° split) is the most natural configuration in a triangular lattice. As light enters the junction, it experiences an optical volume that is slighty larger than that of the constituent waveguides and expands laterally (fig. 8a). This mode expansion is analogous to that in a multi-mode interference (MMI) coupler and leads to the excitation of higher-order modes at the output port. This issue can be addressed by placing a smaller hole at the centre of the junction. The optical volume is then reduced, the mode cannot expand and the excitation of higher order modes is suppressed (fig. 8b). As a result, efficient transmission over a wide bandwidth is observed [13].

This effect has also been observed experimentally. We used a configuration similar to fig. 8b) and placed two smaller holes of increasing size at the centre of the junction in order to achieve a graded transition (fig. 9a). The results are very encouraging, with a good balance between the two ports and a total

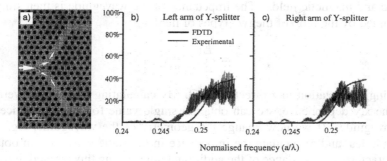

Figure 9. SEM Micrograph and performance of a Y-junction with additional holes at the centre to avoid mode expansion. The vertical scale in b), c) represents transmission relative to a W1 waveguide. The oscillations in the experimental curves are due to Fabry-Perot resonances between the cleaved facets with superimposed resonances arising from reflections at the junction. The fact that the 2D FDTD agrees well with the 3D experiment indicates low out-of plane losses. The total transmission is between 70-80% over a relatively broad spectral range (30-40 nm @ 1300 nm). The bandwidth is limited by the bend rather than the junction.

transmission around 80% being observed (fig. 9b,c). Two of these Y-junctions mounted "back-to back" constitute a Mach-Zehnder interferometer, which has also been operated successfully (passive, no tuning) with 28% transmission observed relative to a W1 waveguide. The design of the interferometer is extremely compact, with overall dimensions of the order of 10 x 10 μm^2.

3.1 Impedance matching

Each change in waveguide geometry presents a discontinuity. Even the symmetric, adjusted Y-junction of fig. 8b has a transmission that is theoretically limited to 44.4% per branch, the remainder of the light being reflected back. In order to improve the transmission further, one can take measures to match the junction to the incoming mode. The general idea is based on the concept of impedance matching that is very familiar to micowave engineers: if the incoming mode is presented with the same impedance in every part of the circuit, it will propagate without reflections. If the impedances of the different circuit elements are different, matching sections have to be incorporated in order to transform the impedances. One such impedance transformer that can be realised in the photonic crystal platform is a "double-stub tuner". A single stub is analogous to a $\lambda/4$ impedance transformer, also known as an antireflection coating in optics. A double stub offers more flexibility and generally features a broader bandwidth. Before discussing the design of such a double stub tuner, we need to understand the concept of impedances in photonic crystal waveguides. A mode of a given polarisation propagating in a waveguide is described by its electric and magnetic fields. The impedance of the waveguide is then simply determined by the ratio of the two types of field, i.e.

$$Z = \frac{E(x, y)}{H(x, y)}, \tag{1}$$

assuming propagation in z-direction [15]. By calculating a weighed average over the x,y dependence, we can obtain a single value for the impedance of the waveguide. If we now change the geometry of the waveguide, e.g. by adding holes, and recalculate the impedance in the same way, we can obtain the characteristic impedance of the additional hole. Using this method, it turns out that the impedance of an additional hole is mainly inductive, with a small capacitive component, i.e. it behaves like an L-C element. We now have a handle to tuning the circuit by designing holes with the appropriate L-C values to match the incoming waveguide an the junction. An example for a Y-junction design incorporating a double-stub tuner is shown in fig. 10. tuning hole impedance cannot be adjusted independently, 100% transmission is difficult to achieve.

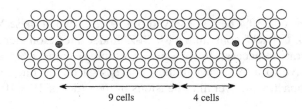

Figure 10.   Y-junction design incorporating a hole at the junction for mode matching and two tuning holes in the waveguide for impedance matching. Simulations using the multiple scattering technique [13] indicate that the tuning holes increase the transmission of the system from 88.8 % to 98.6 %. Because the L-C components of the

4.   Dispersion

Amongst the most exciting aspects of photonic crystals are their dispersive properties that manifest themselves as strong temporal and spatial effects. Spatially dispersive structures are essential for wavelength-selective functions, such as WDM filters and multiplexers, and are discussed in the next section. Temporal dispersion is essential for functions such as optical delay lines and for the compression and dilation of optical pulses ("pulse shaping"). This high degree of dispersion occurs because of the strong interaction between the propagating mode and the lattice. It is a general feature of periodic structures, such as Bragg gratings, that dispersion is high in close proximity to the band-edge. The range of this interaction grows as the index increases, which is the reason why photonic crystals interact strongly with the guided mode over a large frequency range.

The strong dispersion is also one of the key distinctions between photonic crystal waveguides and high index contrast ridge waveguides ("photonic wires"). While photonic wires can confine light equally strongly as photonic crystal waveguides and also guide light sharply around bends, they do not offer the many possibilities offered by the dispersive nature of the photonic lattice.

The reason for the strong dispersion, as mentioned above, is that the propagation characteristics change dramatically at points where the mode strongly interacts with the lattice, e.g. at band-edges, where the mode changes from a propagating to a standing and then an evanescent wave. The same effect also happens at mini-band edges, due to the anticrossing between modes of the similar symmetry, where the mode changes from one type to another, with a large resulting change in propagation constant [16].

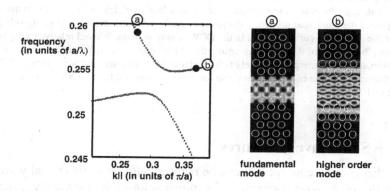

Figure 11. Detail of the anticrossing phenomenon observed in W3 photonic crystal waveguides. The mode profile and thereby the propagation constant changes dramatically near the mini-stopband, which highlights the dispersve nature of the waveguide.

4.1 Dispersion engineering

Coupled cavity waveguides (CCWs) provide a good example of the possibilities offered by photonic crystal waveguides for dispersion engineering. Coupled cavity waveguides (CCWs) [17,18] consist of chains of defects in an otherwise perfect lattice. They allow independent tuning of the spectral position, bandwidth and group velocity dispersion of the waveguide via the shape and size of the individual cavities and via the distance and lattice orientation between cavities. We have studied the propagation of light through a variety of

CCWs and have already found good transmission for closely coupled systems [19]. As an example, fig. 12 illustrates the example of a CCW for optical pulse-compression.

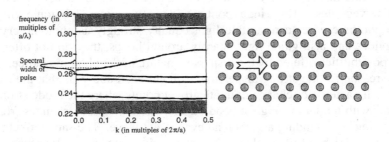

Figure 12.   Example of pulse compression in a coupled cavity waveguide. Assume a 100 fs long pulse at a wavelength of 1.55 μm enters the structure. The spectral width of the pulse corresponds to a frequency of 25 nm, or 0.004 a/λ for a lattice of a=420 nm. The pulse has been dilated, e.g. by travelling through a fibre, to a length of 1ps with the low-frequency ("red") part of the pulse leading. Since the lower frequency experiences a lower group velocity (slope of the dispersion curve, dk/dω) in the CCW, it is now being delayed relative to the high frequency ("blue") end of the pulse. Assuming that the ratio in group velocvity across the pulse is approximately 2, red travels effectively twice as far as blue. In a material with index n=3, 100 μm correspond to a time of 1ps, so the pulse can be recompressed to its original length over a distance of 100 μm.

4.2   Slow wave structures

Another excting concept that can be realised with photonic crystal waveguides is that of slow-wave structures, i.e. propagation of light at greatly reduced speed. Optical delay lines or even optical waveguide memory can be envisaged, as well as electro-optic slow wave modulators that are more or efficient because of the longer interaction time. A recent realisation of slow wave propagation in photonic crystal waveguides has already demonstrated a more than ten-fold reduction in propagation speed [20]. A major issue that needs to be addressed for these devices to become practical, however, is that of impedance or mode matching between the conventional and the slow wave structure, as mismatch inevitably leads to reflection losses.

5.   Spectral Dispersion

One of the remarkable properties of photonic crystals is their high spectral dispersion at selected operating points, which is exploited in devices such as "superprisms" [21]. The operating point of these devices typically lie near a bandedge. Fig. 13 shows such an operating point schematically, calculated for a triangular lattice of holes with a=320 nm and 40% fill-factor, that was realised in a GaAs/AlGaAs heterostructure. The stopband for $\Gamma - M$ direction opens up at $a/\lambda = 0.248$ as indicated. The same bandstructure is shown in fig. 13b as a wavevector diagram, whereby the thick curves represent the dispersion surfaces at $a/\lambda = 0.238$ and 0.254, respectively. Whereas the bandstructure represents the crystal by only showing the major symmetry axes, the wavevector diagram represents the different crystallographic axes as they occur in real space, so repeat every 60°. In other words, the wavevector diagram is obtained by "slicing" the bandstructure horizontally, at a given frequency.

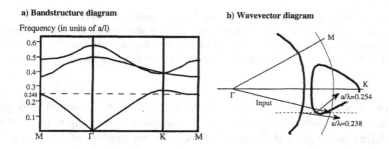

Figure 13. Bandstructure and wavevector diagram for a triangular lattice of holes with 40% air fill-factor. The wavevector diagram highlights the change of the dispersion surface around a/λ=0.24-0.25. The angular dispersion is obtained as follows. The input wave travels in the unpatterned semiconductor waveguide, which is isotropic and therefore represented by a circular dispersion surface. (dotted line). At the interface, the parallel wavevector is conserved, so all possible solutions lie on the dashed line. The intersection with the dispersion surface then determines the direction of the output wavevector. For a/λ=0.238, the output is almost collinear with the input wavevector and there is little change in the propagation direction. For a/λ=0.254, the shape of the dispersion surface has changed dramatically, with a corresponding large change in the output direction.

What is then the origin of the superprism phenomenon is the fact that the dispersion surface, i.e. the shape of the wavevector diagram, changes dramatically with a change in frequency. In the example, the dispersion surface changes from a near circular shape at $a/\lambda = 0.238$ to a set of triangles located at each K-point, because at $a/\lambda = 0.254$, a stopband has opened up in $\Gamma - M$ direction and the only remaining solutions are clustered around the K-point. This change of shape of the dispersion surface yields a large change in output direction as illustrated in fig. 13b.

Following the demonstration of the superprism effect in the autocloned 3D system [21], we have now also been able to observe it in a planar geometry [22,23]. Fig. 14 shows the experimental layout and resulting output as a

a) Layout   b) Output facet

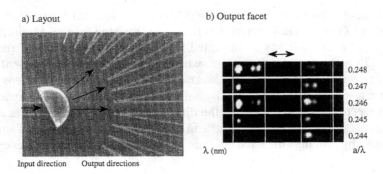

Input direction   Output directions

Figure 14.   Experimental geometry used to observe the superprism effect. The light is incident from the left, impinges on the photonic lattice and is dispersed into different output waveguides. The guides then carry the light to the output facet and are imaged onto a vidicon camera. b) is a montage of the signal on two subsequent output waveguides as a function of wavelength.

function of wavelength. At 1290nm, the light is mainly coupled into guide A. It then swings across to guide B as the wavelength is changed to 1310 nm. Considering the angular difference between the waveguides of $10^\circ/20nm$, the wavelength swing corresponds to a dispersion of $10^\circ/20nm$. For comparison, the wavelength swing predicted from fig. 14b is approximately $50^\circ/80nm$, which is of similar magnitude.

5.1   The mechanism of high spectral resolution

In order to understand the mechanism of this high dispersion, it is instructive to compare the mode of operation with that of a regular grating. For simplicity, let us assume a 1-D grating of the same periodicity, i.e. a=320 nm.

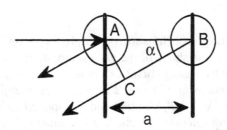

Figure 15.   Resolution of a grating of similar period as the superprism. The two circles at A and B represent the positions of the holes in an equivalent photonic lattice and are shown for illustrative purposes only.

In order for constructive interference to occur, the pathlength difference between the beam reflecting directly from point A and that reflecting from point B should be equal to a wavelength (or integer multiples), so $AB + BC = \lambda \Rightarrow \lambda = a + a \cos \alpha$. Inserting the same parameters as those used to calculate the superprism resolution ($\lambda_1 = 1290nm, \lambda_2 = 1310nm$, a=320 nm and using n=2.5 for the average refractive index of the lattice), we find that $\alpha_1 = 52.23°$ and, $\alpha_2 = 50.39°$, so the angular resolution is approximately $2°/20nm$ or 5 times lower than that of the superprism.

How is it possible that the superprism photonic lattice, which is effectively a 2-D grating, has a significantly higher resolution than a regular 1-D grating ? The answer is related to the group velocity, and is best understood via time-of-flight considerations.

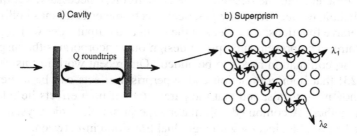

Figure 16. Resolution of a cavity and a superprism. High resolution in a cavity is achieved via a high number of roundtrips that effectively slow the light down. Similarly, the light in the superprism experiences many reflections and is also slowed down, leading to a low grup velocity.

Fundamental quantum mechanics postulates that

$$\Delta p \Delta x \geq \frac{\hbar}{2}, and \Delta E \Delta t \geq \frac{\hbar}{2} \qquad (2)$$

(Heisenberg uncertainty principle). In terms of a normal grating, this tells us that if the grating is large (large Δx), we can resolve a small Δp, which is equivalent to a small Δk or a large $\Delta \lambda$, so large gratings have a high wavelength resolution (because the light can sample many periods). In the case of a cavity (fig. 16a), the $\Delta p \Delta x$ explanation is more difficult, as the cavity is small (small Δx) yet we can achieve high resolution (small Δk), which is an apparent contradiction. This contradiction can be resolved by considering the $\Delta E \Delta t$ pair, i.e. high resolution (small ΔE) is obtained via a long time-of-flight (large Δt). Light goes around the cavity many times, so has a long pathlength and therefore high resolution. The superprism works in a similar way. Light inside the lattice is reflected many times and is thereby slowed down, so the high resolution (small

ΔE) is again achieved via a long interaction time (large Δt). It is no surprise, therefore, that the best operating point of the superprism is at the band-edge, where group velocity is lowest (fig. 13a). so long interaction times are achieved.

5.2   Limitations

While the fundamental superprism phenomenon is fascinating and has now been demonstrated in both 3-D and 2-D geometries, there are several design aspects that need to be addressed before superprism multiplexers can find practical applications. Firstly, the picture shown in fig. 13b is a simplification because it only shows a single propagating mode being excited in the lattice. In reality, multiple modes exist due to the fact that the construction line intersects multiple dispersion surfaces (not shown in fig. 13b).Maximising the coupling efficiency to the desired mode is therefore a necessary, yet difficult design challenge. Secondly, any practical spectrometer combines collimating and dispersive functions, e.g. a lens at the input and output sides with a grating at the centre. The present superprism design only incorporates the dispersive function, so collimation needs to be added. Our recent work in this area has shown [23] that both collimation and superprism effects can be achieved in planar photonic crystals, and that the parameters of both effects lie relatively closely together. A combined collimator-superprism photonic crystal should therefore be possible, ideally with a gradual transition in-between.

6.   Discussion and Conclusion

Several examples of miniaturised photonic crystal circuit elements have now been demonstrated, such as sharp bends, Y-junctions and cavity waveguides. While this makes a clear argument in favour of photonic crystal circuits, miniaturisation alone may not be sufficient to ensure their major impact. Take WDM components as an example; only if miniaturisation can be achieved with comparable or better performance than already offered by existing devices will photonic crystal components succeed. State-of-the-Art Arrayed-waveguide gratings (AWGs), for example, offer 50 GHz channel spacing with a channel accuracy of 2.5GHz and a cross-talk in excess of 30dB. These are benchmarks that are presently well out of reach of photonic crystal circuits based either on superprisms or cavities. Rather than simply copying existing functionalties, we need to look for applications that come natural or offer solutions that cannot otherwise be achieved. On the other hand, miniaturisation offers redundancy, i.e. many circuits can be manufactured on the same piece with only the ones required for a given application being utilised, thus offering simplicity and flexibility at the same time.

If photonic crystal circuits are to succeed, we need to make more use of the inherent properties of the materials out of which we fabricate them. These materials are typically semiconductors, and we use them because of their high index contrast. The fact that the material properties can be adjusted with current injection and electric field effects, however, is mainly ignored in photonic crystal research. These effects offer tunability, but also gain and absorption (for amplifiers amplifier and detectors, respectively). A generic application would therefore be a modulator with integrated dispersion control, or a detector with dispersion compensation at the input side. Alternatively, the effective light-matter interaction in a modulator could be increased via slow wave structures, i.e. photonic crystal waveguides that operate in a low group velocity regime. This should allow modulators to be realised with effective electro-optic coefficients that scale with the reduction in group velocity and be much shorter. Another aspect is that present-day integrated optical devices rely amost exclusively on thermal tuning, which is rather slow and scales badly due to thermal cross-talk and overall power budget.

Dispersion is another fascinating property of photonic crystals, the main attraction being that it can be engineered to suit a given application. Photonic crystal fibres have been made, for example, with either very low dispersion at unusual wavelengths or very high dispersion for "Supercontinuum generation". This high dispersion can be exploited in photonic crystal waveguides, for example in integrated dispersion compensation functions such as the compression and dilation of optical pulses or for slow wave propagation in optical delay lines.

Nonlinearities present further opportunities. The strong confinement achieved in photonic crystal waveguides and cavities offers itself for applications that require high peak fields for nonlinear phenomena, such as all-optical switching, but also wavelength conversion with the phase-matching provided by the periodic structure. While these prospects are very exciting, they are currently held back by the fact that typical nonlinearities are weak on the lengthscale of photonic crystals, and losses are too high. The obvious solution is to reduce propagation losses, a point already made repeatedly. Alternatively, one could investigate new materials; the search for compounds with more suitable electro-optic and nonlinear coefficients has only just started in a field that is almost exclusively concerned with III-V's and silicon-based materials. Whereas photonic crystals offer miniaturisation of passive components, it is the material coefficients of the hosts that need to provide the strong light-matter interaction that will allow us to create both passive and active functionalities on a small lengthscale. Much work remains to be done at this frontier.

Finally, the high Q cavities that may lead to integrated quantum information processing devices rely on another unique property of photonic crystals, their ability to localise light. Although very high Qs in excess of 10,000 have already been predicted for planar photonic crystals [24], higher values may require 3-dimensional lattices. Such lattices would have the additional benefit of full three-dimensional control over spontaneous emission and finally enable the dream of no-threshold, no-noise lasing that motivated the pioneers in this field.

In conclusion, we have shown some examples of the favourable properties of photonic crystals that can be usefully exploited in communications, namely miniaturisation and dispersion. Once the losses have been reduced to more acceptable levels, which appears realistic, and the electro-optic properties of the constituent materials are being utilised, we can indeed expect to see high density integrated photonic crystal circuits.

Acknowledgments

I would like to acknowledge the fruitful discussions and experimental results provided by T.J. Karle, L.J. Wu, M. Mazilu and R. Wilson of St. Andrews University. We are indebted to the Nanoelectronics Research Centre at Glasgow University for access to the e-beam writer and dry etching facility, as well as their general technical support. The work is funded by the EU-IST "PICCO" programme, the UK-EPSRC "Ultrafast Photonics Collaboration" and the Royal Society via a University Research Fellowship.

References

[1] W. Bogaerts, V. Wiaux, D. Taillaert, S. Beckx, B. Luyssaert, P. Bienstman and R. Baets, Fabrication of photonic crystals in silicon-on-insulator using 248-nm deepUV lithography, IEEE J. Select. Topic. Quantum Electron. **8**, 928-934 (2002).

[2] T.F. Krauss, R.M. De la Rue and S. Brand, Two-dimensional photonic-bandgap structures operating at near infrared wavelengths, Nature **383**, 699-702 (1996).

[3] C. Youtsey and I. Adesida, A comparative study of Cl2 and HCl gases for the chemically assisted ion beam etching of InP, J. Vac. Sci. Technol B. **13**, 2360 (1995).

[4] G.R. Hadley, Out-of-plane losses of line-defect photonic crystal waveguides, IEEE Photonics Technol. Lett. **14**, 642-644 (2002).

[5] H. Benisty, C. Weisbuch, D. Labilloy, M. Rattier, C.J.M. Smith, T.F. Krauss, R.M. De la Rue, R. Houdre, U. Oesterle, C. Jouanin and D. Cassagne, Optical and confinement properties of two-dimensional photonic crystals, J. Lightwave Technol. **17**, 2063-2077 (1999).

[6] W. Bogaerts, P. Bienstman P, D. Taillaert, R. Baets and D. De Zutter, Out-of-plane scattering in 1-D photonic crystal slabs, Opt. Quantum Electron. **34**, 195-203 (2002).

[7] O. Toader and S. John, Square spiral photonic crystals: Robust architecture for microfabrication of materials with large three-dimensional photonic band gaps, Phys. Rev. E. **66**, art. No. 016610 Part 2 (2002).

[8] P. Lalanne, Electromagnetic analysis of photonic crystal waveguides operating above the light cone, IEEE J. Quantum Electron. **38**, 800-804 (2002).

[9] K.K. Lee, D.R. Lim, L.C. Kimerling, J. Shin, J and F. Cerrina, Fabrication of ultralow-loss Si/SiO2 waveguides by roughness reduction, Opt. Lett. **26**, 1888-1890 (2001).

[10] J. Arentoft, T. Sondergaard, M. Kristensen, A. Boltasseva, M. Thorhauge and L. Frandsen, Low-loss silicon-on-insulator photonic crystal waveguides, Electron. Lett. **38**, 274-275 (2002.

[11] M. Notomi, A. Shinya, K. Yamada, J. Takahashi, C. Takahashi and I. Yokohama, Optimization of the Q factor in photonic crystal microcavities, IEEE J. Quantum Electron. **38**, 736-742 (2002).

[12] S.Y. Lin, E. Chow, J. Bur, S.G. Johnson and J.D. Joannopoulos, Low-loss, wide-angle Y splitter at similar to 1.6-μm wavelengths built with a two-dimensional photonic crystal, Opt. Lett. **27**, 1400-1402 (2002).

[13] S. Boscolo, M. Midrio and T.F. Krauss, Y junctions in photonic crystal channel waveguides: high transmission and impedance matching, Opt. Lett. **27**, 1001-1003 (2002).

[14] R. Wilson, T. Karle and T.F. Krauss, Efficient Y-Splitter design using Photonic Crystals, submitted for publication.

[15] S. Boscolo, C. Conti, M. Midrio and C.G. Someda, Numerical analysis of propagation and impedance matching in 2-D photonic crystal waveguides with finite length, J. Lightwave Technol. **20**, 304-310 (2002).

[16] S. Olivier, H. Benisty, C.J.M. Smith, M. Rattier, C. Weisbuch, and T.F. Krauss, Transmission properties of two-dimensional photonic crystal channel waveguides, Opt. Quantum Electron. **34**, 171-181 (2002).

[17] N. Stefanou and A. Modinos, Impurity bands in photonic insulators, Phys. Rev. B. **57**, 12127-12133 (1998).

[18] M. Bayindir, B. Temelkuran and E. Ozbay, Propagation of photons by hopping: A waveguiding mechanism through localized coupled cavities in three-dimensional photonic crystals, Phys. Rev. B **61**, 11855-11858 (2000).

[19] T.J. Karle, D.H. Brown, R. Wilson, M. Steer and T.F. Krauss, IEEE J. Select. Topic.Quantum Electron. **8**, 909 (2002).

[20] M. Notomi, K. Yamada, A. Shinya, J. Takahashi, C. Takahashi and I. Yokohama Extremely large group-velocity dispersion of line-defect waveguides in photonic crystal slabs, Phys. Rev. Lett. **87**, art. no. 253902 (2001).

[21] H. Kosaka, T. Kawashima,A. Tomita, M. Notomi, T. Tamamura, T. Sato and S. Kawakami, Superprism phenomena in photonic crystals: Toward microscale lightwave circuits, J Lightwave Technol. **17** 2032-2038 (1999).

[22] L.J. Wu, M. Mazilu, T. Karle and T.F. Krauss, Superprism phenomena in planar photonic crystals, IEEE J. Quantum Electron. **38**, 915-918 (2002).

[23] L.J.Wu,M. Mazilu and T.F. Krauss, Beam steering in planar photonic crystals: From superprism to supercollimator, accepted for IEEE J. Lightwave Technol.

[24] J. Vuckovic and A. Scherer, Optimization of the Q factor in photonic crystal microcavities, IEEE J. Quantum Electron. **38**, 850-856 (2002).

PHYSICS AND APPLICATIONS OF DEFECT STRUCTURES IN PHOTONIC CRYSTALS

Ekmel Ozbay and Mehmet Bayindir
Department of Physics, Bilkent University
Bilkent, 06533 Ankara, Turkey
ozbay@fen.bilkent.edu.tr

Abstract   We propose and demonstrate a new type of propagation mechanism for electromagnetic waves in photonic band gap materials. Photons propagate through coupled cavities due to interaction between the highly localized neighboring cavity
modes. We report a novel waveguide, which we called coupled-cavity waveguide
(CCW), in three-dimensional photonic structures. By using CCWs, we demonstrate lossless and reflectionless waveguide bends, efficient power splitters, and
photonic switches. We also experimentally observe the splitting of eigenmodes
in coupled-cavities and formation of defect band due to interaction between the
cavity modes. The tight-binding (TB) approach, which is originally develop for
the electronic structures, is applied to the photonic structures, and compared to
the experimental results. Our achievements open a new research area, namely
physics and applications of coupled-cavities, in photonic structures. We think
that our results are very important for constructing future all-optical components
on a single chip.

1.   Introduction

The artificially created 3-dimensional (3D) periodic structures inhibit the
propagation of electromagnetic (EM) waves in a certain range of frequencies
in all directions [1, 2]. In analogy with electronic band gaps in semiconductors, these structures are called photonic band gap (PBG) materials or photonic
crystals[3, 4]. The initial interest in this area came from the proposal to use
PBG crystals to control spontaneous emission in photonic devices[1]. However, the technological challenges restricted the experimental demonstrations
and relevant applications of these crystals to millimeter wave and microwave
frequencies[5, 6]. Recently, Fleming and Lin reported a photonic crystal with
a band gap at optical frequencies[7, 8]. With this breakthrough, initially proposed applications like thresholdless semiconductor lasers[9] and single-mode
light-emitting diodes[10] became feasible.

A.S. Shumovsky and V.I. Rupasov (eds.), Quantum Communication and Information Technologies, 273–297.
© 2003 *Kluwer Academic Publishers.*

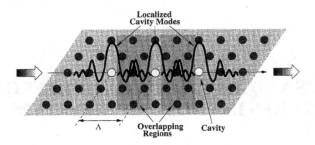

Figure 1.   Schematics of a coupled-cavity (white circles) structure in a two-dimensional photonic crystal (black circles). *Tightly* confined cavity mode interacts *weakly* with the neighboring cavity modes, and therefore the electromagnetic waves can propagate through coupled cavities.

Analogy between the Schrödinger equation and Maxwell's equations allows us to use many important tools which were originally developed for the electronic systems. As an example, it is well known that the TB method has proven to be very useful to study the electronic properties of solids[11]. Recently, the classical wave analog of the TB picture[12] has successfully been applied to the photonic structures[13, 14, 15, 16, 17, 18]. Sterke investigated the properties of the one-dimensional optical superlattices within the TB approximation [13]. Lidorikis *et al.* obtained matrix elements of the TB Hamiltonian for two-dimensional photonic crystals, with and without defects, and tested the TB model by comparing it to the corresponding *ab initio* results[15]. Observation of the normal mode splitting in quartz polystyrene was well explained by the TB photon approach[16].

By using direct implications of the TB picture, a novel propagation mechanism for photons along localized coupled cavity modes in photonic crystals was theoretically proposed[14, 17], and experimentally demonstrated[18]. In these structures, photons can hop from one *tightly* confined mode to the neighboring one due to the *weak* interaction between them (See Fig. 1).   Stefanou and Modinos obtained the cosine-like dispersion relation for their coupled-defect waveguides and waveguide bends with and without disorder[14]. Later, Yariv *et al.* reformulated the same phenomenon in a simple way, and more importantly proposed various applications based on coupled-cavity structures[17].

In the last few years, we proposed and demonstrated various applications based on coupled-cavity structures in photonic crystals. We experimentally observed the eigenmode splitting, and explained by using the TB picture [18, 19]. Guiding and bending of EM wave [20], heavy photons [21], and EM-beam splitting and switching effect [22] were experimentally demonstrated in three-dimensional photonic crystals at microwave frequencies. We also reported observation of directional coupling in coupled photonic crystal waveguides [23], and dropping of photons via cavity and waveguide coupling [24, 25]. In addition, we investigated one-dimensional (1D) coupled optical microcavity (CMC) structures [26, 27], and observed the strong enhancement of spontaneous emission throughout the cavity band in such structures [28].

Very recently, the coupled-cavity structures in photonic band gap materials have inspired considerable attentions [29, 30, 31, 32, 33, 34]. For instance, Lan *et al.* numerically proposed a switching mechanism by changing the positions of the sharp edges of the coupled-cavity band of 1D PBG structures [31], and delay lines for ultrashort optical pulses [32, 33]. Olivier *et al.* reported 2D CCWs at optical wavelengths [29]. The coupled-mode theory is applied to the coupled-cavity structure by Reynolds and his co-workers [30].

2. Localized Coupled-Cavity Modes in Photonic Crystals

In this section, we give a detail analysis of the TB picture in photonic band gap structures. We first investigate splitting of eigenmodes of coupled cavities, photonic molecules, by using the TB approach [18]. Then, we derive simple expressions for physical quantities such as dispersion relation, group velocity, photon lifetime, and dispersion.

Consider a strongly localized mode $\mathbf{E}_\Omega(\mathbf{r})$ corresponding to a single cavity that satisfies simplified version of the Maxwell equations

$$\nabla \times [\nabla \times \mathbf{E}_\Omega(\mathbf{r})] = \epsilon_0(\mathbf{r})(\Omega/c)^2 \mathbf{E}_\Omega(\mathbf{r}) , \qquad (1)$$

where $\epsilon_0(\mathbf{r})$ is the dielectric constant of the single cavity, Ω is the frequency corresponding cavity mode, and c is the speed of light. In order to derive the foregoing equations, we assumed that $\mathbf{E}_\Omega(\mathbf{r})$ is real, nondegenerate and orthonormal:

$$\int d\mathbf{r} \epsilon_0(\mathbf{r}) \mathbf{E}_\Omega(\mathbf{r}) \cdot \mathbf{E}_\Omega(\mathbf{r}) = 1 . \qquad (2)$$

2.1 Photonic molecules: eigenmode splitting

When two localized cavity modes are brought in contact, the corresponding eigenmode can be obtained from superposition of the individual evanescent cavity modes

$$\mathbf{E}_\omega(\mathbf{r}) = A\mathbf{E}_\Omega(\mathbf{r}) + B\mathbf{E}_\Omega(\mathbf{r} - \Lambda\hat{x}) . \qquad (3)$$

The eigenmode $\mathbf{E}_\omega(\mathbf{r})$ also satisfies Eq. (1) where $\epsilon_0(\mathbf{r})$ is replaced with the dielectric constant of the coupled system $\epsilon(\mathbf{r}) = \epsilon(\mathbf{r} - \Lambda\hat{x})$, and Ω is replaced with eigenfrequency ω of the coupled cavity mode.

Inserting $\mathbf{E}_\omega(\mathbf{r})$ into Eq. (1), and multiplying both sides from the left first by $\mathbf{E}_\Omega(\mathbf{r})$ and then by $\mathbf{E}_\Omega(\mathbf{r} - \Lambda\hat{x})$ and spatially integrating the resulting equations, we obtain the splitting

$$\omega_{1,2} = \Omega\sqrt{\frac{1 \pm \beta_1}{1 \pm \alpha_1}} , \tag{4}$$

where α_1 and β_1 are the first order coupling parameters which are given by

$$\alpha_1 = \int \mathrm{d}\mathbf{r}\,\epsilon(\mathbf{r})\mathbf{E}_\Omega(\mathbf{r}) \cdot \mathbf{E}_\Omega(\mathbf{r} - \Lambda\hat{x}) , \tag{5}$$

and

$$\beta_1 = \int \mathrm{d}\mathbf{r}\,\epsilon_0(\mathbf{r} - \Lambda\hat{x})\mathbf{E}_\Omega(\mathbf{r}) \cdot \mathbf{E}_\Omega(\mathbf{r} - \Lambda\hat{x}) . \tag{6}$$

Corresponding photonic modes are given by

$$\mathbf{E}_{\omega_{1,2}}(\mathbf{r}) = \frac{\mathbf{E}_\Omega(\mathbf{r}) \pm \mathbf{E}_\Omega(\mathbf{r} - \Lambda\hat{x})}{\sqrt{2}} , \tag{7}$$

From Eq. 4, we observe that the single cavity mode splits into two distinct photonic modes due to interaction between the cavities. This splitting is analogous to the splitting in the diatomic molecules, for example H_2^+, in which the interaction between the two atoms produce a splitting of the degenerate atomic levels into *bonding* and *antibonding* orbitals [12]. Recently, the bonding/antibonding mechanism in a photonic crystal was theoretically proposed by Antonoyiannakis and Pendry [35, 36]. The splitting phenomenon was experimentally observed in quartz polystyrene [16] and in photonic molecules [37].

Similarly, in the case of three coupled-cavities, the eigenmode can be written as

$$\mathbf{E}_\omega(\mathbf{r}) = A\mathbf{E}_\Omega(\mathbf{r}) + B\mathbf{E}_\Omega(\mathbf{r} - \Lambda\hat{x}) + C\mathbf{E}_\Omega(\mathbf{r} - 2\Lambda\hat{x}) . \tag{8}$$

In this case, the single cavity mode splits into three distinct modes which are given by

$$\Gamma_2 = \Omega\sqrt{\frac{1 - \beta_2}{1 - \alpha_2}} ,$$

$$\Gamma_{1,3} = \Omega\sqrt{\frac{1 \pm \sqrt{2}\beta_1 + \beta_2/2}{1 \pm \sqrt{2}\alpha_1 + \alpha_2/2}} , \tag{9}$$

where α_2 and β_2 are the second nearest neighbor coupling terms which are given by

$$\alpha_2 = \int d\mathbf{r}\, \epsilon(\mathbf{r}) \mathbf{E}_\Omega(\mathbf{r}) \cdot \mathbf{E}_\Omega(\mathbf{r} - 2\Lambda\hat{x}) \,, \tag{10}$$

and

$$\beta_2 = \int d\mathbf{r}\, \epsilon_0(\mathbf{r} - 2\Lambda\hat{x}) \mathbf{E}_\Omega(\mathbf{r}) \cdot \mathbf{E}_\Omega(\mathbf{r} - 2\Lambda\hat{x}) \,. \tag{11}$$

After finding the coefficients in Eq. 8, the corresponding eigenmodes are given by

$$\mathbf{E}_{\Gamma_{1,3}}(\mathbf{r}) = \frac{\mathbf{E}_\Omega(\mathbf{r}) \pm \sqrt{2}\, \mathbf{E}_\Omega(\mathbf{r} - \Lambda\hat{x}) + \mathbf{E}_\Omega(\mathbf{r} - 2\Lambda\hat{x})}{2} \,, \tag{12}$$

$$\mathbf{E}_{\Gamma_2}(\mathbf{r}) = \frac{\mathbf{E}_\Omega(\mathbf{r}) - \mathbf{E}_\Omega(\mathbf{r} - 2\Lambda\hat{x})}{\sqrt{2}} \,. \tag{13}$$

2.2 A novel waveguiding mechanism

When we consider an array of cavities in which each cavity interacts weakly with neighboring cavities, a defect band [1] is formed (See Fig. 1). The eigenmode of this waveguiding band can be written as a superposition of the individual cavity modes which is analogous to the linear combination of atomic orbitals in solid state physics[11]:

$$\mathbf{E}(\mathbf{r}) = E_0 \sum_n e^{-ink\Lambda}\, \mathbf{E}_\Omega(\mathbf{r} - n\Lambda\hat{x}) \,, \tag{14}$$

where the summation over n includes all the cavities, and k is the wavevector. The dispersion relation for this structure can be obtained from Eqs. (1) and (14) keeping only the first two coupling terms

$$\omega(k) = \Omega \left(\frac{\beta_1 \cos(k\Lambda) + 2\beta_2 \cos(2k\Lambda) + 1/2}{\alpha_1 \cos(k\Lambda) + 2\alpha_2 \cos(2k\Lambda) + 1/2} \right)^{1/2} \,. \tag{15}$$

---

[1]This band can also be considered as the waveguiding band.

2.3   Dispersion relation and group velocity

In certain cases, we can safely ignore the second nearest neighbor terms, and this approximation leads to a simpler expression for the dispersion relation

$$\omega(k) = \Omega \left[1 + \kappa \cos(k\Lambda)\right] . \tag{16}$$

Here $\kappa = \beta_1 - \alpha_1$ is a TB parameter which can be obtained from the splitting of the eigenmodes of two coupled cavities. After obtaining Ω, ω_1, and ω_2 from measurements or simulations, one can determine β_1 and α_1 values by using Eq. 4. The bandwidth of the waveguiding band is proportional to the coupling constant and single cavity frequency, which is given by

$$\Delta\omega = 2\kappa\Omega_0 . \tag{17}$$

Group velocity of photons along the coupled cavities can be determined from the dispersion relation as

$$v_g(k) = \nabla_k \omega_k = -\kappa\Lambda\Omega \sin\left(k\Lambda\right) . \tag{18}$$

or the group velocity as a function of frequency is given by

$$v_g(\omega) = \Lambda\Omega \left(\kappa^2 - \left(\frac{\omega}{\Omega} - 1\right)^2\right)^{1/2} . \tag{19}$$

Figure 2 displays the calculated dispersion relation and the group velocity as a function of frequency ω. Due to flat dispersion, the group velocity vanishes at the band edges, i. e., $k = 0$ and $k = \pi/\Lambda$. This property can be used in various applications. For instance, the efficiency of nonlinear processes can be enhanced due to small group velocity at the band edges [17, 38].

2.4   Photon lifetime and dispersion

The net phase difference $\Delta\varphi$ is related with the wave vector k of the crystal

$$kL - k_0 L = \Delta\varphi , \tag{20}$$

where L is the total crystal thickness, $k_0 = 2\pi\omega/c$, and c is the speed of the light in vacuum. Combining Eqs. (16) and (20) along with the definition of $\tau_p = \partial(\Delta\varphi)/\partial\omega$, we obtain a formula for delay time, photon lifetime, as a function of frequency

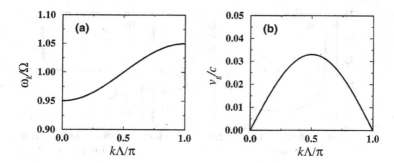

Figure 2. Calculated (a) dispersion relation and (b) group velocity as a function of wavevector k for parameter $\kappa = -0.05$, $\Lambda = 2$ cm, and $\Omega = 10$ GHz. v_g tends towards to zero at band edges $k = 0$ and π/Λ.

$$\tau_p(\omega) = \frac{L/\Lambda}{\Omega\sqrt{\kappa^2 - (\omega/\Omega - 1)^2}} - 2\pi L/c \,. \tag{21}$$

The dispersion is given by

$$D(\omega) = \frac{d}{d\omega}\left(\frac{1}{v_g}\right) = \frac{1}{L}\frac{d\tau_p}{d\omega} = \frac{\omega/\Omega - 1}{\Lambda\Omega^2\left(\kappa^2 - (\omega/\Omega - 1)^2\right)^{3/2}} \,. \tag{22}$$

We calculated the photon lifetime by using Eq.21, and plotted in Fig. 3(a). The photon lifetime goes to the infinity at the band edges, i. e., $\omega = (1 + \kappa)\Omega$ and $\omega = (1 - \kappa)\Omega$. This means that photons move along the coupled cavities very slowly, and therefore we can introduce the *heavy photon* concept in such photonic structures. The heavy photons at the coupled-cavity band edges are analogous to the electrons in semiconductors having energies near the band edges [12].

We also calculated the dispersion D as a function of normalized frequency ω/Ω. As shown in in Fig. 3(b), we observed that $D \to -\infty$ for $\omega = (1 + \kappa)\Omega$ and $D \to +\infty$ for $\omega = (1 - \kappa)\Omega$. Based on this important observation, we can use the coupled-cavity structures as dispersion compensators.

3. Coupled-Cavities in 3D Photonic Crystals

In this section, we investigate experimentally and theoretically the coupling between localized cavity modes in a dielectric-based layer-by-layer 3D photonic crystal within the TB framework. We also report on the observation of the

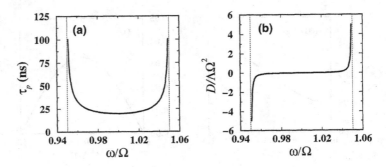

Figure 3.   Calculated (a) delay time and (b) dispersion as a function of normalized frequency ω. The delay time increases rapidly as we approach to the waveguiding band edges. The dispersion goes to $-\infty$ at the lower band edge and $+\infty$ at the upper band edge.

eigenmode splitting in the coupled cavities. Moreover, we demonstrate a new type of waveguiding through localized defect modes.

3.1   3D layer-by-layer photonic crystals

A layer-by-layer dielectric based photonic crystal [39, 40, 7] was used to construct the coupled-cavity structures as shown in Fig. 4(a). The crystal consists of square shaped alumina rods having a refractive index 3.1 at the microwave frequencies and dimensions 0.32 cm×0.32 cm×15.25 cm. A center-to-center separation between the rods of 1.12 cm was chosen to yield a dielectric filling ratio of ~ 0.26. The unit cell consists of 4 layers having the symmetry of a face centered tetragonal (fct) crystal structure. The crystal exhibits a three dimensional photonic band gap extending from 10.6 to 12.8 GHz.

The experimental set-up consists of a HP 8510C network analyzer and microwave horn antennas to measure the transmission-amplitude and transmission-phase spectra [Fig. 4(b)]. The defects were formed by removing a single rod from each unit cell of the crystal. Removing a single rod from an otherwise perfect crystal leads to confined modes with high Q-factors, quality factor defined as the center frequency divided by the full width at half maximum, around 1000. The electric field polarization polarization vector of the incident EM field was parallel to the rods of the defect lines for all measurements.

3.2   Mode splitting

By using the aforementioned experimental setup, we first measured the transmission amplitude through a crystal with a single defective unit cell. This resulted in a localized defect mode within the PBG which is analogous to acceptor

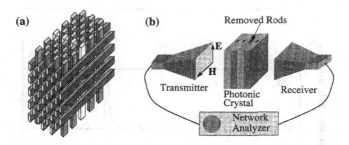

Figure 4. (a) Schematic drawing of the coupled-cavity structures in a layer-by-layer photonic crystal. A single rod (white rods) removed from each unit cell. (b) The experimental setup for measuring the transmission characteristics of the coupled-cavity structures in three-dimensional photonic crystals. The electric field polarization is directed along the removed rods.

impurity state in semiconductor physics [41]. The defect mode occurred at a resonance frequency of $\Omega = 12.150$ GHz with a Q-factor (quality factor, defined as center frequency divided by the peak's full width at half-maximum) of ~ 1000 [Fig. 5(a)].

Next, we measured the transmission through the crystal that contains two consecutive single rod removed unit cells. The intercavity distance for this structure was $\Lambda = 1.28$ cm, which corresponds to single unit cell thickness in the stacking direction. We observed that the mode in the previous case splitted into two resonance modes at frequencies $\omega_1 = 11.831$ GHz and $\omega_1 = 12.402$ GHz [Fig. 5(b)]. The TB parameters found to be $\kappa = \beta - \alpha = -0.047$.

Figure 5(c) shows the transmission characteristics of a crystal having three consecutive defective cells, where the resonant modes were observed at frequencies $\Gamma_1 = 11.708$ GHz, $\Gamma_2 = 12.153$ GHz and $\Gamma_3 = 12.506$ GHz.

	Measured [GHz]	Calculated [GHz]
Γ_1	11.708	11.673
Γ_2	12.153	12.150
Γ_3	12.506	12.492

Table 1. The measured and calculated values of resonant frequencies for the crystal with three defective unit cells.

Table 1 compares the resonance frequencies, which were calculated by inserting TB parameters α_1 and β_1 into the Eq. (9), with the values obtained from the experiment [Fig. 5(c)]. The experimentally measured three splitted modes coincide well with the theoretically expected values. This excellent agreement shows that the classical wave analog of TB formalism is valid for our structure.

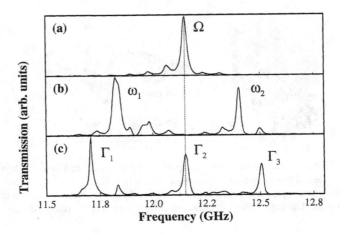

Figure 5.   Transmission characteristics along the stacking direction of the photonic crystal: (a) For single defect with resonance frequency Ω. (b) For two consecutive defects resulting in two splitted modes at resonance frequencies ω_1 and ω_2 with intercavity distance $\Lambda = 1.28$ cm. (c) For three consecutive defects with resonance frequencies Γ_1, Γ_2, and Γ_3.

3.3   Waveguide and waveguide bends

In this section, we demonstrate the observation of guiding and bending of the EM wave through highly localized defect modes in a 3D photonic crystal. The most important feature of this new waveguides, which we called coupled-cavity waveguides (CCWc), is the possibility of constructing lossless and reflectionless bends. This ability has a crucial role to overcome the problem of guiding light around very sharp corners in the optical circuits.

The guiding or bending of EM waves through the localized defect modes via hopping is fundamentally different from previously proposed photonic crystal waveguides [42, 43, 6, 44]. Although, the structural imperfections such as misalignment of rods during the fabrication process affected the efficiency of the CCWs, we have observed nearly 100 percent transmission for various CCWs throughout the entire waveguiding band. Our observation differs from that of Lin *et al.* [43] in which the unity bending efficiency[2] can be obtained only at certain frequencies.

3.4   Transmission measurements

We first measured the transmission characteristics of a straight waveguide which consists of 11 unit cell fct crystal. The defect array was created by

---

[2]In this reference, the bending efficiency is obtained by normalizing the transmission spectrum through the bend to that of the straight waveguide.

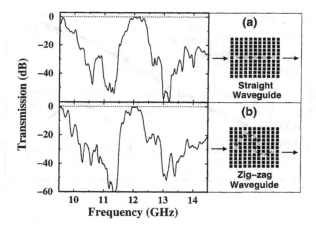

Figure 6. (a) Transmission amplitude as a function of frequency for a straight waveguide geometry which is shown in the right panel. The gray squares represent the missing rods. A full transmission was observed throughout the entire waveguiding band ranging from 11.47 to 12.62 GHz. (b) Transmission characteristics of a zig-zag shaped waveguide which is formed by removing randomly chosen rods while keeping the distance between adjacent defects constant. In all cases, nearly 100 percent transmission amplitudes were measured.

removing a single rod from the first layer of each unit cell with a periodicity of $\Lambda = 1.28$ cm. As shown in left panel of Fig. 6(a), a defect band (guiding band) was formed within the photonic band gap analogous to the impurity bands in the disordered semiconductors. The width of this guiding band can be adjusted by changing the coupling strength between the cavities[3]. For this waveguide structure, nearly a complete transmission of the EM wave was observed within a frequency range extending from 11.47 to 12.62 GHz. It is interesting to note that when we placed one of the removed rods into its original position, we observed almost vanishing transmission amplitude throughout the above frequency range. This result is expected since the second nearest-neighbor coupling amplitude is negligibly small in our structures.

To develop an optical circuit, the problem of the guiding light around sharp corners must be addressed. Conventional dielectric or metallic waveguides have large scattering losses when sharp bends are introduced. Within the TB approximation, we can guide or bend the EM waves along an arbitrarily shaped path by connecting the defects. To verify this idea experimentally, we constructed a zig-zag shaped waveguide while keeping the distance between the consecutive cavities constant. In this waveguide, the propagation direction of photons was randomly changed. As shown in Fig. 6(b), we observed full transmission

---

[3]For instance, the coupling increases when the distance between adjacent defects decreases.

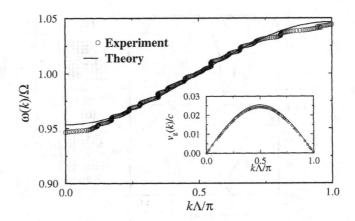

Figure 7.   Dispersion diagram of the waveguiding band predicted from the transmission-phase measurements (○ symbols) and calculated by using tight-binding formalism (solid line) with $\kappa = -0.047$. Inset: The normalized group velocity diagrams calculated by the theory (solid line) and obtained from the experimental data (○ symbols) agree well and both vanish at the guiding band edges, where c is the speed of light.

similar to the results obtained from the straight waveguide. Our results clearly indicate that the sharp corners have no influence on the propagation of EM waves in CCWs. By using CCWs one can achieve the bending of light around a sharp corner without any radiation losses. Therefore, this novel method may have great practical importance in certain applications.

As shown in Fig. 6, the band edges of CCWs are very sharp compared to the PBG edges. This property is important for the switching applications. Conventionally, the switching mechanism is achieved by dynamical shifting of the photonic band gap edges [45] or the position of defect frequency [46] via the nonlinear processes. In our case, the on-off modes of the switch can be achieved by shifting the waveguiding band edges. Therefore, the efficiency of a photonic crystal based switch can be enhanced by using the CCWs.

3.5   Phase measurements

The dispersion relation for the waveguiding band can be obtained from the transmission-phase measurement. Figure 7 shows the comparison of the measured (○ symbols) and calculated (solid line), by using $\omega_k = \Omega[1 + \kappa \cos(k\Lambda)]$, dispersion relations. As shown in Fig 7., TB calculation gives good agreement with the measured result, and the deviations between the experiment and the theory is more pronounced around the edges of the waveguiding band. We expect this discrepancy to vanish as the number of unit cells used in the experiment is increased.

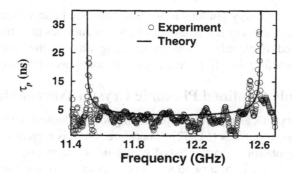

Figure 8. The measured and calculated delay time for the single rod removed CCW. The photon lifetime increases also drastically at the waveguiding band edges.

The inset in Fig. 7 shows the comparison of the theoretical (solid line) and experimental (○ symbols) variation of the group velocity, $v_g(k) = \mathrm{d}\omega(k)/\mathrm{d}k \simeq -\Omega\kappa a \sin(ka)$, of the waveguiding band as a function of wave vector k. The experimental curve is obtained by taking the derivative of the best fitted cosines function to the experimental data. Notice that the group velocity vanishes at the waveguiding band edges [47]. It is important to note that, in the stimulated emission process, the effective gain is inversely proportional to the group velocity [48]. The group velocity can be made smaller if one can reduce the amplitude of the parameter κ.

3.6 Heavy photons at coupled-cavity band edges

Heavy photons or photons with a extremely low group velocity play a critical role in enhancing the efficiency of nonlinear processes, [10, 48, 17] and the gain enhancement in the photonic band edge laser. [10] Moreover, the spontaneous emission rate can be increased since the effective gain is proportional to $1/v_g$. [48] The low group velocity was proposed near the band edges of one-dimensional photonic band-gap structures. [49, 10] Recently, Vlasov *et al.* have observed that the optical pulses significantly slow down at the 3D photonic crystal band edges in visible spectrum. [50]

We measured the delay time of a defect structure where the whole rod is removed from each unit cell of the crystal. [18] We measured the delay time corresponding to a ten unit cell coupled cavity waveguide. The localization volume is bigger than the previous case, and therefore the photon lifetime for a single cavity is around 5 ns. As shown in Fig. 8, the delay time increases drastically at the CCW band edges and agrees well with the calculated result by using Eq. (3). In this case, the group velocity also approaches to zero at the band edges (See the inset in Fig. 7).

Physically, the heavy photons in the photonic band gap structures are analogous to the electrons in semiconductors having energies near the band edges. The corresponding eigenfunctions are standing waves rather than the running waves, and therefore the effective mass of electrons becomes very large. [12]

4.   Highly Confined Photonic Crystal Waveguides

Photonic crystals also provide a promising tool to control the flow of light in integrated optical devices [51, 52]. Therefore, there is a great deal of interest in developing photonic crystal based waveguides where one can confine and efficiently guide the light around sharp corners in two-dimensional (2D) [42, 43, 53, 54, 55, 56] or three-dimensional (3D) [6, 18, 57, 58, 59] photonic crystals. Guiding the light without losses, and even through sharp corners using two-dimensional (2D) photonic crystals was first proposed theoretically by Mekis *et. al.* [42]. Later, researchers have reported experimental observation of waveguiding in 2D photonic crystals first at microwave [43], and then at optical frequencies [53, 54, 55, 56]. However, to avoid the leakage problem in 2D structures, either one has to extend the size of the photonic crystal in the vertical direction, or use a strong index-guiding mechanism in the vertical direction [56, 60]. A way to eliminate the leakage is to use a three-dimensional (3D) photonic crystal. Recently, full confinement of the EM waves utilizing a 3D layer-by-layer photonic crystal structure has been theoretically studied [57, 58]. Although Noda *et al.* reported the fabrication of a 3D sharp bend waveguide at optical wavelengths, they have not reported any optical measurements on this sharp-bend structure [59].

In this section, we demonstrate that the guiding, bending, and splitting of EM waves could be achieved in highly confined waveguides which are constructed by removing a single rod from a perfect 3D layer-by-layer photonic crystal as shown in Fig. 9 [61]. The experimental results are in good agreement with a theory based on tight-binding (TB) approximation [18, 62, 14], and the simulation results of Chutinan and Noda [58]. We also present dropping of EM waves via cavity-waveguide coupling in such structures.

4.1   Experimental setup

A layer-by-layer dielectric based photonic crystal [39, 40, 7] was used in our experiments. We used an HP 8510C network analyzer and microwave horn antennas to measure the transmission-amplitude and the transmission-phase properties of the straight waveguides, the bended waveguide, the power splitter, and WDM structures. We constructed the straight waveguide by removing a single rod from a single layer of a 6 unit cell (24 layer) photonic crystal, so that we had 12 layers at the top and 11 layers at the bottom of the removed rod. The polarization vector **e** of the incident EM wave was kept parallel to the

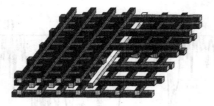

Figure 9.   Schematic drawing of a new type of waveguide structure in 3D layer-by-layer photonic crystals. The EM waves are confined along with the vacancy of a single removed rod (white rod).

stacking direction of the layers. We did not observe any guided modes for the other polarization polarization in any of our experiments.

4.2   Waveguides and waveguide bends

First, we tested the guiding of EM waves by measuring the transmission through a single missing rod. We observed full transmission of the EM waves for certain frequencies within the photonic stop band as shown in Fig. 10(a). The full transmission within the waveguiding band was a proof of how well the wave was confined and guided without losses. The guiding band started from 11.37 GHz and ended at 12.75 GHz. In Fig. 10(a), we also plotted the transmission spectra (dotted line) of the perfect crystal for comparison. The guiding was limited with the photonic band gap of the crystal, for which the crystal had the property of reflecting the EM waves in all directions.

We tested the bending of light through sharp corners in a waveguide structure shown in inset of Fig. 10(b). This structure was constructed by removing part of a single rod from 11th layer, and part of another rod from 12th (adjacent) layer. The resulting vacancies of the missing parts of rods form a 90-degree sharp bend waveguide. The incident wave propagated along the first waveguide (missing portion of the rod on the 11th layer), and successfully coupled to the second waveguide on the 12th layer, which was perpendicular to the propagation direction of the incident EM wave.

As shown in Fig. 10(b), we observed a waveguiding band (extending from 11.55 to 12.87 GHz), for which the frequency range of the band was similar to the straight waveguide. The high transmission-amplitudes, which reached unity around certain frequencies, showed that the EM waves were coupled and guided through the waveguide that contained a sharp bend. This observation is consistent with the TB picture, since the EM waves can propagate through coupled cavities without losses irrespective of change in the propagation direction.

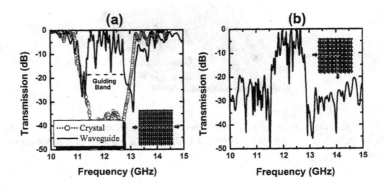

Figure 10.   (a) Transmission amplitude measured from a single rod removed waveguide structures (solid line). Nearly full transmission was achieved for certain frequencies. Transmission through a perfect crystal was plotted for comparison (dotted line). Inset: The schematics of the waveguide structure. (b) Measured transmission characteristics of the 90-degree bended waveguide. A 100% transmission was obtained for certain frequencies throughout the guiding band. Inset: The schematics of the waveguide bend structure.

At this point, we would like to compare our bended waveguide results with the simulations previously reported for this structure by Chutinan and Noda [58]. Based on their simulation, the waveguiding band for sharp bend structure covers 67% of stop band of the photonic crystal which was very close to our experimental value of 68%.

4.3   Power splitters

The power splitters are important for designing photonic crystal based optical components. Previously, the power splitters were theoretically investigated in 2D photonic structures [63, 64, 65], and were experimentally demonstrated in 3D photonic crystals [22]. To test the splitting idea in highly confined waveguides, we removed one rod from the 12th layer partially, and a whole rod from the 13th layer with a crossed configuration [See the inset in Fig. 11]. Figure 11 shows the measured transmission characteristics of this structure. The EM wave inside the input waveguide channel was efficiently coupled to the output channels. Although we have achieved high transmission ($\sim$ 50%) for certain frequencies, the detected powers at each output port were mostly unequal throughout the waveguiding band.

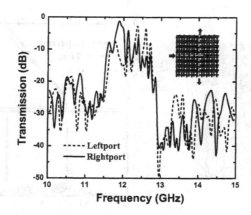

Figure 11. Measured transmission spectra for the splitter structure. The electromagnetic power in the input port splits into the two output ports throughout the guiding band.

4.4 Theoretical model: Tight-binding analysis

In order to understand the underlying physics behind this single rod removed waveguide, we need to closely look at structure of the waveguide. Each vacancy just below the removed rod behaves as a box-like cavity (See the left panel in Fig. 12). The coupling between these localized cavity modes allow propagation of photons by hopping through the vacancy of the missing rod.

Based on this observation, we can obtain the dispersion relation by measuring the transmission-phase characteristics and by using the TB approximation. Within the TB scheme [18], the dispersion relation is given by $\omega(k) = \Omega[1 + \kappa \cos(k\Lambda)]$. Here $\Omega = 12.07\,\text{GHz}$ is the single defect frequency, $\Lambda = 1.12\,\text{cm}$ is the intercavity distance, k is wavevector, and $\kappa = -0.057$ is a TB parameter which was derived from the experimental data ($|\kappa| = \Delta\omega/2\Omega$). As shown in Fig. 12(a), the measured dispersion relation is in good agreement with the TB prediction. We also measured and calculated the photon lifetime, i. e. delay time, of the single rod removed waveguide. Within the TB scheme, the photon lifetime can be written as $\tau_p(\omega) = \Lambda/v_g(\omega) + 2\pi\Lambda/c$ [21]. Here $v_g = \nabla_k\omega(k)$ is group velocity of the guided mode. As displayed in Fig. 12(b), experimental and theoretical results show that the photon lifetime increases drastically, $\tau_p \to \infty$, at the waveguiding band edges.

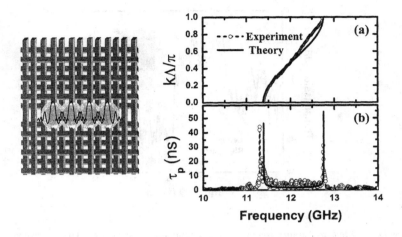

Figure 12.   [Left Panel] Schematic drawing of coupled-cavities in highly confined waveguides. [Right Panel] (a) The dispersion relation obtained from the transmission-phase measurements and the tight-binding approximation. (b) The measured and calculated delay time as a function of frequency. Photon lifetime (τ_p) increases drastically at the waveguiding band edges.

4.5   Dropping of electromagnetic waves through localized modes

Photonic band gap structures can also be used to construct the optical add-drop filters which can be used effectively in WDM applications. The first photonic crystal based WDM structure was proposed by Fan *et. al.* by using resonant tunneling phenomena between two waveguides via cavities [66]. Kosaka *et. al.* reported WDM filters by using superprism phenomena [51]. Noda and his co-workers proposed and experimentally demonstrated trapping and dropping of photons via cavity-waveguide coupling in 2D photonic crystal slabs [59]. Nelson *et. al.* reported wavelength separation by using 1D dielectric multilayer stacks [67]. Recently, various types of WDM structures in 2D photonic crystals have also been reported [68, 69, 70, 71].

In this section, we proposed and demonstrated a new method for dropping photons via the cavity-waveguide coupling in 3D layer-by-layer photonic crystals [24].

In order to demonstrate the demultiplexing phenomena, we designed a structure that consisted of a HCW and cavities [See Fig. 13(a)]. The highly localized defect modes, with quality-factors ($Q = \omega_0/\Delta\omega$) around 1000, were generated either by removing some portion of a rod or by adding additional materials to

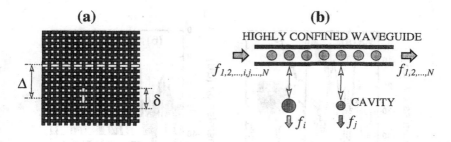

Figure 13.   Dropping of electromagnetic waves in 3D photonic crystal structures. (a) Proposed configuration for the demultiplexing applications in photonic crystals. Some portion, δ, of a single rod was removed to construct an acceptor-like defect mode. (b) Schematic drawing of the mechanism for dropping photons via coupling between a highly confined waveguide and localized cavity modes.

the crystal[4]. Due to coupling between the guided mode inside the waveguide and the localized cavity modes, the EM waves at resonance frequencies of the cavities were dropped from the waveguide mode [See Fig. 13(b)].

The waveguide was constructed by removing a single rod from the 11th layer of the crystal which contains 5 unit cells along the stacking direction. The cavity was formed by cutting some portion of a rod at the upper layer (12th layer) of the HCW layer. The defect volume, and therefore the cavity frequency, was varied by changing δ. The distance between the defect and the waveguide was fixed $\Delta = 4a$.

The measured transmission characteristics were plotted in Fig. 14(a) for the parameter $\Delta = 3.5a$ and $\delta = 0.75$ cm. The EM waves having the resonance frequency $f = 0.4464c/a$ was filtered from the waveguide mode. We also observe a corresponding dip in the HCW spectrum at the same frequency. The tunability of the dropping mode was presented in Fig. 14(b). We measured the transmission spectra by varying the parameters Δ and δ. As shown in Fig. 14(b), the resonance frequency was shifted by changing the parameters Δ and δ.

Based on this observation, we concluded that the photons at the resonance frequency of the defect mode was first trapped in the cavity and then emitted from the cavity. The quality factor of the dropping mode can be adjusted by changing the distance Δ. Since the resonance frequency depends on the defect volume, we can tune the dropping frequency by increasing or decreasing the distance δ. In order to demonstrate tunability of our WDM structures, we measured the transmission spectra for three different values of $\delta = 0.9, 1.3, 1.6$ cm.

---

[4]One can call these modes acceptor- or donor-like photonic modes which are reminiscent to the acceptor and donor states in a semiconductor [41, 72].

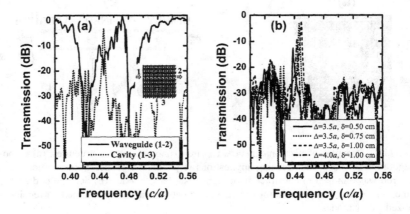

Figure 14.   (a) Measured transmission characteristics of the demultiplexing structure in Fig. 13(b). The EM waves having frequency $f = 0.4464c/a$ was dropped from the HCW mode. There appears a corresponding drop in the transmission spectrum of the waveguide mode at the same frequency. Inset: Schematic drawing of the demultiplexing geometry. (b) Measured transmission spectra corresponding to the various values of the parameters Δ and δ.

As shown in Fig. 14(b), the drop frequency can be tuned by changing the value of δ.

5.   Summary

In this work, we proposed and demonstrated a new type of propagation mechanism for electromagnetic waves in photonic band gap materials. Photons propagate through highly localized cavity modes due to coupling between them. Based on this mechanism, we reported a novel waveguide, which we called coupled-cavity waveguide (CCW), in photonic crystal structures. By using CCWs, we demonstrated various applications such as lossless and reflectionless waveguide bends, efficient power splitters, photonic switches, directional couplers, add-drop filters, and dispersion compensators. We successfully applied the tight-binding (TB) approach, which is originally developed for the electronic structures, to the photonic structures. The excellent agreement between the measured, simulated, and the TB results is an indication of potential usage of TB approximation in photonic structures. We experimentally observed the eigenmode splitting and defect-band formation in coupled localized photonic modes in photonic band gap materials. These observations were explained by using the TB picture. Since the Maxwell's equations have no fundamental length scale, we expect that our microwave results can be extended to the optical frequencies. These results encourage the usage of the layer-by-layer photonic crystals in the design of future ultra-small optoelectronic integrated circuits.

Acknowledgments

This work was supported by NATO Grant No. SfP971970, National Science Foundation Grant No. INT-9820646, Turkish Department of Defense Grant No. KOBRA-001, and Thales JP8.04.

References

[1] E. Yablonovitch, "Inhibited spontaneous emission in solid-state physics and electronics," *Phys. Rev. Lett.*, vol. 58, pp. 2059–2062, 1987.

[2] S. John, "Strong localization of photons in certain disordered dielectric superlattices," *Phys. Rev. Lett.*, vol. 58, pp. 2486–2489, 1987.

[3] J. D. Joannopoulos, R. D. Meade, and J. N. Winn, *Photonic Crystals: Molding the Flow of Light*. Princeton, NJ: Princeton University Press, 1995.

[4] C. M. Soukoulis, ed., *Photonic Crystals and Light Localization in the 21ˢᵗ Century*. Dortrecht: Kluwer, 2001.

[5] M. C. Wanke, O. Lehmann, K. Muller, Q. Wen, and M. Stuke, "Laser rapid prototyping of photonic band-gap microstructures," *Science*, vol. 275, pp. 1284–1286, 1997.

[6] B. Temelkuran and E. Ozbay, "Experimental demonstration of photonic crystal based waveguides," *Appl. Phys. Lett.*, vol. 74, pp. 486–488, 1999.

[7] S. Y. Lin, J. G. Fleming, D. L. Hetherington, B. K. Smith, R. Biswas, K. M. Ho, M. M. Sigalas, W. Zubrzycki, S. R. Kurtz, and J. Bur, "A three-dimensional photonic crystal operating at infrared wavelength," *Nature (London)*, vol. 394, pp. 251–253, 1998.

[8] J. G. Fleming and S.-Y. Lin, "Three-dimensional photonic crystal with a stop band from 1.35 to 1.95 μm," *Opt. Lett.*, vol. 24, pp. 49–51, 1999.

[9] P. R. Villenevue, S. Fan, J. D. Joannopoulos, K.-Y. Lim, G. S. Petrich, L. A. Kolodziejski, and R. Reif, "Air-bridge microcavities," *Appl. Phys. Lett.*, vol. 67, pp. 167–169, 1995.

[10] J. P. Dowling, M.Scalora, M. J. Bloemer, and C. M. Bowden, "The photonic band edge laser: A new approach to gain enhancement," *J. Appl. Phys.*, vol. 75, pp. 1896–1899, 1994.

[11] W. A. Harrison, *Electronic Structure and the Properties of Solids*. San Francisco: Freeman, 1980.

[12] C. Kittel, *Introduction to Solid State Physics*, p. 75. New York: John Wiley and Sons, 7ᵗʰ edition ed., 1996.

[13] C. M. de Sterke, "Superstructure gratings in the tight-binding approximation," *Phys. Rev. E*, vol. 57, pp. 3502–3509, 1998.

[14] N. Stefanou and A. Modinos, "Impurity bands in photonic insulators," *Phys. Rev. B*, vol. 57, pp. 12127–12133, 1998.

[15] E. Lidorikis, M. M. Sigalas, E. N. Economou, and C. M. Soukoulis, "Tight-binding parametrization for photonic band gap materials," *Phys. Rev. Lett.*, vol. 81, pp. 1405–1408, 1998.

[16] T. Mukaiyama, K. Takeda, H. Miyazaki, Y. Jimba, and M. Kuwata-Gonokami, "Tight-binding photonic molecule modes of resonant bispheres," *Phys. Rev. Lett.*, vol. 82, pp. 4623–4626, 1999.

[17] A. Yariv, Y. Xu, R. K. Lee, and A. Scherer, "Coupled-resonator optical waveguide: a proposal and analysis," *Opt. Lett.*, vol. 24, pp. 711–713, 1999.

[18] M. Bayindir, B. Temelkuran, and E. Ozbay, "Tight-binding description of the coupled defect modes in three-dimensional photonic crystals," *Phys. Rev. Lett.*, vol. 84, pp. 2140–2143, 2000.

[19] M. Bayindir, I. Bulu, E. Cubukcu, and E. Ozbay, "Investigation of localized coupled-cavity modes in two-dimensional photonic band gap structures," *IEEE J. Quantum Electron.*, vol. xx, p. yyyy, 2002.

[20] M. Bayindir, B. Temelkuran, and E. Ozbay, "Propagation of photons by hopping: A waveguiding mechanism through localized coupled-cavities in three-dimensional photonic crystals," *Phys. Rev. B*, vol. 61, pp. R11855–R11858, 2000.

[21] M. Bayindir and E. Ozbay, "Heavy photons at coupled-cavity waveguide band edges in a three-dimensional photonic crystal," *Phys. Rev. B*, vol. 62, pp. R2247–R2250, 2000.

[22] M. Bayindir, B. Temelkuran, and E. Ozbay, "Photonic crystal based beam splitters," *Appl. Phys. Lett.*, vol. 77, pp. 3902–3904, 2000.

[23] M. Bayindir and E. Ozbay, "Observation of directional coupling between photonic crystal waveguides," *Phys. Rev. B*, 2002 [Submitted].

[24] M. Bayindir and E. Ozbay, "Dropping of electromagnetic waves through localized modes in three-dimensional photonic band gap structures," *Appl. Phys. Lett.*, 2002 [Submitted].

[25] M. Bayindir and E. Ozbay, "Dropping of photons in two-dimensional photonic band gap structures," *Opt. Express*, 2002 [Submitted].

[26] M. Bayindir, S. Tanriseven, and E. Ozbay, "Propagation of light through localized coupled-cavity modes in one-dimensional photonic band-gap structures," *Appl. Phys. A: Mater. Sci. Process*, vol. 72, pp. 117–119, 2001.

[27] M. Bayindir, C. Kural, and E. Ozbay, "Coupled optical microcavities in one-dimensional photonic band gap structures," *J. Opt. A: Pure and Appl. Opt.*, vol. 3, pp. 184–189, 2001.

[28] M. Bayindir, E. Cubukcu, I. Bulu, and E. Ozbay, "Photonic band gap effect, localization, and waveguiding in two-dimensional penrose lattice," *Phys. Rev. B*, vol. 63, p. 161104(R), 2001.

[29] S. Olivier, C. Smith, M. Rattier, H. Benisty, C. Weisbuch, T. Krauss, R. Houdre, and U. Oesterle, "Miniband transmission in a photonic crystal coupled-resonator optical waveguide," *Opt. Lett.*, vol. 26, pp. 1019–1021, 2001.

[30] A. L. Reynolds, U. Peschel, F. Lederer, P. J. Roberts, T. F. Krauss, and P. J. I. de Maagt, "Coupled defects in photonic crystals," *IEEE Trans. Microwave Theory Tech.*, vol. 49, pp. 1860–1867, 2001.

[31] S. Lan, S. Nishikawa, and O. Wada, "Leveraging deep photonic band gaps in photonic crystal impurity bands," *Appl. Phys. Lett.*, vol. 78, pp. 2101–2103, 2001.

[32] S. Lan, S. Nishikawa, H. Ishikawa, and O. Wada, "Design of impurity band-based photonic crystal waveguides and delay lines for ultrashort optical pulses," *J. Appl. Phys.*, vol. 90, pp. 4321–4327, 2001.

[33] S. Lan, S. Nishikawa, Y. Sugimoto, N. Ikeda, K. Asakawa, and H. Ishikawa, "Analysis of defect coupling in one- and two-dimensional photonic crystals," *Phys. Rev. B*, vol. 65, p. 165208, 2002.

[34] M. M. Sigalas and C. A. Flory, "Microwave measurements of stub tuners in two-dimensional photonic crystal waveguides," *Phys. Rev. B*, vol. 65, p. 125209, 2002.

[35] M. I. Antonoyiannakis and J. B. Pendry, "Mie resonances and bonding in photonic crystals," *Europhys. Lett.*, vol. 40, pp. 613–618, 1997.

[36] M. I. Antonoyiannakis and J. B. Pendry, "Electromagnetic forces in photonic crystals," *Phys. Rev B*, vol. 60, pp. 2363–2374, 1999.

[37] M. Bayer, T. Gutbrod, J. P. Reithmaier, A. Forchel, T. L. Reinecke, P. A. Knipp, A. A. Dremin, and V. D. Kulakovskii, "Optical modes in photonic molecules," *Phys. Rev. Lett.*, vol. 81, pp. 2582–2585, 1998.

[38] Y. Xu, R. K. Lee, and A. Yariv, "Propagation and second-harmonic generation of electromagnetic waves in a coupled-resonator optical waveguide," *J. Opt. Soc. Am. B*, vol. 17, pp. 387–400, 2000.

[39] K. M. Ho, C. T. Chan, C. M. Soukoulis, R. Biswas, and M. M. Sigalas, "Photonic band gaps in three dimensions: New layer-by-layer periodic structures," *Solid State Commun.*, vol. 89, p. 413, 1994.

[40] E. Ozbay, "Layer-by-layer photonic crystals from microwave to far-infrared frequencies," *J. Opt. Soc. Am. B*, vol. 13, pp. 1945–1955, 1996.

[41] E. Yablonovitch, T. J. Gmitter, R. D. Meade, A. M. Rappe, K. D. Brommer, and J. D. Joannopoulos, "Donor and acceptor modes in photonic band structure," *Phys. Rev. Lett.*, vol. 67, pp. 3380–3383, 1991.

[42] A. Mekis, J. C. Chen, I. Kurland, S. Fan, P. R. Villeneuve, and J. D. Joannapoulos, "High transmission through sharp bends in photonic crystal waveguides," *Phys. Rev. Lett.*, vol. 77, pp. 3787–3790, 1996.

[43] S. Y. Lin, E. Chow, V. Hietala, P. R. Villeneuve, and J. D. Joannopoulos, "Experimental demonstration of guiding and bending of electromagnetic waves in a photonic crystal," *Science*, vol. 282, pp. 274–276, 1998.

[44] J. C. Knight, J. Broeng, T. A. Birks, and P. S. J. Russell, "Photonic band gap guidance in optical fibers," *Science*, vol. 282, pp. 1476–1479, 1998.

[45] M. Scalora, J. P. Dowling, C. M. Bowden, and M. J. Bloemer, "Optical limiting and switching of ultrashort pulses in nonlinear photonic band gap materials," *Phys. Rev. Lett.*, vol. 73, pp. 1368–1371, 1994.

[46] P. R. Villeneuve, D. S. Abrams, S. Fan, and J. D. Joannopoulos, "Single-mode waveguide microcavity for fast optical switching," *Opt. Lett.*, vol. 21, pp. 2017–2019, 1996.

[47] K. Sakoda and K. Ohtaka, "Optical response of three-dimensional photonic lattices: Solutions of inhomogeneous Maxwell's equations and their applications," *Phys. Rev. B*, vol. 54, pp. 5732–5741, 1996.

[48] K. Sakoda, "Enhanced light ampli cation due to group-velocity anomaly peculiar to two- and three-dimensional photonic crystals," *Opt. Express*, vol. 4, pp. 167–176, 1999.

[49] J. P. Dowling and C. M. Bowden, "Anomalous index of refraction in photonic bandgap materials," *J. Mod. Opt.*, vol. 42, p. 345, 1994.

[50] Y. A. Vlasov, S. Petit, G. Klein, B. Honerlage, and C. Hirlimann, "Femtosecond measurements of the time of flight of photons in a three-dimensional photonic crystal," *Phys. Rev. E*, vol. 60, pp. 1030–1035, 1999.

[51] H. Kosaka, T. Kawashima, A. Tomita, M. Notomi, T. Tamamura, T. Sato, and S. Kawakami, "Photonic crystals for micro lightwave circuits using wavelength-dependent angular beam steering," *Appl. Phys. Lett.*, vol. 74, pp. 1370–1372, 1999.

[52] A. de Lustrac, F. Gadot, S. Cabaret, J.-M. Lourtioz, T. Brillat, A. Priou, and A. E. Akmansoy, "Experimental demonstration of electrically controllable photonic crystals at centimeter wavelengths," *Appl. Phys. Lett.*, vol. 75, pp. 1625–1627, 1999.

[53] T. Baba, N. Fukaya, and J. Yonekura, "Observation of light propagation in photonic crystal optical waveguides with bends," *Electron. Lett.*, vol. 35, pp. 654–655, 1999.

[54] M. Tokushima, H. Kosaka, A. Tomita, and H. Yamada, "Lightwave propagation through a 120° sharply bent single-line-defect photonic crystal waveguide," *Appl. Phys. Lett.*, vol. 76, pp. 952–954, 2000.

[55] M. Loncar, D. Nedeljkovic, T. Doll, J. Vuckovic, A. Scherer, and T. P. Pearsall, "Waveguiding in planar photonic crystals," *Appl. Phys. Lett.*, vol. 77, pp. 1937–1939, 2000.

[56] S. Y. Lin, E. Chow, S. G. Johnson, and J. D. Joannopoulos, "Demonstration of highly efficient waveguiding in a photonic crystal slab at the 1.5 μm wavelength," *Opt. Lett.*, vol. 25, pp. 1297–1299, 2000.

[57] M. M. Sigalas, R. Biswas, K.-M. Ho, C. M. Soukoulis, D. Turner, B. Vasiliu, S. C. Kothari, and S. Lin, "Waveguide bends in three-dimensional layer-by-layer photonic bandgap materials," *Micro. Opt. Tech. Lett.*, vol. 23, pp. 56–59, 1999.

[58] A. Chutinan and S. Noda, "Highly confined waveguides and waveguide bends in three-dimensional photonic crystal," *Appl. Phys. Lett.*, vol. 75, pp. 3739–3741, 1999.

[59] S. Noda, K. Tomoda, N. Yamamoto, and A. Chutinan, "Full three-dimensional photonic bandgap crystals at near-infrared wavelengths," *Science*, vol. 289, pp. 604–606, 2000.

[60] S. G. Johnson, P. R. Villeneuve, S. Fan, and J. D. Joannopoulos, "Linear waveguides in photonic-crystal slabs," *Phys. Rev. B*, vol. 62, pp. 8212–8222, 2000.

[61] M. Bayindir, E. Ozbay, B. Temelkuran, M. M. Sigalas, C. M. Soukoulis, R. Biswas, and K.-M. Ho, "Guiding, bending, and splitting of electromagnetic waves in highly confined photonic crystal waveguides," *Phys. Rev. B*, vol. 63, p. 081107(R), 2001.

[62] N. W. Ashcroft and N. D. Mermin, *Solid State Physics*. Philadelphia: Saunders, 1976.

[63] J. Yonekura, M. Ikeda, and T. Baba, "Analysis of finite 2-D photonic crystals of columns and lightwave devices using the scattering matrix method," *J. Lightwave Technol.*, vol. 17, pp. 1500–1508, 1999.

[64] R. W. Ziolkowski and M. Tanaka, "FDTD analysis of PBG waveguides, power splitters and switches," *Opt. Quant. Electron.*, vol. 31, pp. 843–855, 1999.

[65] T. Sondergaard and K. H. Dridi, "Energy flow in photonic crystal waveguides," *Phys. Rev. B*, vol. 61, pp. 15688–15696, 2000.

[66] S. Fan, P. R. Villeneuve, J. D. Joannopoulos, and H. A. Haus, "Channel drop tunneling through localized states," *Phys. Rev. Lett.*, vol. 80, pp. 960–963, 1998.

[67] B. E. Nelson, M. Gerken, D. A. B. Miller, R. Piestun, C.-C. Lin, and J. S. Harris, "Use of a dielectric stack as a one-dimensional photonic crystal for wavelength demultiplexing by beam shifting," *Opt. Lett.*, vol. 25, pp. 1502–1504, 2000.

[68] S. S. Oh, C.-S. Kee, J.-E. Kim, H. Y. Park, T. I. Kim, I. Park, and H. Lim, "Duplexer using microwave photonic band gap structure," *Appl. Phys. Lett.*, vol. 76, pp. 2301–2303, 2000.

[69] M. Koshiba, "Wavelength division multiplexing and demultiplexing with photonic crystal waveguide couplers," *J. Lightwave Technol.*, vol. 19, pp. 1970–1975, 2001.

[70] A. Sharkawy, S. Shi, and D. W. Prather, "Multichannel wavelength division multiplexing with photonic crystals," *Appl. Opt.*, vol. 40, pp. 2247–2252, 2001.

[71] C. Jin, S. Han, X. Meng, B. Cheng, and D.Zhang, "Demultiplexer using directly resonant tunneling between point defects and waveguides in a photonic crystal," *J. Appl. Phys.*, vol. 91, pp. 4771–4773, 2002.

[72] E. Ozbay, G. Tuttle, M. M. Sigalas, C. M. Soukoulis, and K. M. Ho, "Defect structures in a layer-by-layer photonic band-gap crystal," *Phys. Rev. B*, vol. 51, pp. 13961–13965, 1995.

COHERENCE AND SUPERFLUIDITY IN ATOMIC GASES

M.P. Tosi

NEST-INFM and Scuola Normale Superiore Piazza dei Cavalieri 7, I-56126 Pisa, Italy

Abstract   The achievement of Bose-Einstein condensation Bose-Einstein condensation in ultracold vapours of alkali atoms has given great impulse to the study of dilute atomic gases in condensed quantum states inside magnetic and optical traps. After the characterization of the thermodynamic and dynamic properties of Bose-condensed gases, a vast set of studies have been addressed to their phase coherence and superfluidity, both in magnetic traps and in optical lattices. Similar cooling and trapping trapping methods are being used to drive gases of fermionic alkali atoms and boson-fermion mixtures deep into the quantum degeneracy regime, with the ultimate aim of realizing novel superfluids from fermion pairing.

1.   Introduction

The phenomenon of Bose-Einstein condensation (BEC) was predicted by Einstein in 1925 by extending to material particles Bose's work on the statistics of photons: there is no restriction on the occupancy of a single quantum state by bosons. boson For an ideal gas in a box at temperature $T = 0$ the bosons can all condense in the zero-momentum state and the macroscopic occupation of this state starts at a critical temperature T_c given by

$$n\lambda_{dB}^3|_c = 2.612 \tag{1}$$

for bosons boson of spin $S = 0$. Here n is the particle number density and dB is the thermal de Broglie wavelength. The relation (1) derives from the vanishing of the chemical potential at T_c. Since $\lambda_{dB} \propto T^{-1/2}$, at $T \leq T_c$ the condensate fraction n_0/n is

$$n_0/n = 1 - (T/T_c)^{3/2}. \tag{2}$$

BEC was associated by Penrose and Onsager with the emergence of off-diagonal long-range order, the condensate density in a homogeneous Bose-condensed fluid being the asymptotic value of the one-body density matrix. density matrix Macroscopic occupation of a single state implies macroscopic quantum coherence.

A.S. Shumovsky and V.I. Rupasov (eds.), Quantum Communication and Information Technologies, 299–328.
© 2003 Kluwer Academic Publishers.

Historically, London in 1938 proposed that the transition from a normal to a superfluid state in the liquid of the bosonic boson isotope of He is associated to Bose-Einstein condensation. While just below its boiling point liquid ^{4}He behaves as an ordinary fluid with low viscosity, it undergoes a liquid-liquid transition at 2.17 K to a superfluid phase phase (HeII), the transition being signalled by a specific-heat anomaly whose shape has led to the name "$\lambda -line$" for the coexistence curve of the two liquid phases. Non-Newtonian flow behaviour and propagation of heat waves are the macroscopic manifestations of quantum mechanics in HeII. These can be explained by viewing HeII as if it were a "mixture" of two fluids, a normal fluid which possesses Newtonian viscosity and a superfluid which is capable of frictionless flow through capillaries or past obstacles. The normal component flows from the source to the sink of heat while the superfluid flows in the opposite direction to keep the total density constant: a periodic supply of heat makes the two components oscillate in antiphase.

The presence of a condensate in HeII has been studied in experiments of neutron inelastic scattering at large momentum transfer, where it was expected to lead to a narrow peak in the differential scattering cross-section owing to the absence of Doppler broadening. However, the liquid is relatively dense and the strong atom-atom interactions deplete the condensate to the point that the measured value of n_0/n is only of order 10% at the lowest temperatures. Almost pure Bose-Einstein condensates Bose-Einstein condensation have instead been realized in 1995 in monatomic alkali gases (^{87}Rb, ^{23}Na, and ^{7}Li) inside magnetic traps. At the time of writing, condensation has also been achieved in spin-polarized atomic H, in atomic He in a metastable state, and in atomic K via "sympathetic cooling" in a gaseous mixture with atomic Rb. The condition (1) for condensation is extremely severe: it requires working with a metastable gas at $T \approx 0.1~\mu$K and $n \approx 10^{12} - 10^{14}$ atoms/cc. BEC inside a trap occurs simultaneously in momentum and coordinate space, and a common method of observation is by measuring the optical density of the cloud after turning off the magnetic trap and allowing it to ballistically expand for a controlled length of time. Accounts of many observations and experiments can be found in the books edited by Inguscio et al.[1] and by Kaiser et al.[2].

In these gases the de Broglie wavelength and the mean interatomic distance are much larger than the range of the interatomic forces. The gas is so dilute that only binary collisions matter and can be described by a contact (delta-function) pseudopotential for s-wave atom-atom scattering: typical values may be $\lambda_{dB} \approx 50A$ against an s-wave scattering length a $a \approx 50A$, and the diluteness parameter is $na^3 \approx 10^{-5}$. Interest in these gases arises from two main perspectives: (1) they are new quantum fluids on which to study many-body physics within the simple model of contact interactions specified by a single (and tunable) parameter, the scattering length; and (2) they are coherent assemblies on which to study the optics of matter waves.

A periodic potential can be superposed onto the harmonic trap by means of a detuned standing wave of laser laser light. Such a configuration allows a variety of novel experiments, relating to macroscopic coherence and superfluidity and to the mechanisms by which decoherence decoherence arises and superfluidity is lost. There also is a parallel effort devoted to the study of ultracold gases of fermions and of gaseous fermion-boson mixtures, with the ultimate aim of realizing new superfluids from fermion pairing.

This short review is mainly aimed at presenting at a tutorial level the theoretical ideas that are relevant to understand the collective behaviour of dilute atomic gases in the deep quantum regime. It is divided into three main chapters: (1) Bose-condensed gases under harmonic confinement, (2) condensates inside optical lattices, and (3) Fermi gases and boson-fermion mixtures.

2. Bose-condensed Gases of Alkali Atoms Inside Harmonic Traps

The magnetic confinement is usually well represented as an external harmonic potential $V(\mathbf{r})$. Real traps are anisotropic, either pancake-shaped or cigar-shaped, but for the sake of simplicity we shall refer to a spherical trap, $V(\mathbf{r}) = m\omega^2 r^2/2$ with the trap frequency ω being typically of order $2\pi \times 100$ Hz. The spread of the ground-state single-particle orbital for non-interacting bosons boson is given by the harmonic-oscillator length, $a_{ho} = \sqrt{(\hbar/m\omega)} \approx 1\mu m$.

The confinement modifies the thermodynamics of the Bose-condensed gas in an essential way [3]. If we denote by L the linear size of a cloud of N atoms in a spherical trap, the critical temperature may be estimated by equating the thermal energy to the confinement energy, $k_B T_c \approx m\omega^2 L^2$, and by using the BEC criterion in Eq. (1) in the form $(N/L^3)\lambda_{dB}|_c \approx 2.6$: we get $k_B T_c \approx \hbar\omega N^{1/3}$. Hence, Eq. (2) is replaced by

$$n_0/n = 1 - (T/T_c)^3. \tag{3}$$

In fact, although the gas is very dilute the atom-atom interactions usually cannot be neglected [4]. We estimate the size of the cloud by equating the confinement energy to the mean potential energy, which for repulsive contact interactions is $(4\pi\hbar^2 a/m)(N/L^3)$: we get $(L/a_{ho}) \approx (8\pi Na/a_{ho})^{1/5}$ or $L \approx 10\mu m$ for a million atoms in the trap. The coupling strength enters in the ratio between the mean potential and kinetic energies,

$$< E_{pot} > / < E_{kin} > \approx (m\omega^2 L^2/2)/[\hbar^2/(2mL^2)] \propto N^{4/5} \tag{4}$$

and this ratio increases rapidly with the number of atoms. Condensates with $N \approx 10^7$ atoms are routinely available in a number of laboratories. In practice, the kinetic energy density becomes appreciable only in the outer regions of the condensed cloud.

As already remarked, interactions are included through two-body collisions which need to be treated, however, to all orders within T-matrix theory. Partial waves with angular momentum Angular momentum $\ell \neq 0$ are negligible at the temperatures of interest [5]. The asymptotic form of the two-particle scattering state with $\ell = 0$ at fixed momentum $\mathbf{k}$ is $\Phi_{0\mathbf{k}}(r) = sin[kr + \delta_0(k)]$ and the s-wave scattering length a is obtained from the phase phase shift $\delta_0(k)$ by the requirement that outside the range of the scattering potential the wave function should vanish at $r = a$, i.e. $a = -\delta_0(k)/k|_{k\to 0}$. The effective interaction between the two atoms, under appropriate conditions and with appropriate qualifications [5], may then be written as a contact potential

$$V(\mathbf{r}) = g\delta(\mathbf{r}) \tag{5}$$

where $g = 4\pi\hbar^2 a/m$. The value of a can be tuned magnetically or optically, and even changed in sign by bringing the system through a Feshbach resonance.

2.1   Bogoliubov Theory of the homogeneous Bose fluid

A gas of neutral bosons boson interacting via a weak Fourier-transformable repulsion is the simplest example of a bosonic superfluid. The theory of this model was given by Bogoliubov. His basic assumption was that a macroscopic number N_0 of particles have condensed into a single quantum state. Even at $T = 0$, N_0 does not necessarily coincide with the total number N of particles (although Bogoliubov's theory implies that N_0/N is not far from unity): as a consequence of quantum kinetic correlations, the interactions modify the ground state ground state of the ideal Bose gas by also causing a depletion of the condensate. Thus the whole of a sample of liquid He is superfluid at $T = 0$, but the condensate fraction N_0/N is only about 10% as previously noted.

Starting from the above assumption, the Hamiltonian Hamiltonian of the homogeneous Bose fluid can be diagonalized to quadratic terms by a canonical transformation into the form

$$H = E_0 + \sum_{k\neq 0} \hbar\omega_{\mathbf{k}} b_{\mathbf{k}}^+ b_{\mathbf{k}} \tag{6}$$

where the b's are operators which create or annihilate an excitation with momentum $\hbar\mathbf{k}$, while the operator $b_{\mathbf{k}}^+ b_{\mathbf{k}}$ counts the number of excitations that are present in the fluid and has expectation value zero in the ground state ground state. In Eq. (6) E_0 is a constant and $\hbar\omega_k$ is the energy of an excitation, given within Bogoliubov's theory by the relation

$$\hbar\omega_k = \sqrt{c^2 k^2 + \hbar^2 k^4/4m^2}. \tag{7}$$

Here, c is the speed of sound, which is equal to $\sqrt{ng/m}$ for contact interactions.

The crucial point of Eq. (7) is that excitations out of the condensate have an energy-momentum dispersion relation which becomes linear at low momenta, instead of a spectrum given by the single-particle kinetic energies as in the ideal Bose gas. That is, the long-wavelength excitations are phonons, i.e. quantized density waves propagating through the fluid. An important feature of the theory is the close relationship between single-particle excitations and collective density excitations, which may be understood by noticing that the emission of a boson from the condensate is accompanied by the excitation of a density wave. It is in fact a theorem, associated with the names of Gavoret and Nozières, that the long-wavelength spectra of quasiparticles and of density fluctuations become identical in the limit of zero temperature.

Related to the above result the Bose fluid was shown in the Bogoliubov theory to have the following properties:

(i) the structure factor S(k) has a linear behaviour at small k, $S(k)|_{k\to0} = \hbar k/2mc$;

(ii) the momentum distribution n(k) is divergent at small k, $n(k)|_{k\to0} = mc/2\hbar k$: a relation between n(k) (a one-body property) and S(k) (a two-body property) being a manifestation of the equivalence between single-particle and collective excitations; and

(iii) the dynamic structure factor $S(k,\omega)$ has a single peak at each value of k, at a frequency $\omega_k = \hbar k^2/2mS(k)$. This expression has become familiar from Feynman's theory for the single-mode excitations of HeII. However, the Bogoliubov dispersion relation in Eq. (7) does not include rotons.

The effects of temperature are easily included in Bogoliubov's theory by appropriate insertions of the Bose thermal distribution function, allowing an evaluation of the condensate density n_0 and of the superfluid density n_s as functions of temperature. The superfluid fraction n_s/n decreases with increasing temperature in the Bogoliubov theory by the Landau mechanism of flow dissipation through sound-wave emission. Given the assumption that a single-particle state $\psi_0(\mathbf{r},t)$ is macroscopically occupied, the conceptual basis for superfluidity is simple (see e.g. Leggett [6]). We write $\psi_0(\mathbf{r},t) = \sqrt{n_0(\mathbf{r},t)}exp[i\varphi(\mathbf{r},t)]$ and define the superfluid velocity $\mathbf{v}_s(\mathbf{r},t)$ by the prescription

$$\mathbf{v}_s(\mathbf{r},t) = (\hbar/m)\nabla\varphi(\mathbf{r},t). \tag{8}$$

This embodies the property $\nabla \times \mathbf{v}_s(\mathbf{r},t) = 0$, i.e. the superfluid flow is irrotational. Also, since no entropy entropy is associated with a single quantum state, the entropy is entirely carried by the particles occupying states other than ψ_0 (the "normal" component of the two-fluid model). Furthermore, from the fact that the phase of ψ_0 must be single-valued modulo 2π, one obtains the Onsager-Feynman quantization condition on superfluid circulation along a closed circuit. Considering HeII inside a rotating bucket, the normal component undergoes rotation with the vessel by being dragged from friction against its walls, whereas the superfluid can only experience vortex motion: at sufficiently high rotational velocities, discrete vortex lines thread the fluid and the superfluid rotates round each line with an angular momentum Angular momentum which is quantized according to

$$\oint \mathbf{v}_s.d\mathbf{l} = nh/m \tag{9}$$

where n is an integer. The integral in Eq. (9) is known as the circulation and its value in a superfluid can only be an integer multiple of a quantum of circulation given by h/m.

Let us quote some properties of the Bogoliubov ground state ground state [7] before proceeding. An expansion in the diluteness parameter na^3 leads to the expression

$$E_0/N = \frac{1}{2}gn[1 + (128/15\sqrt{\pi})(na^3)^{1/2} + ...] \tag{10}$$

for the ground-state energy, where the leading term is the result of mean-field theory and the leading correction comes from second-order perturbation theory. The corresponding result for the condensate depletion at $T = 0$ is

$$n_0/n = 1 - (8/3\sqrt{\pi})(na^3)^{1/2} \tag{11}$$

. These results can be taken to justify the widespread use of mean-field theory and the neglect of condensate depletion at $T = 0$ in the very dilute gases of alkali atoms under confinement inside magnetic traps or in optical lattices.

2.2 The Gross-Pitaevskii approximation

The simplest approximation for the wave function of an inhomogeneous Bose gas is the symmetrized product of single-particle orbitals, which are to be determined variationally. This mean-field, Hartree-Fock theory leads to the Gross-Pitaevskii equation (GPE) for confined BEC systems at $T = 0$. Considering first the case of a stationary ground-state condensate, the theory introduces the order parameter $\Psi(\mathbf{r})$ (the "wave function of the condensate") and asks that its value should minimize the energy functional

$$E[\Psi(\mathbf{r})] = \int d\mathbf{r}[\frac{\hbar^2}{2m}|\nabla\Psi(\mathbf{r})|^2 + V(\mathbf{r})|\Psi(\mathbf{r})|^2 + \frac{1}{2}g|\Psi(\mathbf{r})|^4] \tag{12}$$

where $V(\mathbf{r})$ is the confining potential and the last term in the brackets is for contact interactions. Minimization of the functional yields the GPE,

$$-\frac{\hbar^2}{2m}\nabla^2\Psi(\mathbf{r}) + V(\mathbf{r})\Psi(\mathbf{r}) + g|\Psi(\mathbf{r})|^2\Psi(\mathbf{r}) = \mu\Psi(\mathbf{r}) \tag{13}$$

where μ is the chemical potential. This is to be determined from the condition that $n(\mathbf{r}) = |\Psi(\mathbf{r})|^2$ should integrate to the total number N of particles.

As already noted, in the bulk of the confined alkali gases with repulsive interactions (g > 0) the kinetic energy is small relative to the potential energy. Neglecting, therefore, the Laplacian term in Eq. (13) yields the Thomas-Fermi approximation for the Bose gas:

$$\Psi_{TF}(\mathbf{r}) = g^{-1}\sqrt{\mu - V(\mathbf{r})} \qquad (14)$$

for $\mu > V(\mathbf{r})$ and zero otherwise. For isotropic harmonic confinement Eq. (14) gives a density profile in the shape of an inverted parabola, which is much broader than the Gaussian profile of a condensate of non-interacting bosons. In practice, some spill-out of particles beyond the Thomas-Fermi radius $R = \sqrt{(2\mu/m\omega^2)}$ occurs on account of the kinetic energy. This defines a coherence (healing) length $\xi = \hbar/\sqrt{(2mng)} = (8\pi na)^{-1/2}$ within which an inhomogeneity in the particle distribution is healed.

In the case of attractive interactions ($g < 0$) the condensate is liable to collapse as the number of particles increases. However, it is stabilized by the zero-point energy if it contains a limited number of particles and its wave function can then be represented as a narrow Gaussian. This is the case for ^{7}Li condensates, where the critical number of particles is of order 10^3.

From this introduction to the GPE we proceed to describe how it is used in evaluating (i) the thermodynamics of a Bose-condensed gas at finite temperature, and (ii) the dynamics of a condensate at zero temperature.

2.2.1   Internal energy of a trapped Bose gas.   Experiments on a spin-polarized gas of 87Rb by Ensher et al. [8] have reported the internal energy of the gas as a function of temperature. The data show a pronounced increase of the internal energy over that of an ideal boson boson gas below T_c, while above T_c there is no evidence in the data for deviations from ideality. These data have been accounted for quantitatively by a suitable extension of the mean-field theory [9].

At $T \neq 0$ the particle density is $n(\mathbf{r}) = |\Psi(\mathbf{r})|^2 + \tilde{n}\mathbf{r}$, where the second term is the density of particles that are thermally excited out of the condensate. The condensate wave function obeys the GPE supplemented by the interactions with the thermal cloud,

$$-\frac{\hbar^2}{2m}\nabla^2\Psi(\mathbf{r}) + V(\mathbf{r})\Psi(\mathbf{r}) + g|\Psi(\mathbf{r})|^2\Psi(\mathbf{r}) + 2g\tilde{n}(\mathbf{r})\Psi(\mathbf{r}) = \mu\Psi(\mathbf{r}) \quad (15)$$

the factor 2 in the fourth term being a consequence of exchange for Bose particles. The Thomas-Fermi approximation is again useful if T is not too close to the critical temperature. The thermal cloud, on the other hand, is an extremely rarefied and weakly interacting gas that can be treated in a semiclassical approximation as an ideal Bose gas in an effective external potential [3]. That is,

$$\widetilde{n}(\mathbf{r}) = (2\pi\hbar)^{-3} \int d^3p \{exp[(p^2/2m + V_{eff}(\mathbf{r}) - \mu)/k_BT] - 1\}^{-1} \quad (16)$$

where $V_{eff}(\mathbf{r}) = V(\mathbf{r}) + 2g|\Psi(\mathbf{r})|^2 + 2g\widetilde{n}(\mathbf{r})$. The chemical potential is fixed from the total number N of particles according to

$$N = \int d\mathbf{r}|\Psi(\mathbf{r})|^2 + \int dE\rho(E)\{exp[(E - \mu)/k_BT] - 1\}^{-1} \quad (17)$$

and the density of state density of state $\rho(E)$ is obtained from the semiclassical expression

$$\rho(E) = [(2m)^{3/2}/(4\pi^2\hbar^3)] \int d^3r\sqrt{E - V_{eff}(\mathbf{r})}|_{V_{eff}(\mathbf{r})<E} \quad (18)$$

Under appropriate conditions and with appropriate qualifications this approach has proved useful to treat both Bose gases and boson-fermion mixtures at $T \neq 0$.

2.2.2   Dynamical solutions of the GPE.   Let us now examine the collective dynamical behaviour of a condensate at $T = 0$. As already indicated in § 2.1, we write the space and time-dependent order parameter as

$$\Psi(\mathbf{r}, t) = \sqrt{n_0(\mathbf{r}, t)}exp[i\varphi(\mathbf{r}, t)] \quad (19)$$

where $n(\mathbf{r}, t)$ is the condensate density and the gradient of $\varphi(\mathbf{r}, t)$ gives the superfluid velocity by Eq. (8). The condensate wave function obeys the time-dependent GPE,

$$-i\hbar\partial_t\Psi(\mathbf{r}, t) = [-(\hbar^2/2m)\nabla^2 + V(\mathbf{r}) + g|\Psi(\mathbf{r})|^2]\Psi(\mathbf{r}), \quad (20)$$

giving back Eq. (13) for $\Psi(\mathbf{r}, t) = \Psi(\mathbf{r})exp(-i\mu t/\hbar)$. Using Eq. (19), Eq. (20) can be rewritten as the combination of a continuity equation

$$\partial_t n_0(\mathbf{r}, t) = -\nabla.[n_0(\mathbf{r}, t)\mathbf{v}_s(\mathbf{r}, t)] \quad (21)$$

and a collisionless quantum Navier-Stokes equation,

$$m\partial_t\mathbf{v}_s(\mathbf{r}, t) = -\nabla.\{[2m\sqrt{n_0(\mathbf{r}, t)}]^{-1}\nabla^2 n_0(\mathbf{r}, t) + V(\mathbf{r}) + gn_0(\mathbf{r}, t) - \mu\}. \quad (22)$$

The first term on the RHS of Eq. (22) comes from the kinetic energy and can be dropped in the Thomas-Fermi approximation in the case $g > 0$.

Let us consider the case of small dynamical distortions of the ground-state density $n_0(\mathbf{r})$. We set $n_0(\mathbf{r}, t) = n_0(\mathbf{r}) + \delta n(\mathbf{r}, t)$ and to linear terms we find

$$\partial_t^2 \delta n(\mathbf{r}, t) = \nabla.\{[gn_0(\mathbf{r})/m]\nabla\delta n(\mathbf{r}, t))\}. \tag{23}$$

In the homogeneous limit this equation yields $\omega^2 \delta n(\mathbf{q}, \omega) = (gn_0 q^2/m)\delta n(\mathbf{q}, \omega)$, i.e. we recover the Bogoliubov sound waves. In the harmonically confined gas, assuming for simplicity a spherical confinement, one can take $\delta n(\mathbf{r}, t) = \delta n(\mathbf{r})exp(i\omega t)$ with $\delta n(\mathbf{r}) = r^{\ell}F_{n\ell}(r^2)Y_{\ell m}(\theta, \phi)$ and find a set of discrete modes in correspondence to integer values of the quantum numbers n and ℓ. Their frequencies are given at strong coupling by

$$\omega_{n\ell} = \omega\sqrt{2n^2 + 2n\ell + 3n + \ell} \tag{24}$$

in terms of the bare trap frequency . A full discussion of this and other applications of the GPE can be found in the review of Dalfovo et al. [10].

The quadrupolar modes corresponding to $n = 0$ and $\ell = 2$ have been observed in anisotropic traps, where the degeneracy associated with the z component of the angular momentum Angular momentum is split. Taking for example the case of a condensate inside a pancake-shaped trap [11], the $m = 0$ quadrupolar excitation is driven by a weak modulation of the radial spring constant of the trap, while the $m = \pm 2$ excitations are driven by breaking the symmetry with elliptical contours rotating in either direction (we shall later refer to the use of these quadrupolar modes in revealing vortices). The quadrupolar mode frequencies and their damping have been followed with increasing temperature towards T_c [12].

Notice also from Eq. (24) that the frequency of the dipolar mode corresponding to $n = 0$ and $\ell = 1$ is equal to that of the trap. This is a general consequence of its character as a bodily oscillation of the condensate inside the harmonic bowl provided by the trap. We shall later see the role of this mode in connection with the superfluidity of a condensate being driven through an optical lattice by a harmonic force.

2.3   Coherence

The process of formation and growth of a condensate by evaporative cooling and elastic collisions from a gas of cold atoms inside a magnetic trap can in itself be viewed as the generation of a mesoscopic matter wave, by a process which is in some sense akin to the stimulated emission of coherent light in an optical laser. laser The intrinsic dynamics of condensate growth towards equilibrium was followed experimentally by Miesner et al. [13] and shown to be in quantitative agreement with a model which includes bosonic stimulation factors entering the collision probability from occupation of the final state.

A condensate is a coherent system because it is described by a macroscopic wave function and a common phase exists for the whole condensed cloud at equilibrium. A coherent state coherent state can be viewed as built from a linear combination of number states, with the implication that there is no definite value for the number of atoms in the condensate. If the phase is measured, the detection process itself causes the condensate to evolve from a number state into a coherent state with a definite phase (see e.g. [14]).

The high degree of spatial and temporal coherence of a Bose condensate has been confirmed in several experiments. For example, the long coherence time introduces strong correlations between successive events of Rayleigh light scattering by the atomic cloud, leading to highly directional scattering of light and atoms [15]. This allows extraction of a directional beam of recoiling atoms by matter-wave amplification.

2.3.1   Interference between condensates and condensate interferometry.   Given a precise phase difference between two condensates, an interference pattern will be observed if they are brought to overlap with each other. Such an experiment on Bose condensates was carried out by Andrews et al. [16], in a form which is analogous to the interference of two independent optical laser beams where each measurement shows an interference signal but the phase phase is random for each experimental realization. In their experiment two independent condensates were produced by evaporatively cooling atoms in a double-well trap created by splitting a magnetic trap in half with an off-resonant laser beam. After the trap was switched off, the falling atom clouds expanded ballistically and overlapped: an interference pattern was observed in the region of overlap by tomographic absorption imaging. The pattern is determined by the de Broglie wavelength associated with the relative velocity of the two condensates and consists of straight fringes with a huge spacing of about $15 \mu m$.

These data have been accounted for by Röhrl et al. [17], who studied from the GPE the evolution of the coherent superposition $\psi(\mathbf{r}) = \psi_1(\mathbf{r}) + \psi_2(\mathbf{r}) exp(i\varphi)$ of two independent condensate wave functions with an a priori arbitrary relative phase. These authors also studied the transition from interference of two independent condensates to interference of two coupled condensates, which is comparable with the Josephson effect in superconductors and leads to buildup and broadening of the central peak in the interference pattern.

More generally, condensate interferometry implies the use of a condensate that has split into two parts with a definite initial phase relationship between them, these parts being then brought into overlap and interference as for interference from a double slit illuminated by a single optical-laser beam. Coherent splitting of condensates has been achieved by optically induced Bragg diffraction [18]. A number of ingenious techniques have been developed to extract a collimated beam of atoms from a Bose condensate: we mention (i) the application of an rf field inducing spin flips between trapped and untrapped states [19]; (ii) the use of optical Raman processes driving transitions between trapped and untrapped magnetic sublevels [20]; and (iii) the splitting off of an atomic wave packet by diffraction against an optical standing wave [21].

2.3.2   Josephson-type effects in a two-state system.   As an example of a Josephson-type effect that can be met in Bose-Einstein condensates, Bose-Einstein condensation let us consider with Leggett [5] the split trap of the experiment of Andrews et al. [16] on the interference between two condensates, but now in the case where the laser laser power creating a double-well trap has been lowered so that tunneling of bosons between the two wells becomes non-negligible. The single-particle tunneling matrix element may be estimated to be $\approx \hbar\omega exp(-B)$, where ω is an in-well frequency and B is an appropriate WKB exponent. The case $B \gg 1$ corresponds to a weak link between the two wells: the tunneling is weak and the gas of condensed bosons can be described as a system with two accessible states given by the self-consistent ground states ground state on the two sides of the barrier. This is the problem treated in the original work of Josephson, which considered a weak-link junction between two superconductors containing bosonic Cooper pairs.

In the Josephson weak-link regime, and assuming that the number of bosons is large enough that phase fluctuations are negligible (from $\delta N.\delta\varphi \approx 1$ and $\delta N/N \approx N^{-1/2}$ we get $\delta\varphi \approx N^{-1/2}$, the system can be mapped into a pendulum: a free pendulum oscillating between the two states on opposite sides of the barrier if no bias is applied but there is some initial unbalance in their populations, or a driven pendulum if a bias is applied across the barrier. We go back to Eq. (20), neglecting for simplicity all space dependences, and add a coupling between the two states from boson boson tunneling. The equations of motion for the two macroscopic wave functions $\psi_i(t)(i = 1, 2)$ are

$$i\hbar\partial_t\psi_1 = \mu_1\psi_1 + \varepsilon\psi_2, i\hbar\partial_t\psi_2 = \mu_2\psi_2 + \varepsilon\psi_1 \qquad (25)$$

, where μ_i are the chemical potentials and ε is the coupling. These yield

$$\partial_t n_1 = -\partial_t n_2 = (2\varepsilon/\hbar)\sqrt{n_1 n_2}\sin\varphi, \partial_t\varphi = \Delta\mu/\hbar \qquad (26)$$

with $\Delta\mu = \mu_1 - \mu_2$, so that an external current can be driven through the junction in the absence of a bias across the junction. More generally, allowing for a periodic driving potential $\nu(t) = U\cos(\omega t + \vartheta)$ applied across the barrier, a multimode analysis yields the current through the weak link in the form

$$I(t) = I_0 \sum_{n=-\infty}^{\infty} (-1)^n J_n(U/\hbar\omega)\sin[(\Delta\mu - n\hbar\omega)t/\hbar - n\vartheta + \varphi_0] \qquad (27)$$

where $J_n(x)$ are the Bessel functions of order n. We return to Eq. (27) in the context of the motion of a condensate being driven by external forces through an optical lattice

2.4   Superfluidity and vortices

As already remarked in § 1, a characteristic behaviour of liquid HeII is the presence of two hydrodynamic modes at finite frequency, i.e. ordinary sound waves and so-called second sound - the latter being associated with propagation of heat waves through out-of-phase oscillations of the normal and superfluid components. In a Bose-condensed gas at finite temperature, the superfluid component corresponds to the condensate atoms and the normal component to the thermal cloud. The large coefficient of thermal expansion of the gas results in a decoupling of the motions of the condensate from those of the thermal cloud: near the critical temperature the in-phase oscillations involve mainly the latter, while the out-of-phase oscillations are mainly confined to the condensate [22]. The collective excitations of a dilute Bose gas were probed in an experiment by Stamper-Kurn et al. [23]: hydrodynamic oscillations of the thermal cloud were observed as well as an out-of-phase dipolar oscillation of the thermal cloud and the condensate. The visible separation between the normal and the superfluid component in these experiments directly confirms the two-fluid model of superfluids.

A clear demonstration of the transition from normal to superfluid behaviour is afforded by the observation of the scissors mode as a function of temperature [24]. This mode of oscillation is excited by a sudden rotation of an anisotropic trap. In the normal fluid such oscillations occur at two frequencies, with the lower one corresponding to the rigid-body moment of inertia. Well below the critical temperature this low-frequency mode disappears and the condensate is seen to oscillate at a single undamped frequency.

In a further experiment Onofrio et al. [25] studied the hydrodynamic flow in a Bose-Einstein-condensed fluid stirred by a blue-detuned laser beam acting as a macroscopic moving object. A density-dependent critical velocity for the onset of a distortion in the density distribution around the moving object was observed, the distortion being associated with a pressure gradient arising from a drag force between the beam and the condensate. A model for the drag force is that it arises from the periodic shedding of vortex lines at a rate that increases with velocity. We shall return on dissipation of superfluidity in the case of a condensate being driven through an optical lattice.

A number of direct observations of quantized vortices in confined condensates have been reported [26, 27, 28, 29, 30] and an exhaustive review article on this topic is available from Fetter and Svidzinsky [31]. The presence of a single vortex line inside a condensate breaks time-inversion symmetry, leading to a splitting of the quadrupolar surface modes with azimuthal quantum number $m = \pm 2$: the time derivative ∂_t in the equation of motion for the condensate is replaced by $\partial_t + \mathbf{v}.\nabla$ where $\mathbf{v}$ is the velocity field. A slow precession is induced in a quadrupolar distortion of a condensate containing a vortex at its centre, with a relative angular velocity $(\omega_+ - \omega_-)/\omega_+ \cong |m|(a_{ho\perp}/R_\perp)^2$, and the precession of the axes of the quadrupolar mode can be used to reveal the formation of vortices in an unambiguous manner [27]. Stirring of the condensate by a laser beam leads to the generation of one or more vortices above a critical value of the stirring frequency, and the measured angular momentum Angular momentum just above the critical frequncy is , corresponding to a single quantum of circulation. Ordered triangular arrays of vortices [29] and the phase phase singularity associated with a vortex [30] have also been observed.

3.   Condensates in Optical Lattices

Linear optical lattices are realized by means of a detuned standing wave of laser laser light and provide a periodic potential where the excitations and the transport behaviour of cold atomic gases and Bose-Einstein condensates Bose-Einstein condensation can be investigated. The use of optical lattices in the study of ultracold atoms [32] has allowed to demonstrate transport phenomena like Bloch oscillations, Landau-Zener tunnelling, and Wannier-Stark ladders, and to realize a paradigm of chaotic dynamics, namely the periodically kicked rotator.

The use of Bose condensates inside optical lattices has been proposed as a tool for developing quantum computing Quantum computing devices and for studying the transition from classical to quantum chaotic behaviour and that from solitonic transport to dynamical localization (for a short review see [33]). Experiments have concerned the generation of coherent matter-wave pulses [34], laser cooling methods [35], and the realization of an atom-laser outcoupler [20]. Moving optical lattices have been used to induce Bragg diffraction from a condensate [18] and for momentum spectroscopy [36]. Through the combined use of a harmonic force and an optical lattice, a condensate has been driven across a local velocity threshold into a gradual loss of superfluidity via sound wave emission [37].

3.1   Ground state and band excitations in a 1D lattice

The field of two counter-propagating beams of detuned laser light acts on an atomic gas as a periodic quasi-1D potential of the form $U_L(z) = U_0 \sin^2(\pi z/d)$, with a period d equal to one-half the light wavelength U_0 and in the range 0.5 - 20 recoil energies $E_R = \hbar^2/(2md^2)$. The theory of the dynamics of a condensate in such a lattice involves concepts which are well known in solid state theory for particles in a 1D crystalline potential. The energy - momentum dispersion relation is a multi-valued periodic function: the "energy bands" $E_n(q)$ inside the Brillouin zone $(-\pi/d < q \leq \pi/d)$. Two sets of Bloch orbitals having definite symmetry under time reversal, namely $Z_{nq\pm}(x) = 2^{-1/2}[u_{nq}(x) \pm v_{nq}(x)]$, need to be built from the functions $u_{nq}(x)$ and $v_{nq}(x)$ entering Bogoliubov's canonical transformation and obeying the Bogoliubov - des Gennes equations [38]. Of main interest for the experiments is the lowest energy band starting at $q = 0$ from the chemical potential (the band $n = 0$, say).

It is convenient to express the Bloch orbitals in terms of Wannier functions centred on the lattice sites [39], namely

$$Z_{nq\pm}(x) = N^{-1/2} \sum_\ell exp(i\ell qd) w_{n\pm}(x - \ell d) \qquad (28)$$

This expansion allows one to demonstrate a number of exact properties of a periodic 1D condensate, regarding the phase-density nature of the excitations in the various bands:

- (i)since the condensate at equilibrium occupies the lowest-energy state at $q = 0$ in the $n = 0$ band, the whole of the $n = 0$ band corresponds to pure modulations of the phase of the condensate;

- (ii)the higher bands derive from the superposition of localized excitations with definite phase relations and the density in each band reflects that of an isolated-well state.

Interestingly, the axial form of the optical potential reproduces the "Mathieu problem" for electrons in a 1D crystal that was studied by Slater in an early work on the band-structure theory of solids. Slater was able to specify the conditions under which the Wannier functions for the lowermost energy bands of this problem can be approximated by on-site tight-binding orbitals of harmonic oscillator form, with Gaussian widths determined by the curvature of the bottom of each potential well. In this situation the relationships that exist between the low-lying excitations of a condensate in a lattice and those of a condensate in a single harmonic well are easily visualized.

Various pump-probe methods which could be used to explore condensate excitations inside an optical lattice under currently attainable experimental conditions have been numerically analyzed from the GPE [40]. In a first method the condensate is put into a state of flow through the lattice with a velocity $v = \hbar q/m$ and the velocity spectrum is probed, e.g. by measuring the momentum distribution at different times and evaluating the time-Fourier transform of the average velocity. In a second method a resonant parametric driving of the lattice is performed for a variable length of time and the density spectrum of the condensate is determined from the analysis of its dynamics. The results of these two methods agree with each other and with those obtained by a third method, in which the state of a phase-modulated condensate is propagated in imaginary time and the excitations are probed through their interband and intraband relaxations.

A further consequence of the Wannier-function expansion is that the current carried by a periodic condensate moving in the n = 0 band under a force F(t) is given by

$$\overline{p}(t) = \frac{\hbar}{d} \frac{\sum_{\ell \geq 1}[-2\partial g_0(\ell d)/\partial \ell] \sin[\ell q(t)d]}{1 + 2\sum_{t \geq 1} g_0(\ell d) \cos[\ell q(t)d]} \tag{29}$$

where $g_0(\ell d) \equiv \int dk |f_0(k)|^2 \cos[\ell k d]$ and $f_0(k)$ is the Fourier transform of . Eq. (29) has been derived within a semiclassical approximation, in which the quasi-momentum q(t) evolves in time according to $\hbar \dot{q}(t) = F(t)$. It shows that the average current driven by a constant force is a periodic function of time, with a structure arising from the relative phases between the lattice wells. The details of the periodic potential and of the interparticle coupling enter through the weighting function $g_0(\ell d)$.

3.2 Coherent drop emission under gravity

Let us consider some examples of coherent transport behaviour which can be analyzed with the aid of Eq. (29). The simplest case is a condensate under the action of a constant force F: the quasimomentum $q(t)$ increases linearly with time and the average current is periodic with the period T_B of Bloch oscillations, given by $T_B = h/(Fd)$.

The first observation of Bloch oscillations in a condensate is due to Anderson and Kasevich [33] and was obtained by pouring the condensate into a vertical optical lattice where it is acted upon by the force of gravity. In essence, the condensate starts out from its ground state ground state and can coherently climb the energy dispersion curve as if it were a single quasi-particle. In the conditions of the experiment, as the condensate reaches the edge of the Brillouin zone a part of it tunnels out into vacuum as a coherent drop of atomic matter, while the rest is Bragg-reflected across the Brillouin zone to return from there down into the lowest-energy state and again to climb up the energy dispersion curve. The measured time-interval in the emission of matter drops is 1.1 ms, against a theoretical value of 1.09 ms for the Bloch period.

Computer simulation models based on the GPE [33] reproduce the drop-emission period and demonstrate its constancy for varying strengths of the interactions and of the barrier height. The drop emission is found to be a threshold process, the size of the drops being a universal function of the governing parameters.

3.3 Josephson-type oscillations

We turn to the case where the external force acting on the condensate inside the optical lattice is the sum of a constant force F and of a harmonic term $-m\omega^2 x$ (here, x is the displacement of the centre-of-mass coordinate and we are for the moment assuming that its value is small enough for the condensate to be in a non-dissipative regime). We have

$$\hbar q(t) = Ft + m\omega A \sin(\omega t + \varphi_0) \tag{30}$$

with $\sin(\varphi_0)$ and A to be determined from the initial conditions. The expression for the average current is obtained by substituting Eq. (30) in Eq. (29) and by performing a multimode expansion of both the numerator and the denominator, through the use of the Bessel functions $J_n(\ell A d/a_{ho}^2)$ [41].

It turns out that there exists a range of realistic system parameters for which the term $\ell = 1$ in the numerator of Eq. (29) is dominant and the denominator can be replaced by unity. In this case the average current takes the expression

$$\bar{p}(t) = p_0 \sum_{n=-\infty}^{\infty} (-1)^n J_n(Ad/a_{ho}^2) \sin[(\omega_B - n\omega)t - n\varphi_0], \tag{31}$$

where $p_0 = -2\hbar\partial g_0(d)/\partial d$. The condensate is driven to perform multimode oscillations with frequencies $\omega_n = \omega_B - n\omega$ and amplitudes determined by $J_n(Ad/a_{ho}^2)$, where $\omega_B = |F|d/\hbar$ is the angular frequency of Bloch oscillations.

The simplified equation (31) is formally equivalent to the expression for the current passing through a Josephson weak link when a constant voltage drop $\Delta\mu = Fd$ and an oscillating voltage of amplitude $U = m\omega^2 Ad$ is applied to it (see Eq. (27)). We can expect, therefore, that coherent phenomena typical of Josephson junctions between superconductors, such as multiple resonances and multimode oscillations, should become observable when a superfluid condensate is driven by external forces through an optical lattice. A number of different cases are reported below, for 10^4 ^{87}Rb atoms moving through a lattice with $U_0 = 1.5E_R$ and $\lambda = 795nm$ [41].

As a first example, we consider the case when the constant force F is tuned to match the resonance condition $\omega_B = n\omega$. At each resonance the pattern of the current from Eq. (29) suddenly changes, displaying a number n of sub-oscillation peaks within a fundamental period $T^* = (0.9\omega/2\pi)^{-1}$ with an effective mass correction as determined from the GPE. If anharmonicity were present, the pattern would change continuously from each resonance to the next, though the number of peaks would be preserved.

This resonant behaviour could thus be used to measure the shift of the harmonic-oscillator levels due to mean-field or lattice interactions, by tuning the constant force. Conversely, an unknown constant force could be measured by tuning the driving frequency. The observation of the resonances would be possible through a measure of the momentum distribution at different times. The strength of the forces and the range of the harmonic-oscillator levels which can be probed is limited by the need to stay below a dissipation threshold: this requires low condensate velocities (see below).

As a second example we consider the case where the BEC is subjected only to a harmonic force making it oscillate back and forth through the optical lattice. Here, multimode oscillations could be observed for suitable choices of system parameters and could provide a tool to tailor the pulsed emission of coherent matter waves. An important parameter in this situation is the intensity of the harmonic force, or equivalently the maximum velocity v with which the condensate passes through the bottom of the harmonic bowl. Both from Eq. (29) and from a numerical solution of the GPE it is found that for $v/v_{BZ} < 0.5$, with $v_{BZ} \equiv \hbar\pi/(md)$ being the velocity corresponding to the quasi-momentum at the Brillouin zone boundary, the condensate current executes monochromatic oscillations with period T^*. Terms with $|n| > 1$ start significantly contributing for $v/v_{BZ} > 0.5$ and terms up to $|n| = 3$ determine a flattening of the maximum current which becomes very evident at $v/v_{BZ} = 0.70$. The numerical solution of the GPE shows that throughout this dynamical range the condensate remains coherent for the system parameters used in these calculations. The simplified Eq. (31) gives a quantitative account of all these data at low values of the maximum velocity, but tends to overemphasize the weight of higher multimode components as v/v_{BZ} is increased.

3.4   Decoherence

Decoherence decoherence comes in at higher centre-of-mass velocities, prior to the onset of Bragg scattering at $v/v_{BZ} = 1$. The break-up of quasi-particle dynamical behaviour is seen from the GPE as a fragmentation of the condensate into mutually incoherent parts. Above a threshold velocity the condensate breaks up into an incoherent assembly of subsystems randomly moving in the various lattice wells, and at the same time its momentum distribution crumbles away and spreads over the whole Brillouin zone [42].

This behaviour as obtained from the numerical solution of the GPE is reminiscent of dynamical localization and quantum chaotic behaviour [43]. The Hamiltonian which governs the system can be traced back to that of a periodically kicked rotator, where a simple oscillator comes into resonance with an external driving force and the chaotic behaviour sets in after the activation of the many degrees of freedom of the system.

Of special interest for testing fundamental concepts as well as for applications is the issue of quantum chaos. chaos The question arises whether a suitable set of experimental parameters or set-up arrangements could be found to observe the chaotic behaviour predicted by the GPE before the condensation excitations become dominant. More work is needed along these lines to develop an appropriate quantum mapping of the problem and to define the distinctive observables to be examined.

3.5   Superfluidity and dissipation in a real condensate

In a study combining experimental and simulational methods [37] a ^{87}Rb condensate was placed inside a static magnetic trap with a superimposed optical lattice and its dynamical evolution was controlled by displacing the centre of the magnetic trap by a variable amount Δx in the longitudinal direction (parallel to the lattice). The displacement was very rapid compared with the fundamental period of the magnetic trap. Different dynamical regimes were observed, depending on the magnitude of Δx.: different values of Δx correspond to different velocities of transit of the condensate through the bottom of the trap.

For small displacements the condensate performs undamped oscillations in the harmonic potential of the magnetic trap and feels the lattice only through a shift in its oscillation frequency. In this small-amplitude regime only the states in the lowest band near the Brillouin zone centre are being explored, so that the frequency shift can be interpreted as an increase in effective mass (by some 18 %). This behaviour is a clear manifestation of superfluidity: the coherent condensate is being accelerated through band states as if it were a single quasiparticle. In this regime there is quantitative agreement between the predictions from the GPE and the experiments on a real condensate.

At larger initial displacements, corresponding to maximum velocities higher than about 3 mm/s for the specific values of the system parameters, the onset of dissipative processes is observed. The amplitude of the oscillations is more strongly attenuated as the initial displacement Δx increases and at this level the GPE qualitatively agrees with the experimental findings. However, in regard to the dynamical behaviour of the condensate shape, the fragmentation predicted from the GPE is in disagreement with the observed density profiles of the real system.

The measured density profiles can be fitted by a two-component distribution, since the central part of the atomic cloud proceeds along the trajectory which it would have in the superfluid state while the rest lags behind. This suggests an interpretation in terms of a two-fluid model. From the study of homogeneous superfluids one expects that coherence and superfluid behaviour will be lost when the velocity of flow is sufficient for the spontaneous emission of elementary excitations. The essentially 1D dynamics of the sample in these experiments implies an important role for longitudinal phonon emission, with a spectrum of critical velocities because of the inhomogeneity of the sample. In a simplified picture the lattice may be viewed as a medium with microscopic roughness, producing local compressions of the gas and friction forces that damp the oscillatory motions in favour of the formation of a non-superfluid thermal cloud.

The experiments in the dissipative regime can be interpreted in terms of an inhomogeneous superfluid having a density-dependent local critical velocity for destruction of superfluidity. This x-dependent critical velocity is calculated as the local speed of sound,

$$c(x) = [(n(x)/m)(\delta\mu/\delta n)]^{1/2}, \tag{32}$$

where μ is the chemical potential and $\delta\mu/\delta n$ gives the 1D elastic stiffness constant of the condensate against longitudinal stretching. For the condensate under study this formula yields a maximum speed of sound, corresponding to the maximum density of the condensate, as $c_{max} \cong 5.2$ mm/s, to be compared with the measured value $v_{crit} \cong 5$ mm/s for the critical velocity at which the superfluid state is completely destroyed in the experiment. Assuming, therefore, that the critical velocity $v_{crit}(x)$ for local disappearance of superfluidity in the inhomogeneous condensate coincides with the local speed of sound $c(x)$, one has $v_{crit}(x) \propto \sqrt{n(x)}$: this relation gives a fully quantitative and parameter-free account of the data [37].

3.6 Condensates in a 3D optical lattice

Optical lattice potentials of higher dimensionalities have been created for ultracold atoms and Bose-Einstein condensates. Bose-Einstein condensation In the experiments of Greiner et al. [44] a spherically symmetric condensate of ^{87}Rb atoms is created inside a magnetic trap and a 3D lattice potential is superposed by means of three optical standing waves, which are aligned orthogonally to each other and have their crossing point positioned at the centre of the condensate. The 3D optical potential for the atoms is proportional to the sum of the intensities of the three standing waves, leading to a simple cubic geometry of the lattice. In order to test the phase phase coherence between different lattice sites after ramping up the lattice potential, the combined trapping potentials are suddenly turned off and the atomic wave functions are thus allowed to expand freely and to interfere with one another. Two different behaviours are observed, depending on the value of the lattice potential depth.

In the superfluid regime, where all atoms are delocalized over the lattice with equal relative phases between lattice sites, a high-contrast 3D interference pattern is observed as expected for a periodic array of phase coherent matter-wave sources. As the lattice well depth is increased, the strength of the higher-order interference maxima increases, due to the tighter localization of the atomic wave functions at a single lattice site. Above a critical strength, an incoherent background of atoms gains strength until no interference pattern remains. This transition from coherence to incoherence is reversible.

The interpretation of these observations that Greiner et al. offer is by establishing a parallelism with a quantum phase transition from a coherent conductor to a correlation-induced Mott insulator within a Bose-Hubbard model [45]. In the superfluid phase each atom is spread out over the whole lattice with long-range phase coherence. In the insulating state instead equal numbers of atoms are localized at individual lattice sites and phase coherence across the lattice is lost. The reader may refer to Sachdev [46] for a review of phase transitions that may be driven in a quantum system at $T = 0$ by varying some parameter in its Hamiltonian.

4.   Confined Fermi Gases and Boson-fermion Mixtures

Atomic vapours of fermionic ^{6}Li and ^{40}K atoms have been trapped and cooled down to degeneracy in several experiments using techniques similar to those employed to realize Bose-Einstein condensation Bose-Einstein condensation. At very low temperatures collisions between atoms are essentially limited to the s-wave channel, but for fermions in the same hyperfine state these are forbidden by the Pauli exclusion principle. A dilute one-component gas of Fermi atoms in a fully spin-polarized state under magnetic confinement thus is a close laboratory realization of the ideal Fermi gas. In a two-component Fermi gas, on the other hand, s-wave scattering is operative between pairs of atoms in different hyperfine states: while the scattering length is small and repulsive for ^{40}K, it is strong and attractive for ^{6}Li so that s-wave pairing in a mixture of ^{6}Li atoms is the simplest mechanism that may yield a novel superfluid. The ineffectiveness of the evaporative cooling method, caused by the suppression of s-wave collisions, can be partly circumvented by recourse to "sympathetic cooling", either in a two-component Fermi mixture or in a boson-fermion mixture such as ^{6}Li-^{7}Li. At the time of writing several groups have reached by these methods temperatures as low as $0.2TF_F$ in the Fermi cloud.

4.1   One-component Fermi gas

Dynamical equations for an ideal Fermi gas in a trapping potential, that are analogous to Eqs. (21) and (22) for a Bose-Einstein condensate, Bose-Einstein condensation can be obtained from the kinetic equation for the one-body density matrix density matrix or equivalently for the Wigner distribution function [47]. Of particular interest is the dynamics of the Fermi cloud in the collisional regime, which in the homogeneous limit corresponds to hydrodynamic sound associated with a local breathing of the Fermi sphere. The local pressure $P(\mathbf{r}, t)$ in the trapped gas is essentially determined by the kinetic energy density and in a local-density model is

$$P(\mathbf{r}, t) = (2A/5)[n(\mathbf{r}, t)]^{5/3} \tag{33}$$

at $T = 0$, where $n(\mathbf{r}, t)$ is the particle number density and $A = (6\pi^2)^{2/3}(\hbar^2/2m)$. The corresponding Thomas-Fermi result for the equilibrium density profile is

$$n_{TF}(\mathbf{r}) = \theta(\varepsilon_F - V(\mathbf{r}))[(\varepsilon_F - V(\mathbf{r}))/A]^{3/2} \qquad (34)$$

and the linearized equation of motion for density fluctuations in a spherical trap yields a frequency spectrum labelled by the principal and angular-momentum numbers n and ℓ,

$$\omega(n, \ell) = \omega\sqrt{\ell + (4n/3)(n + \ell + 2)} \qquad (35)$$

ω being the trap frequency [48]. The frequencies of the surface modes of the mesoscopic Fermi cloud, corresponding to $n = 0$ in Eq. (34), coincide with those obtained in the Thomas-Fermi approximation for a Bose cloud (see Eq. (24)), since in these modes the fluid is incompressible.

While the above results refer to the normal state of the Fermi gas, the low-lying collective excitations of a trapped Fermi gas in the superfluid state have been studied by Baranov and Petrov [49] in the non-interacting limit. Their theory is formulated in terms of the fluctuations in the phase of the order parameter, but they show that these modes also manifest themselves as density fluctuations. In fact, their equation of motion for the density fluctuations in the superfluid Fermi gas coincides with that obtained by Amoruso et al. [48] for the normal Fermi gas in the collisional regime. Equation (34) thus also gives the frequencies of the superfluid Fermi gas under isotropic harmonic confinement. This is for the trapped fluid the analogue of a well-known property of the homogeneous Fermi fluid: the opening of the superfluid gap in the single-particle excitation spectrum leaves a window inside which compressional waves can propagate. In the homogeneous superfluid this mode is known as the Bogoliubov-Anderson (BA) mode [50] and can be viewed as a coherent oscillation of the order parameter of the condensed phase, whose presence is required by the Goldstone theorem in order to ensure gauge invariance. The speed of propagation of the BA sound wave in the superfluid phase is the same as that of hydrodynamic sound in the normal phase, since the compressibility of the fluid is practically unaffected by the phase transition [51].

4.2 Mixtures of Fermi gases

For a Fermi gas trapped in two different hyperfine levels, the density profiles of the two components are coupled by their mean-field interactions. For sufficiently strong repulsive interactions the gas may phase-separate [52], as a result of the competition between the repulsive interaction energy driving spatial separation and the kinetic energy disfavouring localization of each component in a limited rgion of the trap. Fermion pairing may instead take place for sufficiently strong attractive interactions [53]. In a symmetric mixture (equal numbers of particles in each component, subject to the same harmonic confinement), the phase transitions are estimated to lie at $k_F \approx 0.8$ for phase separation and at $k_B T_{BCS} \approx 0.28\varepsilon_F exp[-\pi/(2k_F|a|)]$ for a BCS transition, with k_F the non-interacting Fermi wave number and a the s-wave scattering length.

The linearized equations of motion for the density and concentration fluctuations in a two-component Fermi gas with repulsive interspecies interactions in the collisional regime have been studied by Amoruso et al. [52]. For a symmetric system at weak coupling the interactions enter the eigenfrequencies through a simple parameter C reflecting the shift of the Fermi energy, yielding back Eq. (34) in the non-interacting case. In the same case the frequencies of the concentration modes become degenerate with those of the total-density modes.

The superfluid state in a two-component Fermi gas is expected to be different from that attained in liquid ^{3}He, which has a p-wave pairing and is not in the dilute regime, as well as from that of conventional superconducting metals where the excitation spectrum is dominated by the Coulomb interactions. An important issue for future experiments is to identify a clear signature of the superfluid transition in an atomic Fermi gas: contrary to the case of Bose-Einstein condensates, Bose-Einstein condensation for fermions the transition affects only slightly the density profile and the internal energy of the Fermi gas. Various ideas have been put forward for this objective, with special regard to observations of the change in dynamical properties across the transition from normal to superfluid. A possible approach is to identify the superfluid phase through a measurement of the bulk excitations of the gas, i.e. excitations with a wavelength smaller than the spatial extension of the atomic cloud. A calculation of the dynamic structure factor as a function of temperature by the Random Phase Approximation [54] has demonstrated a prominent role of the BA mode in the spectrum of the superfluid phase. The calculated spectrum for a homogeneous two-component Fermi gas possesses a continuum of particle-hole excitations above the gap and a peak from the BA phonon lying in the gap below the continuum, the peak being Landau damped at finite T by the interplay with thermal excitations. These results are then adapted to a trapped Fermi cloud by means of a local-density model, which predicts further broadening of the spectrum due to inhomogeneity of the density profile.

An alternative approach to revealing the phase transition is to rely on the character of the dynamical response of a superfluid to a long-wavelength transverse probe, which can only excite the nonsuperfluid component [55]. A test of superfluidity in an inhomogeneous finite system can thus be given by exciting a small-angle oscillation in a plane where the confinement is anisotropic [56], that is the scissors mode already referred to in § 2.4 for a Bose condensate. As noted there, this mode is predicted to show only one frequency component in the deep superfluid state, while it has two components in the normal fluid due to the additional contribution from transverse excitations.

4.3   Fermion-Boson mixtures

Sympathetic cooling of a spin-polarized Fermi gas by elastic scattering against bosons in a fermion-boson mixture seems to minimize the effects of "Pauli blocking", i.e. the final-state occupation factors that limit the process of cooling in clouds containing two fermionic components [57]. At the time of writing several experiments are currently in progress on trapping and cooling various fermion-boson mixtures, i.e. ^{6}Li-^{7}Li, ^{6}Li-^{23}Na and ^{40}K-^{87}Rb. From the theoretical point of view a mixture of condensed bosons and normal fermions is an interesting system, because it can show spatial demixing of the two components in the case of repulsive boson-fermion interactions [58]. On increasing these interactions in the homogeneous gas the total energy is minimized by placing the bosons boson and fermions (or boson-fermion mixtures of different composition) in different regions of space [59], even though this implies a high cost in kinetic energy at the interface. There are important consequences of system finiteness on demixing in trapped mixtures, and in particular the transition is spread out in a finite system and the anisotropy of the trap acts differently on the two types of atoms. Locating the onset of partial demixing is relevant to the practicalities of fermion cooling, since at that point the diminishing overlap between the two clouds will start reducing the effectiveness of the collisional processes between the two species.

The conditions for full demixing of a 50-50 mixture in a spherical trap at $T = 0$ have been set out [60] in terms of two simple system parameters, i.e. the ratio a_{bb}/a_{bf} of the boson-fermion and boson-boson boson scattering lengths and the quantity $k_F a_{bb}$ where k_F is the Fermi wave number of the isolated fermion component. A mixed state is stable if the inequality

$$4k_F a_{bb} \leq 3\pi(a_{bb}/a_{bf})^2 \tag{36}$$

holds, and subsequent studies [61, 62] have confirmed the essential correctness of this criterion. The onset of partial demixing has been determined from the location of the maximum for the overlap interaction energy of the two atomic components, which is given in terms of the numbers of bosons and fermions as

$$a_{bb}/a_{bf}|_{max} = (c_1 N_F^{1/2} N_B^{-2/5} + c_2 N_B^{2/5} N_F^{-1/3})^{-1}. \tag{37}$$

Here the c's are explicitly known in terms of systems parameters (the atomic masses and scattering lengths, and the trap frequencies and anisotropies).

Extensive calculations [61, 62] have identified inside the regime of full demixing several exotic configurations for the two phase-separated clouds, in addition to the simplest ones consisting of a core of bosons enveloped by fermions and vice versa. With increasing temperature the main effect of the growing thermal cloud of bosons is to transform some of the exotic clusters into more symmetric ones, until demixing is ultimately lost. Demixing also occurs between the fermions and the thermally excited bosons at very high values of the fermion-boson boson coupling.

The approach of the mixture to phase separation will be signalled by changes in dynamical properties and in the collective excitation spectrum, in addition to changes in equilibrium properties. In particular, a softening of the frequencies of surface or bulk modes having the appropriate symmetry for a phase-separated configuration is expected.

4.4   The Tonks gas and Boson-Fermion mapping

The geometry of a magnetic trap can be adapted to have cylindrical symmetry with a transverse confinement which may be much weaker or much stronger than the longitudinal confinement. It is thus possible to experimentally generate and study strongly anisotropic atomic fluids effectively approaching dimensionality $D = 2$ or $D = 1$, at very low temperature and with very high purity [63]. Parallel advances in atom waveguide technology [64, 65, 66, 67, 68], with potential applications to atom interferometry and integrated atom optics, especially motivate theoretical studies of dilute systems in a regime where the quantum dynamics becomes essentially quasi-onedimensional (1D).

A gas of N bosons boson interacting via a repulsive delta-function potential and moving in $D = 1$ under periodic boundary conditions was solved exactly by Lieb and Liniger [69, 70], while Yang and Yang [71] studied the equilibrium thermodynamics and the excitation spectrum at finite T. The so-called Tonks gas limit is reached as the scattering length becomes very large and the bosons boson become effectively impenetrable. The dynamics of a condensate in a thin waveguide approaches this limit at very low temperature, high dilution, and large 1D coupling constant [72].

A precise mapping exists in 1D between spin-polarized fermions and impenetrable bosons [73] boson. In both systems each particle behaves as a hard point which has zero probability of being found on top of any other particle, but is at any time free to move on a segment comprised by its two near neighbours and can exchange with either of them. The mapping in the point-like limit thus descends from the fact that a many-body wave function of either system describes free motion of each particle in an external one-body potential between nodes that are determined by the positions of the other particles. On account of the node structure of the fermionic wave function, for the ground state ground state the mapping takes the form of an identity between the wave function of the 1D Bose gas in the Tonks limit and the modulus of the wave function for the 1D Fermi gas,

$$\psi_0^B = |\psi_0^F|. \tag{38}$$

The latter wave function is, of course, a Slater determinant of single-particle orbitals determined by the longitudinal confining potential and therefore the ground state properties of either gas can, at least in principle, be calculated exactly. In fact, the mapping between impenetrable bosons boson and spin-polarized (effectively impenetrable) fermions in 1D extends to the excited states and to time-dependent external potentials. Notice that quantum mechanical exchange still allows each particle to be delocalized over the length of the gas: whereas classical hard particles are unable to get past one another, two identical quantum particles entering a binary collision can undergo backward scattering as well as forward scattering through exchange.

The mapping does not hold for the momentum distributions, since they involve the wave function in the momentum representation and the result of taking a Fourier transform depends on the relative signs of the configurational wave function in the various regions of space. The single-particle momentum distribution in the ground state reflects for the Fermi system a rapid drop as the value of the chemical potential is being crossed; it instead bears trace of Bose-Einstein condensation Bose-Einstein condensation for the Bose system, in combination with long high-momentum tails from the nodes in x-space [74]. In a strict sense there is no condensate in this 1D case: in particular, in the absence of a trap potential the occupation of the lowest single-particle state is known to be only of order $N_{1/2}$ in the many-body ground state ground state of the 1D gas of hard-core bosons. boson However, the momentum distribution still has a strong peak in the neighbourhood of zero momentum and some coherence effects are present as in ordinary BEC. These have been discussed in the literature in regard to dark soliton-like behaviour in response to a phase-imprinting pulse [75] and to Talbot recurrences following an optical lattice pulse [76].

It should be emphasized that an approach based on the use of Hermite polynomials for numerical calculations on mesoscopic clouds under harmonic confinement is severely limited in the number of atoms that it can treat. Brack and van Zyl [77] have developed a more powerful method for non-interacting fermions occupying a set of closed shells under isotropic harmonic confinement in D dimensions. A Green's function Green's function method, which altogether avoids the use of wave functions in favour of the matrix elements of the position and momentum operators, is most powerful in $D = 1$ [78]. This method can in principle be used for any confining potential and is easily extended to systems at finite temperature and to gases under anisotropic confinement in higher dimensionalities. ·

5.   Concluding Remarks

The special conditions of ultralow temperature, ultrahigh purity, and tunable interactions which apply in dilute quantum gases hold the promise of novel developments and of accurate tests of fundamental ideas in physics. In this short review I have started from some basic models of many-body theory, that is the Bogoliubov theory for a dilute Bose gas and the Thomas-Fermi theory for a confined Fermi gas, and tried to cover at an introductory level the very remarkable progress that has been made both in experiment and in theory to advance basic knowledge of confined atomic vapours in the regime of deep quantum degeneracy. In consideration of the topics that are covered by other contributors to the volume, I have paid special attention to the phenomena of coherence and superfluidity and tried to give ample introductory references to the recent literature.

References

[1] Inguscio, M., Stringari, S., and Wieman, C.E. (1999) Bose-Einstein Condensation in Atomic Gases, IOS, Amsterdam.

[2] Kaiser, R., Westbrook, C., and David, F. (2001) Coherent Matter Waves, EDP Sciences, Paris.

[3] Bagnato, V., Pritchard, D.A., and Kleppner, D. (1987) Bose-Einstein condensation in an external potential, Phys. Rev. A **35**, 4354-4358.

[4] Baym, G. and Pethick, C.J. (1996) Ground-state properties of magnetically trapped Bose-condensed Rubidium gas, Phys. Rev. Lett. **76**, 6-9.

[5] Leggett, A.J. (2001) Bose-Einstein condensation in the alkali gases: some fundamental concepts, Rev. Mod. Phys. **73**, 307-356.

[6] Leggett, A.J. (1999) Superfluidity, Rev. Mod. Phys. **71**, S318-S323.

[7] Huang, K. (1987) Statistical Mechanics, Wiley, New York.

[8] Ensher, J.R., Jin, D.S., Matthews, M.R., Wieman, C.E., and Cornell, E.A. (1996) Bose-Einstein condensation in a dilute gas: measurement of energy and ground-state occupation, Phys. Rev. Lett. **77**, 4984-4987.

[9] Minguzzi, A., Conti, S., and Tosi, M.P. (1997) The internal energy and condensate fraction of a trapped interacting Bose gas, J. Phys.: Condens. Matter **9**, L33-L38.

[10] Dalfovo, F., Giorgini, S., Pitaevskii, L.P., and Stringari, S. (1999) Theory of Bose-Einstein condensation in trapped gases, Rev. Mod. Phys. **71**, 463-512.

[11] Jin, D.S., Ensher, J.R., Matthews, M.R., Wieman, C.E., and Cornell, E.A. (1996) Collective excitations of a Bose-Einstein condensate in a dilute gas, Phys. Rev. Lett. **77**, 420-423.

[12] Jin, D.S., Matthews, M.R., Ensher, J.R., Wieman, C.E., and Cornell, E.A. (1997) Temperature-dependent damping and frequency shifts in collective excitations of a dilute Bose-Einstein condensate, Phys. Rev. Lett. **78**, 764-767.

[13] Miesner, H.-J., Stamper-Kurn, D.M., Andrews, M.R., Durfee, D.S., Inouye, S., and Ketterle, W. (1998) Bosonic stimulation in the formation of a Bose-Einstein condensate, Science **279**, 1005-1007.

[14] Castin, Y. and Dalibard, J. (1997) Relative phase of two Bose-Einstein condensates, Phys. Rev. A **55**, 4330-4337.

[15] Inouye, S., Chikkatur, A.P., Stamper-Kurn, D.M., Stenger, J., Pritchard, D.E., and Ketterle, W. (1999) Superradiant Rayleigh scattering from a Bose-Einstein condensate, Science **285**, 571-574.

[16] Andrews, M.R., Townsend, C.G., Miesner, H.-J., Durfee, D.S., Kurn, D.M., and Ketterle, W. (1997) Observation of interference between two Bose condensates, Science **275**, 637-641.

[17] Röhrl, A., Naraschewski, M., Schenzle, A., and Wallis, H., (1997) Transition from phase locking to the interference of indepedent Bose condensates: theory versus experiment, Phys. Rev. Lett. **78**, 4143-4146.

[18] Kozuma, M., Deng, L., Hagley, E.W., Wen, J., Lutwak, R., Helmerson, K., Rolston, S.L., and Phillips, W.D. (1999) Coherent splitting of Bose-Einstein condensed atoms with optically induced Bragg diffraction, Phys. Rev. Lett. **82**, 871-875.

[19] Bloch, I., Hänsch, T.W., and Esslinger, T. (1999) Atom laser with a cw output coupler, Phys. Rev. Lett. **82**, 3008-3011.

[20] Hagley, E.W., Deng, L., Kozuma, M., Wen, J., Helmerson, K., Rolston, S.L., and Phillips, W.D. (1999) A well-collimated quasi-continuous atom laser, Science **283**, 1706-1709.

[21] Inouye, S., Pfau, T., Gupta, S., Chikkatur, A.P., Görlitz, A., Pritchard, D.E., and Ketterle, W. (1999) Phase-coherent amplification of atomic matter waves, Nature **402**, 641-644.

[22] Griffin, A. and Zaremba, E. (1997) First and second sound in a uniform Bose gas, Phys. Rev. A **56**, 4839-4844.

[23] Stamper-Kurn, D.M., Miesner, H.-J., Inouye, S., Andrews, M.R., and Ketterle, W. (1998) Collisionless and hydrodynamic excitations of a Bose-Einstein condensate, Phys. Rev. Lett. **81**, 500-503.

[24] Maragò, O.M., Hopkins, S.A., Arlt, J., Hodby, E., Hechenblaikner, G., and Foot, C.J. (2000) Observation of the scissors mode and evidence for superfluidity of a trapped Bose-Einstein condensed gas, Phys. Rev. Lett. **84**, 2056-2059.

[25] Onofrio, R., Raman, C., Vogels, J.M., Abo-Shaeer, J.R., Chikkatur, A.P., and Ketterle, W. (2000) Observation of superfluid flow in a Bose-Einstein condensed gas, Phys. Rev. Lett. **85**, 2228-2231.

[26] Madison, K.W., Chevy, F., Wohlleben, W., and Dalibard, J. (2000) Vortex formation in a stirred Bose-Einstein condensate, Phys. Rev. Lett. **84**, 806-809.

[27] Chevy, F., Madison, K.W., and Dalibard, J. (2000) Measurement of the angular momentum of a rotating Bose-Einstein condensate, Phys. Rev. Lett. **85**, 2223-2227.

[28] Anderson, B.P., Haljan, P.C., Wieman, C.E., and Cornell, E.A. (2000) Vortex precession in Bose-Einstein condensates: observations with filled and empty cores, Phys. Rev. Lett. **85**, 2857-2860.

[29] Abo-Shaeer, J.R., Raman, C., Vogels, J.M., and Ketterle, W. (2001) Observation of vortex lattices in Bose-Einstein condensates, Science **292**, 476-479.

[30] Inouye, S., Gupta, S., Rosenband, T., Chikkatur, A.P., Görlitz, A., Gustavson, T. L., Leanhardt, A.E., Pritchard, D.E., and Ketterle, W. (2001) Observation of vortex phase singularities in Bose-Einstein condensates, Phys. Rev. Lett. **87**, 080402.1-4.

[31] Fetter, A.L. and Svidzinsky, A.A. (2001) Vortices in a trapped dilute Bose-Einstein condensate, J. Phys.: Condens. Matter **13**, R135-R194.

[32] Grynberg, G. and Robilliard, C. (2001) Cold atoms in dissipative optical lattices, Phys. Reps. **355**, 335-451.

[33] Chiofalo, M.L. and Tosi, M.P. (2001) Coherent and dissipative transport of a Bose-Einstein condensate inside an optical lattice, J. Phys. B **34**, 4551-4560.

[34] Anderson, B.P. and Kasevich, M.A. (1998) Macroscopic quantum interference from atomic tunnel arrays, Science **282**, 1686-1689.

[35] Vuletic, V., Kerman, A.J, Chin, C., and Chu, S. (1999) Observation of low-field Feshbach resonances in collisions of Cesium atoms, Phys. Rev. Lett. **82**, 1406-1409.

[36] Stenger, J., Inouye, S., Chikkatur, A.P., Stamper-Kurn, D.M., Pritchard, D.E., and Ketterle, W. (1999) Bragg spectroscopy of a Bose-Einstein condensate, Phys. Rev. Lett. **82**, 4569-4573.

[37] Burger, S., Cataliotti, F.S., Fort, C., Minardi, F., Inguscio, M., Chiofalo, M.L., and Tosi, M.P. (2001) Superfluid and dissipative dynamics of a Bose-Einstein condensate in a periodic optical potential, Phys. Rev. Lett. **86**, 4447-4450.

[38] Berg-Sørensen, K. and Mølmer, K. (1998) Bose-Einstein condensates in spatially periodic potentials, Phys. Rev. A **58**, 1480-1484.

[39] Chiofalo, M.L., Polini, M., and Tosi, M.P. (2000) Collective excitations of a periodic Bose condensate in the Wannier representation, Eur. Phys. J. D **11**, 371-378.

[40] Chiofalo, M.L., Succi, S., and Tosi, M.P. (2001) Probing the energy bands of a Bose-Einstein condensate in an optical lattice, Phys. Rev. A **63**, 063613.1-5.

[41] Chiofalo, M.L., and Tosi, M.P. (2001) Josephson-type oscillations of a driven Bose-Einstein condensate in an optical latice, Europhys. Lett. **56**, 326-332.

[42] Cardenas, M., Chiofalo, M.L., and Tosi, M.P. (2002) Matter wave dynamics in an optical lattice: decoherence of Josephson-type oscillations from the Gross-Pitaevskii equation, Physica B, in press.

[43] Gardiner, S.A., Jaksch, D., Dum, R., Cirac, J.I., and Zoller, P. (2000) Nonlinear matter wave dynamics with a chaotic potential, Phys. Rev. A **62**, 023612.1-21.

[44] Greiner, M., Mandel, O., Esslinger, T., Hänsch, T.W., and Bloch, I. (2002) Quantum phase transition from a superfluid to a Mott insulator in a gas of ultracold atoms, Nature **415**, 39-44.

[45] Fisher, M.P.A., Weichman, P.B., Grinstein, G., and Fisher, D.S. (1989) Boson localization and the superfluid-insulator transition, Phys. Rev. B **40**, 546-570.

[46] Sachdev, S. (1999) Quantum phase transitions, University Press, Cambridge.

[47] See e.g. Minguzzi, A. and Tosi, M.P. (2001) Collective excitations in confined Fermi gases, Physica B **300**, 27-37.

[48] Amoruso, M., Meccoli, I., Minguzzi, A., and Tosi, M.P. (1999) Collective excitations of a degenerate Fermi vapour in a magnetic trap, Eur. Phys. J. D **7**, 441-447.

[49] Baranov, M.A. and Petrov, D.S. (2000) Low-energy collective excitations in a superfluid trapped Fermi gas, Phys. Rev. A **62**, 041601.1-4.

[50] Anderson, P.W. (1958) Random phase approximation in the theory of superconductivity, Phys. Rev. **112**, 1900-1916.

[51] Leggett, A.J. (1965) Theory of a superfluid Fermi liquid. I. General formulation and static properties, Phys. Rev. **140**, A1869-A1888.

[52] Amoruso, M., Meccoli, I., Minguzzi, A., and Tosi, M.P. (2000) Density profiles and collective excitations of a trapped two-component Fermi vapour, Eur. Phys. J. D **8**, 361-369.

[53] Stoof, H.T.C., Houbiers, M., Sackett, C.A., and Hulet, R.G. (1996) Superfluidity of spin-polarized 6Li, Phys. Rev. Lett. **76**, 10-13.

[54] Minguzzi, A., Ferrari, G., and Castin, Y. (2001) Dynamic structure factor of a superfluid Fermi gas, Eur. Phys. J. D **17**, 49-55.

[55] Forster, D. (1975) Hydrodynamic Fluctuations, Broken Symmetry, and Correlation Functions, Benjamin, Reading, MA.

[56] Minguzzi, A. and Tosi, M.P. (2001) Scissors mode in a superfluid Fermi gas, Phys. Rev. A **63**, 023609.1-4.

[57] DeMarco, B., Papp, S.B., and Jin, D.S. (2001) Pauli blocking of collisions in a quantum degenerate atomic Fermi gas, Phys. Rev. Lett. **86**, 5409-5412.

[58] Mølmer, K. (1998) Bose condensates and Fermi gases at zero temperature, Phys. Rev. Lett. **80**, 1804-1807.

[59] Viverit, L., Pethick, C.J., and Smith, H. (2000) Zero-temperature phase diagram of binary boson-fermion mixtures, Phys. Rev. A **61**, 053605.1-12.

[60] Minguzzi, A. and Tosi, M.P., (2000) Schematic phase diagram and collective excitations in the collisional regime for trapped boson-fermion mixtures at zero temperature, Phys. Lett. A **268**, 142-148.

[61] Akdeniz, Z., Vignolo, P., Minguzzi, A., and Tosi, M.P., (2002) Phase separation in a boson-fermion mixture of lithium atoms, J. Phys. B **35**, L1-L7.

[62] Akdeniz, Z., Minguzzi, A., Vignolo, P., and Tosi, M.P., (2002) Demixing in mesoscopic boson-fermion clouds inside cylindrical harmonic traps: quantum phase diagram and the role of temperature, cond-mat/0203025.

[63] Görlitz, A., Vogels, J.M., Leanhardt, A.E., Raman, C., Gustavson, T.L., Abo-Shaeer, J.R., Chikkatur, A.P., Gupta, S., Inouye, S., and Ketterle, W. (2001) Realization of Bose-Einstein condensates in lower dimensions, Phys. Rev. Lett. **87**, 130402.1-4.

[64] Thywissen, J.H., Westervelt, R.M., and Prentiss, M. (1999) Quantum point contacts for neutral atoms, Phys. Rev. Lett. **83**, 3762-3765.

[65] Müller, D., Anderson, D.Z., Grow, R.J., Schwindt, P.D.D., and Cornell, E.A. (1999) Guiding neutral atoms around curves with lithographically patterned current-carrying wires, Phys. Rev. Lett. **83**, 5194-5197.

[66] Dekker, D.H., Lee, C.S., Lorent, V., Thywissen, J.H., Smith, S.P., Drndic, M., Westervelt, R.M., and Prentiss, M. (2000) Guiding neutral atoms on a chip, Phys. Rev. Lett. **84**, 1124-1127.

[67] Key, M., Hughes, J.G., Rooijakkers, W., Sauer, B.E., and Hinds, E.A. (2000) Propagation of cold atoms along a miniature magnetic guide, Phys. Rev. Lett. **84**, 1371-1373.

[68] Bongs, K., Burger, S., Dettmer, S., Hellweg, D., Arlt, J., Ertmer, W., and Sengstock, K. (2001) Waveguide for Bose-Einstein condensates, Phys. Rev. A **63**, 031602.1-4.

[69] Lieb, E.H. and Liniger, W. (1963) Exact analysis of an interacting Bose gas. I. The general solution and the ground state, Phys. Rev. **130**, 1605-1616.

[70] Lieb, E.H. (1963) Exact analysis of an interacting Bose gas. II. The excitation spectrum, Phys. Rev. **130**, 1616-1624.

[71] Yang, C.N. and Yang, C.P. (1969) Thermodynamics of a one-dimensional system of bosons with repulsive delta-function interaction, J. Math. Phys. **10**, 1115-1122.

[72] Dunjko, V., Lorent, V., and Olshanii, M. (2001) Bosons in cigar-shaped traps: Thomas-Fermi regime, Tonks-Girardeau regime, and in between, Phys. Rev. Lett. **86**,5413-5416.

[73] Girardeau, M. (1960) Relationship between systems of impenetrable bosons and fermions in one dimension, J. Math. Phys. **1**, 516-523.

[74] Minguzzi, A., Vignolo, P., and Tosi, M.P. (2002) High-momentum tails in the Tonks gas under harmonic confinement, Phys. Lett. A **294**, 222-226.

[75] Girardeau, M.D. and Wright, E.M. (2000) Dark solitons in a one-dimensional condensate of hard-core bosons, Phys. Rev. Lett. **84**, 5691-5694.

[76] Rojo, A.G., Cohen. J.L., and Berman, P.R. (1999) Talbot oscillations and periodic focusing in a one-dimensional condensate, Phys. Rev. A **60**, 1482-1490.

[77] Brack, M. and van Zyl, B.P. (2001) Simple analytical particle and kinetic energy densities for a dilute fermionic gas in a d-dimensional harmonic trap, Phys. Rev. Lett. **86**, 1574-1577.

[78] Vignolo, P., Minguzzi, A., and Tosi, M.P. (2000) Exact particle and kinetic-energy densities for one-dimensional confined gases of noninteracting fermions, Phys. Rev. Lett. **85**, 2850-2853.

DIFFERENT SCATTERING REGIMES IN TWO-DIMENSIONAL BOSE-EINSTEIN CONDENSATES

B. Tanatar

Department of Physics,
Bilkent University,
06533 Ankara, Turkey

Abstract   Motivated by the recent efforts to produce low-dimensional condensates, we study the ground-state density profiles of two-dimensional Bose-Einstein condensed atoms at zero temperature within a mean-field theory. The interplay between the tight harmonic confinement in the axial direction and collisional properties of the condensate atoms help identify three distinct regimes of experimental interest. Each regime is described by a different atom-atom coupling which depends on the density of the condensate as the system starts to be influenced by two-dimensional collisions. We trace the regions of experimentally accessible system parameters for which the crossover between different dimensionality behaviors in the scattering properties may become observable.

Keywords: Bose-Einstein condensates, low-dimensional systems

1.   Introduction

The observation of Bose-Einstein condensation Bose-Einstein condensation (BEC) in externally confined atomic vapors[1, 2, 3] has stimulated a big interest in the theoretical and experimental work on interacting systems of bosons. boson The thermodynamic, ground-state static and dynamic properties of condensates are extensively investigated and the main results are compiled in a number of reviews.[4, 5, 6, 7] Most of the excitement stems from the possibility of understanding properties of a macroscopic quantum state. Other than the fundamental physics considerations, the Bose-Einstein condensed systems offer interesting applications of atom laser laser and atom optics.

The BEC phenomenon in low dimensional and particularly in two-dimensional (2D) systems has attracted considerable amount of interest from the point of view of understanding effects of dimensionality. As the homogeneous 2D system of bosons boson would not undergo BEC at a finite temperature, the prospects of observing BEC in systems with an external potential[8, 9] provide a strong motivation for such investigations. It was argued by Mullin[10] that

329

A.S. Shumovsky and V.I. Rupasov (eds.), Quantum Communication and Information Technologies, 329–338.
© 2003 *Kluwer Academic Publishers.*

BEC is not possible for strictly 2D systems even in a trapping trapping potential in the thermodynamic limit. However, by varying the trapping field so that it is very narrow in one direction, it should be possible to separate the single-particle states of the oscillator potential into well-defined bands, and occupying the lowest band should produce an effectively two-dimensional system. Growing number of experiments[11, 12, 13, 14, 15, 16] exploring the possibility of realizing quasi-one-dimensional (Q1D) and quasi-two-dimensional (Q2D) trapped atomic gases culminated in the recent experiments of Görlitz et al.[17] in achieving low-dimensional condensates. and measurements on the BEC transition temperature and other properties are expected to follow.

The studies on the BEC in 2D systems can be broadly divided into two categories. In the first group the interaction effects are treated parametrically without reference to the actual interaction potential or the scattering length which describes it as in the 3D formulation of the interacting boson boson condensates.[18, 19, 20, 21] Included in the same spirit of calculations there has been path integral Monte Carlo simulations[22] at finite temperature to give indications of a BEC transition in 2D systems. In the second group, some effort is made to relate the 2D interaction strength to the 3D scattering length[23, 24, 25] or to solve the scattering problem in strictly 2D to obtain the relevant dependence of the interaction coupling on the scattering length. Kim et al.[26] using the scattering theory in 2D, found that the interaction strength depends logarithmically on the scattering length. Shevchenko[27, 28] in a series of papers studied interacting 2D Bose gas in a nonuniform field arriving at the conclusion even though a Bose-Einstein condensation Bose-Einstein condensation does not take place, the system exhibits superfluidity. Recently, Kolomeisky et al.[29] in their treatment of low-dimensional Bose liquids suggested a modified form for the mean-field description of 2D condensates. Lieb, Seiringer, and Yngvason [30] have rigorously analyzed this and related approximations as applied to the practical cases of interest. Petrov and coworkers[31, 32] using scattering theory arguments obtained a slightly more detailed form of the effective interaction coupling for Q2D systems that distinguishes different density regimes. Similar results are also obtained by Lee et al.[33] within a many-body T-matrix approach.

In this paper we introduce various interaction coupling models to describe the pancake shaped 2D condensates formed in highly anisotropic traps. Our central aim is to calculate the equilibrium density (or wavefunction) profiles of the condensate with different choices of system parameters such as anisotropy in the trap frequencies and scattering length. Our calculations show that with increasing anistropy the condensate first becomes 2D with regard to the confinement and then also in its collisional properties. We find that these different regimes may be identified by measuring the size of the cloud in the radial plane and we characterize the crossover regime in terms of the relevant physical parameters.

2. Theory

The ground-state properties of a condensed system of bosons boson at zero temperature are described by the Gross-Pitaevskii equation which is a nonlinear Schrödinger equation. In the presence of external trap potentials, therefore an inhomogeneous Bose system, it is useful to adopt the local-density approximation which regards the system locally homogeneous. We consider a dilute Bose-condensed gas in anisotropic harmonic confinement characterized by the frequencies $\omega_\perp$ and $\omega_z = \lambda\omega_\perp$ with $\lambda >> 1$ to yield a pancake shaped condensate. The mean-field energy functional in the local-density approximation can very generally be written as[34, 35, 36]

$$E = \int d\mathbf{r} \left[\frac{\hbar^2}{2m} |\nabla\psi|^2 + V_{\text{ext}}(\mathbf{r})|\psi|^2 + \epsilon(\rho)|\psi|^2 \right], \qquad (1)$$

where $\epsilon(\rho) = g\rho/2$ is the ground-state energy (per particle) of the homogeneous system, and $\rho = |\psi|^2$ is the density. The coupling parameter g can depend on the condensate density, $g = g(\psi)$. Variation of the energy functional with respect to ψ^*, subject to the normalization condition $\int d\mathbf{r} \, |\psi|^2 = N$, yields the nonlinear Schrödinger equation

$$-\frac{\hbar^2}{2m}\nabla^2\psi + V_{\text{ext}}(\mathbf{r})\psi + \frac{\partial[\epsilon(\rho)\rho]}{\partial\rho}\psi = \mu\psi, \qquad (2)$$

where μ is the chemical potential. The local-density approximation was used by Fabrocini and Polls[35] and Nunes[34] to study the high density effects in 3D condensates, by making use of the 3D homogeneous hard-sphere Bose gas results from perturbation theory. These studies showed that the modifications to GP equation become important as the number of particles N and the hard-sphere radius a are increased making the system less dilute.

To explore different regimes of 2D condensates we consider various models. [37] In the first approach, we consider a condensate whose third dimension is of the order of the harmonic confinement length $a_z = (\hbar/m\omega_z)^{1/2}$. If the condition $a << a_z$ holds, the system undergoes collisions in three dimensions and the coupling constant we use in the 2D GP equation is $g_{Q3D} = g_3 D|\phi(z = 0)|^2$ where $\phi(z = 0) = (2\pi a_z^2)^{-1/4}$ is the axial ground state ground state wavefunction evaluated at $z = 0$. It takes the form

$$g_{Q3D} = 2\sqrt{2\pi}\frac{\hbar^2}{m}\frac{a}{a_z} \qquad (3)$$

reflecting the geometrical effects of the reduced dimensionality. A number of authors[23, 24, 25] have used this coupling constant to describe 2D condensates.

When the anisotropy further increases and the scattering length a becomes of the order of a_z, the collisions among the atoms start to be influenced by the presence of the trap in the tight $z-$direction. Petrov *et al.*[31, 32] and Lee *et al.*[33] studied the scattering problem in quasi-two-dimension (Q2D) of a system harmonically confined in the $z-$direction and homogeneous in the perpendicular direction, to obtain the following expression for the coupling strength

$$g_{Q2D} = \frac{2\sqrt{2\pi}(\hbar^2/m)(a/a_z)}{1 + \frac{(a/a_z)}{\sqrt{2\pi}}|\ln(g_{Q2D}\rho(2\pi m/\hbar^2)a_z^2)|}. \tag{4}$$

In the Q2D regime the coupling strength depends on the condensate density ρ. Furthermore, g_{Q2D} is given by an implicit relation which has to be determined numerically during the solution of the 2D GP equation. If we make a zero-order approximation for g_{Q2D} appearing in the logarithm, i.e. $g_{Q2D} \approx 2\sqrt{2\pi}(\hbar^2/m)(a/a_z)$, then the expresion for Q2D coupling strength simplifies to

$$g_{Q2D} = \frac{2\sqrt{2\pi}(\hbar^2/m)(a/a_z)}{1 + \frac{(a/a_z)}{\sqrt{2\pi}}|\ln(2(2\pi)^{3/2}\rho a a_z)|}. \tag{5}$$

In our numerical calculations we have found that the above implicit and explicit forms of g_{Q2D} slightly change the wavefunction profiles only for very large values of asymmetry parameter λ.

The strictly 2D regime is attained when the collisional properties are such that $a >> a_z$. harmonic isotropic trapping potential $V_{ext}(\mathbf{r}) = m\omega_\perp^2 r^2/2$. Perturbation theory calculations for a homogeneous system of 2D hard-disk bosons boson yield[38]

$$\epsilon(\rho) = \frac{\hbar^2}{2m} \frac{4\pi\rho}{|\ln\rho a^2|}, \tag{6}$$

for the ground-state energy (per particle). The corresponding mean-field equation for the condensate wave function thus reads

$$-\frac{\hbar^2}{2m}\nabla^2\psi(r) + \frac{1}{2}m\omega_\perp^2 r^2\,\psi(r) + \frac{\hbar^2}{2m}\frac{8\pi}{|\ln\psi^2 a^2|}\,|\psi(r)|^2\psi(r) = \mu\psi, \tag{7}$$

from which the 2D intertaction coupling can be identified

$$g_{2D} = \frac{4\pi\hbar^2}{m}\frac{1}{|\ln(\rho a^2)|} \tag{8}$$

The energy functional corresponding to this equation has also been suggested by Shevchenko[28] and more recently by Kolomeisky *et al.*[29]

We note that the ensuing GP equation in 2D is quite different than its 3D counterpart, in that the dimensionless interaction term g_{Q2D} or g_{2D} not only depends on the hard-disk radius a logarithmically, but it also depends on $|\psi|^2$, making the GP equation highly nonlinear. In 3D the density dependent interactions arise as corrections or modifications to the GP equation, whereas in 2D the mean-field interaction is density dependent.

In the previous applications[34, 35, 39] of the local-density approximation in 3D condensates, effects beyond the mean-field theory (GP equation) has been explored by including higher order terms in the homogeneous energy density $\epsilon(\rho)$. In 2D, the correct mean-field description is given by Eq. (2). To go beyond the mean-field theory one would have to use the higher order terms in the perturbation series for $\epsilon(\rho)$. Another possible correction to the mean-field description has been suggested by Andersen and Haugerud.[40] In their treatment the kinetic energy functional is modified to include a gradient term

$$E_{\text{kin}} = \frac{\hbar^2}{2m} \int dr \left[1 - \frac{2}{3} \frac{1}{|\ln |\psi|^2 a^2|} \right] |\nabla \psi|^2. \tag{9}$$

In our numerical calculations to be presented and discussed in the next section, we have tested the significance of corrections brought by the gradient term. We have found that it has negligible effect on the condensate profile $\psi(r)$ and other physical quantities for the range of N and scattering length values we examined.

3. Results and Discussion

We solve the Gross-Pitaevskii equation given in Eq. (2) for three models of $g(\psi)$ numerically using the steepest descent method[35, 41] which is known to produce accurate results. A further check of our numerical calculations is provided by the virial relation. Under the scaling transformation $\mathbf{r} \to \lambda \mathbf{r}$, so that $\psi(\mathbf{r}) \to \psi(\mathbf{r})/\lambda$, and using the variational nature of the energy we obtain

$$E_{\text{kin}} - E_{\text{ext}} + E_{\text{int}} - \frac{1}{2} \int dr \frac{dg(\psi)}{d\psi} |\psi(\mathbf{r})|^5 = 0. \tag{10}$$

In this form of the virial relation, we have omitted the gradient correction to the kinetic energy, since numerically it is found to be negligibly small. Our solution of the GP equation satisfies the virial relation to a high degree.

In the following we illustrate the ground-state density $|\psi(r)|^2$ predicted by the three models. We first consider the parameters as appropriate for ^{23}Na atoms in the experiment of Görlitz *et al.*[17] taking $N = 5 \times 10^5$, $\lambda = 26.33$, and $a = 2.8\,\text{nm}$. For these parameters the condensate appears as 2D due to its confinement characteristics ($\mu \simeq 2.08\,\hbar\omega_z$) but the collisions have a 3D nature ($a/a_z \simeq 3.8 \times 10^{-3}$). Density profiles in this Q3D regime are displayed in Fig. 1 for all three models. It is observed that Q3D and Q2D interaction models yield almost identical predictions (solid and dotted lines). The fully 2D model,

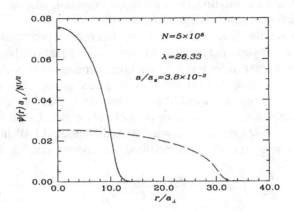

Figure 1. The condensate density $|\psi(r)|^2$ (in units of $N/a_\perp^2$) as a function of the radial distance for a system of $N = 10^5$ atoms, $a/a_z = 3.8 \times 10^{-3}$ and $\lambda = 26.33$ from the numerical solution of GP equation for the three models: Q3D (dotted line), Q2D (solid line), and 2D (dashed line).

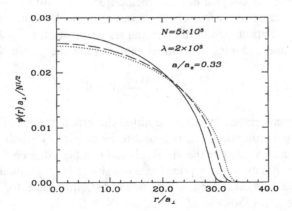

Figure 2. The same as in Fig. 1, for $a/a_z = 0.33$ and $\lambda = 2 \times 10^5$.

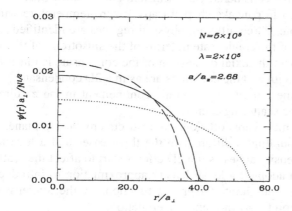

Figure 3. The same as in Fig. 1, for $a/a_z = 2.68$ and $\lambda = 2 \times 10^5$.

on the other hand, produces a quantitatively very different density profile with a much larger chemical potential ($\mu \simeq 17.8\hbar\omega_z$). It is evident that 2D model is not applicaple to the current experimental situation.[17]

In the second case we increase the anisotropy parameter to a much larger value $\lambda = 2 \times 10^5$ and make the choice $a/a_z = 0.33$ for the scattering length. These parameters correspond to the regime of crossover in the scattering properties from 3D to 2D in a condensate where the motion in the third dimension is completely frozen by the confinement effects ($\mu \simeq 0.002\hbar\omega_z$). As shown in Fig. 2 in the crossover regime the three models predict comparable shapes of the condensate cloud indicating the approach to 2D collisions.

Finally, we further increase the value of the scattering length to $a/a_z = 2.68$, keeping the same values of λ and N as the previous case. For this choice of parameters the collisions are truly 2D and the strictky 2D model should be relevant in predicting an accurate profile of the condensate cloud. The other models would overestimate the width of $\psi(r)$. Our results confirming this scenario are shown in Fig. 3.

4.   Summary and Concluding Remarks

In summary we have considered Bose-Einstein condensates Bose-Einstein condensation confined in pancake shaped traps at zero temperature within a mean-field description. Three different physical regimes are identified for the scattering properties of the condensate in terms of the anisotropy of the trap;
(i) a Q3D regime where the axial dimension of the condensate is much larger than the scattering length and the collisions are as in a 3D condensate,
(ii) a Q2D regime where the tight harmonic confinement in the z-direction begins to influence the scattering events,
(iii) a strictly 2D regime where collisions are restricted to the x-y plane.

The interaction coupling is different in the three cases and a logarithmic dependence on the density arises as the 2D effects start to affect the scattering properties. We have adopted a local-density approximation to introduce the appropriate coupling strength into the energy functional of the condensate and to obtain a GP equation for its in-plane wavefunction.

Our results should be relevant to the experimental efforts aiming to produce low-dimensional condensates. The results of our calculations show that the different collisional regimes are reflected in the width of the cloud. Observable properties are expected for flat condensates not only due to the geometric confinement effects (i.e. when $\mu < \hbar\omega_z$) but also when the condition $a > a_z$ holds. The strictly 2D limit can be reached by making the tight confinement very large and by increasing the scattering length. Recent experiments of Cornish *et al.* [42] have demonstrated the feasibility of tuning the scattering length through Feshbach resonances. This opens the possibility of studying the effects of interactions and regimes those described beyond the Gross-Pitaevskii equation more systematically. To assess the importance of interactions and high density effects Monte Carlo simulations would be useful as a test of range of validity of the mean-field and local-density approximations. Our calculations should be the starting point for finite temperature studies in which the role of interactions with the non-condensed thermal particles and phase correlations at increased temperatures may be addressed.

Acknowledgments

This work was partially supported by the Scientific and Technical Research Council of Turkey (TUBITAK), by NATO-SfP, by the Turkish Department of Defense, and by Turkish Academy of Sciences (TUBA). Fruitful discussions with Professor M. P. Tosi, and Drs. A. Minguzzi and P. Vignolo are kindly acknowledged.

References

[1] Anderson M. H., Ensher J. R., Matthews M. R., Wieman C. E., and Cornell E. A., Science **269**, 198 (1995).

[2] Davis K. B., Mewes M.-O., Andrews M. R., van Druten N. J., Durfee D. S., Kurn D. M., and Ketterle W. (1995) Bose-Einstein condensation of Sodium atoms, *Phys. Rev. Lett.* **75**, 3969.

[3] Bradley C. C., Sackett C. A., and Hulet R. G. (1997) Bose-Einstein condensation of Lithium: observation of limited condensate number, *Phys. Rev. Lett.* **78**, 985.

[4] Dalfovo F., Giorgini S., Pitaevskii L. P., and Stringari S. (1999) Theory of Bose-Einstein condensation in trapped gases, *Rev. Mod. Phys.* **71**, 463.

[5] Parkins A. S. and Walls D. F. (1998) The physics of trapped dilute-gas Bose-Einstein condensates, *Phys. Rep.* **303**, 1.

[6] Leggett A. J. (2001) Bose-Einstein condensation in the alkali gases: some fundamental concepts, *Rev. Mod. Phys.* **73**, 307.

[7] Courtieille Ph. W., Bagnato V. S., and Yukalov V. I. (2001) Bose-Einstein condensation of trapped atomic gases, *Laser Phys.* **11**, 659.

[8] Bagnato V. and Kleppner D. (1991) Bose-Einstein condensation in low-dimensional traps, *Phys. Rev. A* **44**, 7439.

[9] De Groot S. R., Hooyman G. J., and ten Seldam C. A. (1950) On the Bose-Einstein condensation, *Proc. R. Soc. London Ser. A* **203**, 266.

[10] Mullin W. J. (1998) A study of Bose-Einstein condensation in a two-dimensional trapped gas, *J. Low Temp. Phys.* **110**, 167.

[11] Hinds E. A., Boshier M. G., and Hughes I. G. (1998) Magnetic waveguide for trapping cold atom gases in two-dimensions, *Phys. Rev. Lett.* **80**, 645.

[12] Gauck H., Hartl M., Schneble D., Schnitzler H., Pfau T., and Mlynek J. (1998) Quasi-2D gas of laser cooled atoms in a planar matter waveguide, *Phys. Rev. Lett.* **81**, 5298.

[13] Vuletić V., Kerman A. J., Chin C., and Chu S. (1999) Observation of low-field Feshbach resonances in collisions of Cesium atoms, *Phys. Rev. Lett.* **82**, 1406.

[14] Bouchoule I., Perrin H., Kuhn A., Morinaga M., and Salomon C. (1999) Neutral atoms prepared in Fock states of a one-dimensional harmonic potential, *Phys. Rev. A* **59**, 8.

[15] Dettmer S., Hellweg D., Ryytty P., Arlt J. J., Ertmer W., Sengstock K., Petrov D. S., Shlyapnikov G. Y., Kreutzmann H., Santos L., and Lewenstein M. (2001) Observation of phase fluctuations in elongated Bose-Einstein condensates, *Phys. Rev. Lett.* **87**, 160406.

[16] Burger S., Cataliotti F. S., Fort C., Maddaloni P., Minardi F., and Inguscio M. (2002) Quasi-2D Bose-Einstein condensation in an optical lattice, *Europhys. Lett.* **57**, 1.

[17] A. Görlitz, J. M. Vogels, A. E. leanhardt, C. Raman, T. L. Gustavson, J. R. Abo-Shaeer, A. P. Chikkatur, S. Gupta, S. Inouye, T. P. Rosenband, D. E. Pritchard, and W. Ketterle, (2001) Realization of Bose-Einstein condensates in lower dimensions, *Phys. Rev. Lett.* **87**, 130402.

[18] Haugset T. and Haugerud H. (1998) Exact diagonalization of the Hamiltonian for trapped interacting bosons in lower dimensions, *Phys. Rev. A* **57**, 3809.

[19] Bayindir M. and Tanatar B. (1998) Bose-Einstein condensation in a two-dimensional, trapped, interacting gas *Phys. Rev. A* **58**, 3134.

[20] Kim J. G. and Lee E. K. (1999) Thermodynamic properties of trapped, two-dimensional interacting Bose gases in Hartree-Fock approximation, *J. Phys. B* **32**, 5575.

[21] Adhikari S. K. (2000), *Phys. Lett.* A **265**, 91.

[22] Heinrichs S. and Mullin W. J. (1998), *J. Low Temp. Phys.* **113**, 231.

[23] Bhaduri R. K., Reimann S. M., Viefers S., Choudhury A. G., and Srivastava M. K. (2000) The effect of interactions on Bose-Einstein condensation in a quasi two-dimensional harmonic trap, *J. Phys.* B **33**, 3895.

[24] Salasnich S., Parola A., and Reatto L. (2002) Effective wave equation for the dynamics of cigar-shaped and disk-shaped Bose condensates, *Phys. Rev.* A **65**, 043614.

[25] Kim J. G., Kang K. S., Kim B. S., and Lee E. K. (2000) Hartree-Fock calculation for trapped, two-dimensional interacting Bose gases of finite size, *J. Phys.* B **33**, 2559.

[26] Kim S.-H., Oh S. D., and Jhe W. (2000) Two-dimensional condensation of dilute Bose atoms in a harmonic trap, *J. Kor. Phys. Soc.* **37**, 665.

[27] Shevchenko S. I. (1991) Theory of two-dimensional superfluidity in a nonuniform external field, *Sov. Phys. JETP* **73**, 1009.

[28] Shevchenko S. I. (1992) On the theory of a Bose gas in a nonuniform field, *Sov. J. Low Temp. Phys.* **18**, 223.

[29] Kolomeisky E. B., Newman T. J., Straley J. P., and Qi X. (2000) Low-dimensional Bose liquids: beyond the Gross-Pitaevskii approximation, *Phys. Rev. Lett.* **85**, 1146.

[30] Lieb E. H., Seiringer R., and Yngvason J. (2001) A rigorous derivation of the Gross-Pitaevskii energy functional for a two-dimensional Bose gas, *Commun. Math. Phys.* **224**, 17.

[31] Petrov D. S., Holzmann M., and Shlyapnikov G. V. (2000) Bose-Einstein condensation in quasi-2D trapped gases, *Phys. Rev. Lett.* **84**, 2551.

[32] Petrov D. S. and Shlyapnikov G. V. (2001) Intertomic collisions in a tightly confined Bose gas, *Phys. Rev.* A **64**, 012706.

[33] Lee M. D., Morgan S. A., Davis M. J., and Burnett K. (2002) Energy-dependent scattering and the Gross-Pitaevskii equation in two-dimensional Bose-Einstein condensates, *Phys. Rev.* A **65**, 043617.

[34] Nunes G. S. (1999) Density functional theory of the inhomogeneous Bose-Einstein condensate, *J. Phys.* B **32**, 4293.

[35] Fabrocini A. and Polls A. (1999) Beyond the Gross-Pitaevskii approximation: local density versus correlated basis approach for trapped bosons, *Phys. Rev.* A **60**, 2319.

[36] Kim Y. E. and Zubarev A. L. (2003) Density-functional theory of bosons in a trap, *Phys. Rev.* A**67**, 015602.

[37] Tanatar B., Minguzzi A., Vignolo P., and Tosi M. P. (2002) DEnsity profile of a Bose-Einstein condensate inside a pancake-shaped trap: observational consequences of the dimensional cross-over in the scattering regime, *Phys. Lett.* A **302**, 131.

[38] Schick M. (1971) Two-dimensional system of hard-core bosons, *Phys. Rev.* A **3**, 1067.

[39] Banerjee A. and Singh M. P. (2001) Ground-state properties of a trapped Bose gas beyond the mean-field approximation, *Phys. Rev.* A **64**, 063604.

[40] Andersen J. O. and Haugerud H. (2001) Ground state of a trapped Bose-Einstein condensate in two dimensions; beyond the mean-field approximation, *Phys. Rev.* A **65**, 033615.

[41] Dalfovo F. and Stringari S. (1996) Bosons in anisotropic traps: ground state and vortices, *Phys. Rev.* A **53**, 2477.

[42] Cornish S. L., Claussen N. R., Roberts J. L., Cornell E. A., and Wiemann C. E. (2000) Stable [85]Rb Bose-Einstein condensates with widely tunable interactions, *Phys. Rev. Lett.* **85**, 1795.

AN INFORMATION-THEORETIC ANALYSIS OF GROVER'S ALGORITHM

Erdal Arikan

Electrical-Electronics Engineering Department,
Bilkent University, 06533 Ankara, Turkey
arikan@ee.bilkent.edu.tr

Abstract   Grover discovered a quantum algorithm for identifying a target element in an
unstructured search universe of N items in approximately $\pi/4\sqrt{N}$ queries to a
quantum oracle. For classical search using a classical oracle, the search complexity is of order $N/2$ queries since on average half of the items must be searched.
In work preceding Grover's, Bennett et al. had shown that no quantum algorithm
can solve the search problem in fewer than $O(\sqrt{N})$ queries. Thus, Grover's
algorithm has optimal order of complexity. Here, we present an information-
theoretic analysis of Grover's algorithm and show that the square-root speed-up
by Grover's algorithm is the best possible by any algorithm using the same quantum oracle.

Keywords: Grover's algorithm, quantum search, entropy.

1.   Introduction

Grover [1], [2] discovered a quantum algorithm for identifying a target element in an unstructured search universe of N items in approximately $\pi/4\sqrt{N}$
queries to a quantum oracle oracle. For classical search using a classical oracle,
the search complexity is clearly of order $N/2$ queries since on average half of the
items must be searched. It has been proven that this square-root speed-up is the
best attainable performance gain by any quantum algorithm. In work preceding
Grover's, Bennett et al. [4] had shown that no quantum algorithm can solve
the search problem in fewer than $O(\sqrt{N})$ queries. Following Grover's work,
Boyer et al. [5] showed that Grover's algorithm is optimal asymptotically, and
that square-root speed-up cannot be improved even if one allows, e.g., a 50%
probability of error. Zalka [3] strengthened these results to show that Grover's
algorithm is optimal exactly (not only asymptotically). In this correspondence
we present an information-theoretic analysis of Grover's algorithm and show
the optimality of Grover's algorithm from a different point of view.

A.S. Shumovsky and V.I. Rupasov (eds.), Quantum Communication and Information Technologies, 339–347.
© 2003 *Kluwer Academic Publishers.*

2.   A General Framework for Quantum Search

We consider the following general framework for quantum search algorithms. We let X denote the state of the target and Y the output of the search algorithm. We assume that X is uniformly distributed over the integers 0 through $N - 1$. Y is also a random variable distributed over the same set of integers. The event $Y = X$ signifies that the algorithm correctly identifies the target. The probability of error for the algorithm is defined as $P_e = P(Y \neq X)$.

The state of the target is given by the density matrix density matrix

$$\rho_T = \sum_{x=0}^{N-1} (1/N)|x\rangle\langle x|, \tag{1}$$

where $\{|x\rangle\}$ is an orthonormal set. We assume that this state is accessible to the search algorithm only through calls to an oracle oracle whose exact specification will be given later. The algorithm output Y is obtained by a measurement performed on the state of the quantum computer at the end of the algorithm. We shall denote the state of the computer at time $k = 0, 1, \ldots$ by the density matrix $\rho_C(k)$. We assume that the computation begins at time 0 with the state of the computer given by an initial state $\rho_C(0)$ independent of the target state. The computer state evolves to a state of the form

$$\rho_C(k) = \sum_{x=0}^{N-1} (1/N)\rho_x(k) \tag{2}$$

at time k, under the control of the algorithm. Here, $\rho_x(k)$ is the state of the computer at time k, conditional on the target value being x. The joint state of the target and the computer at time k is given by

$$\rho_{TC}(k) = \sum_{x=0}^{N-1} (1/N)|x\rangle\langle x| \otimes \rho_x(k). \tag{3}$$

The target state (1) and the computer state (2) can be obtained as partial traces of this joint state.

We assume that the search algorithm consists of the application of a sequence of unitary operators on the joint state. Each operator takes one time unit to complete. The computation starts at time 0 and terminates at a predetermined time K, when a measurement is taken on $\rho_C(K)$ and Y is obtained. In accordance with these assumptions, we shall assume that the time index k is an integer in the range 0 to K, unless otherwise specified.

There are two types of unitary operators that may be applied to the joint state by a search algorithm: oracle oracle and non-oracle. A non-oracle operator is of the form $I \otimes U$ and acts on the joint state as

$$\rho_{TC}(k+1) = (I \otimes U)\,\rho_{TC}(k)\,(I \otimes U)^{\dagger} = \sum_{x}(1/N)|x\rangle\langle x| \otimes U\rho_x(k)U^{\dagger}.$$

$$(4)$$

Under such an operation the computer state is transformed as

$$\rho_C(k+1) = U\rho_C(k)U^{\dagger}. \qquad (5)$$

Thus, non-oracle operators act on the conditional states $\rho_x(k)$ uniformly; $\rho_x(k+1) = U\rho_x(k)U^{\dagger}$. Only oracle oracle operators have the capability of acting on conditional states non-uniformly.

An oracle operator is of the form $\sum_x |x\rangle\langle x| \otimes O_x$ and takes the joint state $\rho_{TC}(k)$ to

$$\rho_{TC}(k+1) = \sum_{x}(1/N)|x\rangle\langle x| \otimes O_x\rho_x(k)O_x^{\dagger}. \qquad (6)$$

The action on the computer state is

$$\rho_C(k+1) = \sum_{x}(1/N)O_x\rho_x(k)O_x^{\dagger}. \qquad (7)$$

All operators, involving an oracle or not, preserve the entropy entropy of the joint state $\rho_{TC}(k)$. The von Neumann entropy Von Neumann entropy of the joint state remains fixed at $S[\rho_{TC}(k)] = \log N$ throughout the algorithm. Non-oracle operators preserve also the entropy of the computer state; the action (5) is reversible, hence $S[\rho_C(k+1)] = S[\rho_C(k)]$. Oracle action on the computer state (7), however, does not preserve entropy; $S[\rho_C(k+1)] \neq S[\rho_C(k)]$, in general.

Progress towards identifying the target is made only by oracle oracle calls that have the capability of transferring information from the target state to the computer state. We illustrate this information transfer in the next section.

3.   Grover's Algorithm

Grover's algorithm can be described within the above framework as follows. The initial state of the quantum computer is set to

$$\rho_C(0) = |s\rangle\langle s| \qquad (8)$$

where

$$|s\rangle = \sum_{x=0}^{N-1}(1/\sqrt{N})|x\rangle. \qquad (9)$$

Since the initial state is pure, the conditional states $\rho_x(k)$ will also be pure for all $k \geq 1$.

Grover's algorithm uses two operators: an oracle operator with

$$O_x = I - 2|x\rangle\langle x|, \tag{10}$$

and a non-oracle operator (called 'inversion about the mean') given by $I \otimes U_s$ where

$$U_s = 2|s\rangle\langle s| - I. \tag{11}$$

Both operators are Hermitian.

Grover's algorithm interlaces oracle calls with inversion-about-the-mean operations. So, it is convenient to combine these two operations in a single operation, called Grover iteration, by defining $G_x = U_s O_x$. The Grover iteration takes the joint state $\rho_{TC}(k)$ to

$$\rho_{TC}(k+1) = \sum_x (1/N)|x\rangle\langle x| \otimes G_x \rho_x(k) G_x^\dagger \tag{12}$$

In writing this, we assumed, for notational simplicity, that G_x takes one time unit to complete, although it consists of the succession of two unit-time operators.

Grover's algorithm consists of $K = (\pi/4)\sqrt{N}$ successive applications of Grover's iteration beginning with the initial state (8), followed by a measurement on $\rho_C(K)$ to obtain Y. The algorithm works because the operator G_x can be interpreted as a rotation of the x–s plane by an angle $\theta = \arccos(1-2/N) \approx 2/\sqrt{N}$ radians. So, in K iterations, the initial vector $|s\rangle$, which is almost orthogonal to $|x\rangle$, is brought into alignment with $|x\rangle$.

Grover's algorithm lends itself to exact calculation of the eigenvalues of $\rho_C(k)$, hence to computation of its entropy. The eigenvalues of $\rho_C(k)$ are

$$\lambda_1(k) = \cos^2(\theta k) \tag{13}$$

of multiplicity 1, and

$$\lambda_2(k) = \frac{\sin^2(\theta k)}{N-1} \tag{14}$$

of multiplicity $N - 1$. The entropy of $\rho_C(k)$ is given by

$$S(\rho_C(k)) = -\lambda_1(k)\log\lambda_1(k) - (N-1)\lambda_2(k)\log\lambda_2(k) \tag{15}$$

and is plotted in Fig. 1 for $N = 2^{20}$. (Throughout the paper, the unit of entropy is bits and log denotes base 2 logarithm.) The entropy $S(\rho_c(k))$ has period $\pi/\theta \approx (\pi/2)\sqrt{N}$.

Our main result is the following lower bound on time-complexity.

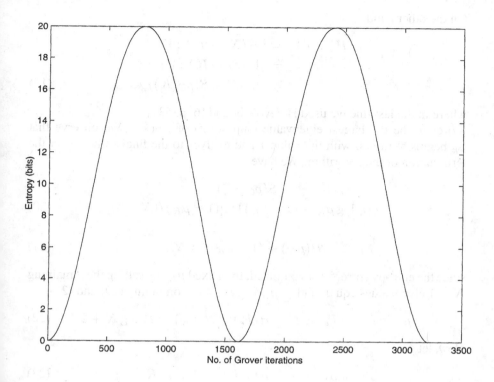

Figure 1. Evolution of entropy in Grover's algorithm.

Proposition 1 *Any quantum search algorithm that uses the oracle calls $\{O_x\}$ as defined by (10) must call the oracle at least*

$$K \geq \left(\frac{1 - P_e}{2\pi} + \frac{1}{\pi \log N}\right) \sqrt{N} \qquad (16)$$

times to achieve a probability of error P_e.

For the proof we first derive an information-theoretic inequality. For any quantum search algorithm of the type described in section 2, we have by Fano's inequality,

$$H(Y|X) \leq \mathcal{H}(P_e) + P_e \log(N - 1) \leq \mathcal{H}(P_e) + P_e \log(N), \qquad (17)$$

where for any $0 \leq u \leq 1$

$$\mathcal{H}(u) = -\delta \log \delta - (1 - \delta) \log(1 - \delta). \qquad (18)$$

On the other hand,

$$
\begin{aligned}
H(X|Y) &= H(X) - I(X;Y) \\
&= \log N - I(X;Y) \\
&\geq \log N - S(\rho_C(K))
\end{aligned}
\tag{19}
$$

where in the last line we used Holevo's bound [6, p. 531].

Let μ_k be the largest eigenvalue (sup-norm) of $\rho_C(k)$. We observe that μ_k begins at time 0 with the value 1 and evolves to the final value μ_K at the termination of the algorithm. We have

$$
S(\rho_C(K))
$$
$$
-\mu_K \log \mu_K - (1 - \mu_K) \log[(1 - \mu_K)/(N - 1)]
\tag{20}
$$

$$
\mathcal{H}(\mu_K) + (1 - \mu_K) \log N.
\tag{21}
$$

since the entropy entropy is maximized, for a fixed μ_K, by setting the remaining $N - 1$ eigenvalues equal to $(1 - \mu_K)/(N - 1)$. Combining (19) and (21),

$$
\mu_K \log N \leq P_e \log N + \mathcal{H}(P_e) + \mathcal{H}(\mu_K) \leq P_e \log N + 2
\tag{22}
$$

Now, let

$$
\Delta = \sup\{|\mu_{k+1} - \mu_k| : k = 0, 1, \ldots, K - 1\}.
\tag{23}
$$

This is the maximum change in the sup norm of $\rho_C(k)$ per algorithmic step. Clearly,

$$
K \geq \frac{1 - \mu_K}{\Delta}.
$$

Using the inequality (22), we obtain

$$
K \geq \frac{1 - P_e + 2/\log N}{\Delta}.
\tag{24}
$$

Thus, any upper bound on Δ yields a lower bound on K. The proof will be completed by proving

Lemma 1 $\Delta \leq 2\pi/\sqrt{N}$.

We know that operators that do not involve oracle calls do not change the eigenvalues, hence the sup norm, of $\rho_C(k)$. So, we should only be interested in bounding the perturbation of the eigenvalues of $\rho_C(k)$ as a result of an oracle call. We confine our analysis to the oracle operator (10) that the Grover algorithm uses.

For purposes of this analysis, we shall consider a continuous-time representation for the operator O_x so that we may break the action of O_x into infinitesimal time steps. So, we define the Hamiltonian

$$H_x = -\pi|x\rangle\langle x| \tag{25}$$

and an associated evolution operator

$$O_x(\tau) = e^{-i\tau H_x} = I + (e^{i\pi\tau} - 1)|x\rangle\langle x|.$$

The operator O_x is related to $O_x(\tau)$ by $O_x = O_x(1)$.

We extend the definition of conditional density to continuous time by

$$\rho_x(k_0 + \tau) = O_x(\tau)\rho_x(k_0)O_x(\tau)^\dagger \tag{26}$$

for $0 \leq \tau \leq 1$. The computer state in continuous-time is defined as

$$\rho_C(t) = \sum_x (1/N)\rho_x(t). \tag{27}$$

Let $\{\lambda_n(t), u_n(t)\}$, $n = 1, \ldots, N$, be the eigenvalues and associated normalized eigenvectors of $\rho_C(t)$. Thus,

$$\rho_C(t)|u_n(t)\rangle = \lambda_n(t)|u_n(t)\rangle, \quad \langle u_n(t)|\rho_C(t) = \lambda_n(t)\langle u_n(t)|,$$

$$\langle u_n(t)|u_m(t)\rangle = \delta_{n,m}. \tag{28}$$

Since $\rho_C(t)$ evolves continuously, so do $\lambda_n(t)$ and $u_n(t)$ for each n.

Now let $(\lambda(t), u(t))$ be any one of these eigenvalue-eigenvector pairs. By a general result from linear algebra (see, e.g., Theorem 6.9.8 of Stoer and Bulirsch [7, p. 389] and the discussion on p. 391 of the same book),

$$\frac{d\lambda(t)}{dt} = \langle u(t)|\frac{d\rho_C(t)}{dt}|u(t)\rangle. \tag{29}$$

To see this, we differentiate the two sides of the identity $\lambda(t) = \langle u(t)|\rho_C(t)|u(t)\rangle$, to obtain

$$
\begin{aligned}
\frac{d\lambda(t)}{dt} &= \langle u'(t)|\rho_C(t)|u(t)\rangle + \langle u(t)|\frac{d\rho_C(t)}{dt}|u(t)\rangle + \langle u(t)|\rho_C(t)|u'(t)\rangle \\
&= \langle u(t)|\frac{d\rho_C(t)}{dt}|u(t)\rangle + \lambda(t)[\langle u'(t)|u(t)\rangle + \langle u(t)|u'(t)\rangle] \\
&= \langle u(t)|\frac{d\rho_C(t)}{dt}|u(t)\rangle + \lambda(t)\frac{d}{dt}\langle u(t)|u(t)\rangle \\
&= \langle u(t)|\frac{d\rho_C(t)}{dt}|u(t)\rangle
\end{aligned}
$$

where the last line follows since $\langle u(t)|u(t)\rangle \equiv 1$.

Differentiating (27), we obtain

$$\frac{d\rho_C(t)}{dt} = \sum_x -(i/N)[H_x, \rho_x(t)] \tag{30}$$

where $[\cdot, \cdot]$ is the commutation operator. Substituting this into (29), we obtain

$$
\begin{aligned}
\left|\frac{d\lambda(t)}{dt}\right| &= \left|\langle u(t)| - \frac{i}{N}\sum_x [H_x, \rho_x(t)] |u(t)\rangle\right| \\
&\leq \frac{2}{N}\left|\sum_x \langle u(t)|H_x\rho_x(t)|u(t)\rangle\right| \\
&\stackrel{(a)}{\leq} \frac{2}{N}\sqrt{\sum_x \langle u(t)|H_x^2|u(t)\rangle}\sqrt{\sum_x \langle u(t)|\rho_x^2(t)|u(t)\rangle} \\
&\stackrel{(b)}{=} \frac{2}{N}\sqrt{\sum_x \pi^2 |\langle u(t)|x\rangle|^2}\sqrt{N\langle u(t)|\rho_C(t)|u(t)\rangle} \\
&= \frac{2\pi}{\sqrt{N}} 1 \cdot \sqrt{\lambda(t)} \\
&\leq \frac{2\pi}{\sqrt{N}}
\end{aligned}
$$

where (a) is the Cauchy-Schwarz inequality, (b) is due to (i) $\rho_x^2(t) = \rho_x(t)$ as it is a pure state, and (ii) the definition (27). Thus,

$$|\lambda(k_0 + 1) - \lambda(k_0)| = \left|\int_{k_0}^{k_0+1} \frac{d\lambda(t)}{dt}dt\right| \leq 2\pi/\sqrt{N}. \tag{31}$$

Since this bound is true for any eigenvalue, the change in the sup norm of $\rho_C(t)$ is also bounded by $2\pi/\sqrt{N}$.

Discussion

The bound (16) captures the $\sqrt{N}$ complexity of Grover's search algorithm. As mentioned in the Introduction, lower-bounds on Grover's algorithm have been known before; and, in fact, the present bound is not as tight as some of these earlier ones. The significance of the present bound is that it is largely based on information-theoretic concepts. Also worth noting is that the probability of error P_e appears explicitly in (16), unlike other bounds known to us.

References

[1] L. K. Grover, 'A fast quantum mechanical algorithm for database search,' *Proceedings, 28th Annual ACM Symposium on the Theory of Computing (STOC)*, May 1996, pp. 212-219. (quant-p/9605043)

[2] L. K. Grover, 'Quantum mechanics helps in searching for a needle in a haystack,' *Phys. Rev. Letters,* 78(2), 325-328, 1997. (quant-ph/9605043)

[3] C. Zalka, 'Grover's quantum searching is optimal,' *Phys. Rev. A,* 60, 2746 (1999). (quant-ph/9711070v2)

[4] C. H. Bennett, E. Bernstein, G. Brassard, and U. V. Vazirani, 'Strength and weaknesses of quantum computing,' *SIAM Journal on Computing,* vol. 26, no. 5, pp. 1510-1523, Oct. 1997. (quant-ph/9701001)

[5] M. Boyer, G. Brassard, P. Hoeyer, and A. Tapp, 'Tight bounds on quantum computing,' *Proceedings 4th Workshop on Physics and Computation,* pp. 36-43, 1996. Also *Fortsch. Phys.* 46(1998) 493-506. (quant-ph/9605034)

[6] M.A. Nielsen and I.L. Chuang, *Quantum Computation and Quantum Information,* Cambridge University Press, 2000.

[7] J. Stoer and R. Bulirsch, *Introduction to Numerical Analysis.* Springer, NY: 1980.

Index